Oxford Lecture Series in
Mathematics and its Applications

Series editors

John Ball Dominic Welsh

OXFORD LECTURE SERIES
IN MATHEMATICS AND ITS APPLICATIONS

1. J. C. Baez (ed.): *Knots and quantum gravity*
2. I. Fonseca and W. Gangbo: *Degree theory in analysis and applications*
3. P. L. Lions: *Mathematical topics in fluid mechanics, Vol. 1: Incompressible models*
4. J. E. Beasley (ed.): *Advances in linear and integer programming*
5. L. W. Beineke and R. J. Wilson (eds): *Graph connections: Relationships between graph theory and other areas of mathematics*
6. I. Anderson: *Combinatorial designs and tournaments*
7. G. David and S. W. Semmes: *Fractured fractals and broken dreams*
8. Oliver Pretzel: *Codes and algebraic curves*
9. M. Karpinski and W. Rytter: *Fast parallel algorithms for graph matching problems*
10. P. L. Lions: *Mathematical topics in fluid mechanics, Vol. 2: Compressible models*
11. W. T. Tutte: *Graph theory as I have known it*
12. Andrea Braides and Anneliese Defranceschi: *Homogenization of multiple integrals*
13. Thierry Cazenave and Alain Haraux: *An introduction to semilinear evolution equations*
14. J. Y. Chemin: *Perfect incompressible fluids*
15. Giuseppe Buttazzo, Mariano Giaquinta and Stefan Hildebrandt: *One-dimensional variational problems: an introduction*
16. Alexander I. Bobenko and Ruedi Seiler: *Discrete integrable geometry and physics*
17. Doina Cioranescu and Patrizia Donato: *An introduction to homogenization*
18. E. J. Janse van Rensburg: *The statistical mechanics of interacting walks. polygons. animals and vesicles*
19. S. Kuksin: *Hamiltonian partial differential equations*
20. Alberto Bressan: *Hyperbolic systems of conservation laws: the one-dimensional Cauchy problem*
21. B. Perthame: *Kinetic formulation of conservation laws*
22. A. Braides: *Gamma-convergence for beginners*
23. Robert Leese and Stephen Hurley: *Methods and algorithms for radio channel assignment*
24. Charles Semple and Mike Steel: *Phylogenetics*
25. Luigi Ambrosio and Paolo Tilli: *Topics on analysis in metric spaces*
26. Eduard Feireisl: *Dynamics of viscous compressible fluids*
27. Antonín Novotný and Ivan Straškraba: *Introduction to the mathematical theory of compressible flow*
28. Pavol Hell and Jarik Nesetril: *Graphs and homomorphisms*
29. Pavel Etingof and Frederic Latour: *The dynamical Yang-Baxter equation, representation theory, and quantum integrable systems*
30. Jorge Ramirez Alfonsin: *The Diophantine Frobenius problem*
31. Rolf Niedermeier: *Invitation to fixed parameter algorithms*
32. Jean-Yves Chemin, Benoit Desjardins, Isabelle Gallagher and Emmanuel Grenier: *Mathematical Geophysics: An introduction to rotating fluids and the Navier-Stokes equations*

Mathematical Geophysics

An introduction to rotating fluids and the Navier–Stokes equations

J.-Y. Chemin

Laboratoire J.-L. Lions, Université Paris 6, Case 187, 4 place Jussieu, 75252 Paris Cedex 05, France

B. Desjardins

C.E.A./D.I.F., B.P. 12, 91680 Bruyères le Châtel, France

I. Gallagher

Institut de Mathématiques de Jussieu, Université Paris 7, Case 7012, 2 place Jussieu 75251 Paris Cedex 05, France

E. Grenier

E.N.S. LYON, 46 Allée d'Italie, 69364 Lyon cedex 7, France

CLARENDON PRESS • OXFORD
2006

This book has been printed digitally and produced in a standard specification in order to ensure its continuing availability

OXFORD
UNIVERSITY PRESS

Great Clarendon Street, Oxford OX2 6DP

Oxford University Press is a department of the University of Oxford.
It furthers the University's objective of excellence in research, scholarship,
and education by publishing worldwide in

Oxford New York

Auckland Cape Town Dar es Salaam Hong Kong Karachi
Kuala Lumpur Madrid Melbourne Mexico City Nairobi
New Delhi Shanghai Taipei Toronto
With offices in
Argentina Austria Brazil Chile Czech Republic France Greece
Guatemala Hungary Italy Japan South Korea Poland Portugal
Singapore Switzerland Thailand Turkey Ukraine Vietnam

Oxford is a registered trade mark of Oxford University Press
in the UK and in certain other countries

Published in the United States
by Oxford University Press Inc., New York

Reprinted 2008

ISBN 978-0-19-857133-9

CONTENTS

PREFACE

This book is divided into four parts. The introduction (Part I) provides the physical background of the geophysical models that are analyzed in this book from a mathematical viewpoint.

Part II is devoted to a self-contained proof of the existence of weak (or strong) solutions to the incompressible Navier–Stokes equations.

Part III deals with the rapidly rotating Navier–Stokes equations, first in the whole space, where dispersion effects are considered. Then the case where the domain has periodic boundary conditions is considered, and finally rotating Navier–Stokes equations between two plates are studied, both in the case of horizontal coordinates in $\mathbf{R}^2$ and periodic.

In Part IV the stability of Ekman boundary layers, and boundary layer effects in magnetohydrodynamics and quasigeostrophic equations are discussed. The boundary layers which appear near vertical walls are presented and formally linked with the classical Prandlt equations. Finally spherical layers are introduced, whose study is completely open.

ACKNOWLEDGEMENTS

This book is the fruit of several years of work on the Navier–Stokes equations and on rotating fluids. During those years we had the opportunity for discussion with many colleagues who played an invaluable role in our understanding of many problems related to these questions. Among those people we wish to thank particularly Laure Saint-Raymond.

A preliminary version of this book was given to read to a number of people who pointed out many mistakes and misprints, and this book is undoubtedly a great improvement thanks to their comments. We are therefore very grateful to Frédéric Charve, Thierry Gallay, Pierre Germain, David Lannes, Fabrice Planchon, Violaine Roussier and Jean Starynkevitch, among others, for their careful proof-reading and their precise and useful remarks.

Finally we acknowledge the hospitality of the Centre International de Recherche Mathématique in Luminy, where part of this work was accomplished.

PART I

General introduction

The aim of this part is to provide a short introduction to the physical theory of rotating fluids, which is a significant part of geophysical fluid dynamics. This chapter contains no rigorous results but rather gives a general overview of classical phenomena occurring in rotating fluids, in particular the propagation of waves and of boundary layers in the neighborhood of horizontal and vertical walls. It also makes links with mathematical results proved in this book and gives some references to the physical literature (very abundant on this subject).

Meteorology and oceanography

Let us begin by simple computations of orders of magnitude. The typical amplitude of velocity in the ocean is a few meters per second (except in strong oceanic currents like the Gulf Stream), and the typical size of an ocean is 5000 kilometers. It therefore takes 50 days for a fluid particle to cross the ocean. Meanwhile the Earth has made 50 rotations. As a consequence, if we want to study oceans at a global level, the Coriolis force cannot be neglected. It is important in magnitude and also for its physical consequences. Therefore all the models of oceanography and meteorology dealing with large-scale phenomena include the Coriolis force. Of course other physical effects are of similar importance, like temperature variations, salinity, stratification, and so on, but a first step in the study of more complex models is to understand the behavior of rotating fluids. Many important features of oceanic circulation can be explained by large rotation, like for instance the intensification of oceanic currents near western coasts (the Gulf Stream near the Gulf of Mexico, Kuroshio near Japan, and so on). It is striking to realize that a model as simple as the incompressible Navier–Stokes equations together with a large Coriolis term, with natural boundary conditions, is sufficient to recover with quite good accuracy the large-scale oceanic circulation and to give a preliminary explanation to strong currents! Of course precise explanations require refined models, including topographical effects, stratification, for instance, to predict the location where the Gulf Stream leaves the American shore.

So a first step is to neglect temperature, salinity, and stratification, and to consider only Navier–Stokes equations for incompressible fluids. Note that at these scales, compressibility effects can be completely dismissed: we are not concerned with sound effects, and the speed of air or water at large scales is so small that the Mach number is almost 0 and the assumption of incompressibility is fully justified. Another way to justify the incompressibility condition is to derive it from so-called "primitive equations" after a change of vertical coordinates (pressure coordinates). We will not detail this point here. Moreover, the speed of rotation of the Earth can be considered as a constant on the time-scales considered (a few months or a few years). In that case, up to terms which we write as gradients, the Coriolis force reduces to $2\omega \wedge u$ where ω denotes the rotation vector, which will be taken along the x_3-direction, and u the velocity of the fluid. The equations are then the so-called Navier–Stokes–Coriolis equations (NSC_ε), written in a non-dimensional way (for the sake of simplicity, we will set in the whole book the characteristic length to one)

$$(\mathrm{NSC}_\varepsilon) \qquad \begin{cases} \partial_t u + u \cdot \nabla u - \nu \Delta u + \dfrac{e^3 \wedge u}{\varepsilon} + \nabla p = f \\ \mathrm{div}\ u = 0, \end{cases}$$

in a domain Ω (with boundary conditions to be made precise later on). In these equations, u denotes the velocity of the fluid, p its pressure, e^3 the unit vector in the x_3-direction, ν the rescaled viscosity and ε^{-1} the rescaled speed of rotation, and f a forcing term (heating, gravity, and so on). The parameter ε is called the Rossby number, and is a small parameter, usually of order 10^{-1} to 10^{-3}. The high rotation limit corresponds to the regime when ε tends to 0, possibly with a link between ν and ε. In physical situations, ν is of order ε and we assume that $\nu = \beta\varepsilon$ where β is fixed. As a matter of fact, the limit ε tends to 0 with fixed ν leads to $u = 0$, which is not very interesting: the fluid is immediately stopped (see the section below on Ekman layers).

In real situations. however, the fluid is turbulent and ν no longer denotes the molecular kinematic viscosity of the fluid, but rather a turbulent viscosity, measured from the speed of diffusion of tracers for instance. This is of course a very crude approximation of turbulent phenomena. In particular, it does not take into account the anisotropy created by large rotation. The Coriolis force creates an asymmetry between horizontal and vertical motions, vertical motion being penalized. This induces an anisotropy in the turbulent behavior, the horizontal turbulence being more important than the vertical turbulence. To take this effect into account, it is usual in meteorology and oceanography to replace the $-\nu\Delta$ term by $-\nu_h\Delta_h - \nu_V\partial_3^2$ where $\Delta_h = \partial_1^2 + \partial_2^2$, ν_h denotes the horizontal viscosity and ν_V the vertical viscosity, and we state $x = (x_h, x_3)$. Here and throughout this text we have denoted $x_h = (x_1, x_2)$ and ∂_i stands for $\partial/\partial x_i$. We take ν_V of order ε, ν_h large compared to ν_V and either ν_h tends to 0 with ε, or ν_h is constant as ε goes to 0.

This is of course a fully unjustified turbulence model. More complicated models can be studied, where the viscosity is linked to the local shear, or where a dynamical model for the energy of the small-scale flow is considered, like the $k-\varepsilon$ model.

Let us now discuss the geometry of the domain. The natural domain would be the oceans or the whole atmosphere. However to isolate the various phenomena it is more interesting to begin with simple geometries like the whole space $\mathbf{R}^3$, the periodic setting $\mathbf{T}^3$, between two plates $\mathbf{R}^2 \times [0,1]$ or $\mathbf{T}^2 \times [0,1]$, or domains with vertical boundaries $\Omega_h \times [0,1]$ where Ω_h is a two-dimensional domain. These domains already cover the stratification effect of rotation, inertial waves, boundary layers, and Ekman pumping. To go further towards refined models and to study the effect of curvature of the Earth, we should consider a spherical shell $R_1 \leq r \leq R_2$ where $R_2 - R_1 \ll R_1$, or a part of it (see [47]). This leads to the so-called β effect and to equatorial singularities. Topographical effects can also be included (see below).

For a more detailed introduction we refer to the monographs of Pedlovsky [103] and Greenspan [67].

Review of physical phenomena

The aim of this section is to describe various physical problems which arise in the high-rotation limit of Navier–Stokes–Coriolis equations (NSC_ε) – in particular, the two-dimensional limit constraint, Poincaré waves, and horizontal boundary layers – and to refer to the corresponding mathematical results and chapters.

Taylor–Proudman columns

The first step in the study of rotating fluids (NSC_ε) is to verify that the only way to control the Coriolis force as ε tends to 0 is to balance it with the pressure gradient term. Hence in the limit, $e^3 \wedge u$ must be a gradient

$$e^3 \wedge u = -\nabla p,$$

which leads to

$$\begin{cases} -u^2 = -\partial_1 p \\ u^1 = -\partial_2 p \\ 0 = -\partial_3 p. \end{cases}$$

In particular, the limit pressure p must be independent of x_3, hence depends only on t and x_h. We see that u^1 and u^2 are also independent of x_3, and that

$$\partial_1 u^1 + \partial_2 u^2 = -\partial_1\partial_2 p + \partial_2\partial_1 p = 0.$$

In particular (u^1, u^2) is a two-dimensional, horizontal, divergence-free vector field. On the other hand,

$$\partial_3 u^3 = -\partial_1 u^1 - \partial_2 u^2 = 0,$$

therefore u^3 is also independent of x_3. Physically, the fluid is limited by rigid (fixed) boundaries or interfaces, from above or from below, which in general leads to $u^3 = 0$ (at least to first order in ε; this will be detailed later). In that case, the fluid has a two-dimensional behavior. Throughout this book we will be working with vector fields which may depend on the vertical variable or not, and which may have two components or three. In order to fix the terminology, we will denote by a "horizontal vector field" any two-component vector field, whereas a two-dimensional vector field will denote a vector field independent of the vertical variable.

In physical cases, all the particles which have the same x_1 and x_2 have the same velocity $(u^1(t,x_h), u^2(t,x_h), u^3(t,x_h))$. The particles of fluid move in vertical columns, called Taylor–Proudman columns. That is the main effect of high rotation and a very strong constraint on the fluid motion. First note that on the boundary of the domain, where the velocity vanishes usually or is prescribed, the x_3 independence is violated. That leads to boundary layers which are investigated below. Second, note that as the fluid is incompressible, the height of Taylor–Proudman columns must be constant as time evolves.

If the domain of evolution is limited by two parallel planes $a \leq x_3 \leq b$ or is periodic in x_3 then columns move freely and in the limit of high rotation the fluid behaves like a two-dimensional incompressible fluid (we forget for a while the boundary layers).

If the domain of evolution is limited by two non-parallel planes (for instance the domain defined by $0 \geq x_3 \geq -x_1$), the motion is even more constrained. First it has to be two-dimensional and horizontal. Second the columns must have a constant height. Therefore the fluid moves in the x_2-direction and the velocity field is of the form $(0, u^2(t,x_h), 0)$. By incompressibility we even have $\partial_2 u^2 = 0$ hence u^2 depends only on t and x_1! Of course such motion is too constrained and is not of much interest.

If the domain of evolution is a sphere or an ellipsoid, then again the fluid particles move along paths of constant height, which are closed circles or closed ellipses. Again the motion is too constrained and not relevant in meteorology and oceanography, but can be easily studied in laboratories (we refer to impressive pictures in [67]).

That conclusion may seem strange at first glance since oceans have a non-negligible topography, both in terms of amplitude (ranging from a few hundred meters near shores to ten kilometers at most) and also through the influence of global circulation. However we must keep in mind that the rotating Navier–Stokes equations are a very crude model since they in particular neglect the effects of stratification and density, of temperature and salinity. As a consequence the effects of topography are amplified and exaggerated, and if we want to keep a reasonable role for topography, we must study variations in heights of the domain of order ε, or else the motion is completely constrained [103]. A typical domain of evolution to study topography is therefore

$$\Omega_\varepsilon = \{-1 + \varepsilon\eta(x_h) \leq x_3 \leq 0\},$$

where η is a smooth function. A domain of the form $\Omega = \{\eta(x_h) \leq x_3 \leq 0\}$ leads to degenerate motions if η is not a constant independent of x_h. Such domains have been studied in [48] and will not be investigated in this book.

In conclusion, if Ω is the whole space $\mathbf{R}^3$ or in the periodic setting, the limit flow is a two-dimensional, horizontal divergence-free flow $u = (u^1, u^2)$. We shall show later on that it simply satisfies the two-dimensional incompressible Euler or Navier–Stokes equations. In the case of an isotropic viscosity $-\nu\Delta$ when ν goes to 0 with ε, the limit equation is simply the incompressible Euler system

$$(E) \quad \begin{cases} \partial_t u + u \cdot \nabla u + \nabla p = 0 \\ \operatorname{div} u = 0, \end{cases}$$

in two-dimensional space. That limit still holds in the case when the viscosity is anisotropic and ν_h goes to 0 with ε. Finally, if ν_h is constant one finds the incompressible Navier–Stokes system in two-dimensional space:

$$\begin{cases} \partial_t u + u \cdot \nabla u - \nu_h \Delta_h u + \nabla p = 0 \\ \operatorname{div} u = 0. \end{cases}$$

The proof will be detailed in Part III and requires the understanding of another important physical phenomenon: the propagation of high-speed waves in the fluid, which we will now detail.

Poincaré waves

In the previous section we have seen that the limit flow is two dimensional. That is not always the case for general initial data (which may depend on x_3) and it remains to describe what happens to the three-dimensional part of the initial data. Let us, in this paragraph, again forget the role of boundaries, and focus on the whole space or on periodic cases. If we omit for a moment the non-linearity of (NSC_ε) and the viscous terms, we end up with the Coriolis system

$$(C_\varepsilon) \quad \begin{cases} \partial_t v + \dfrac{e^3 \wedge v}{\varepsilon} + \dfrac{\nabla p}{\varepsilon} = 0 \\ \operatorname{div} v = 0, \end{cases}$$

which turns out to describe the propagation of waves, called Poincaré waves, or inertial waves. As we will see in the second section of Chapter 5, the corresponding dispersion law relating the pulsation ω to the wavenumber $\xi \in \mathbf{R}^3$ is

$$\pm\omega(\xi) = \varepsilon^{-1} \frac{\xi_3}{|\xi|} \cdot$$

Therefore the two-dimensional part of the initial data evolves according to two-dimensional Euler or Navier–Stokes equations, and the three-dimensional part generates waves, which propagate very rapidly in the domain (with a speed of order ε^{-1}). The time average of these waves vanishes, like their weak limit, but they carry a non-zero energy. Note that the wavenumbers of these waves are bounded as ε tends to 0 and a priori no short wavelengths are created. On the contrary, time frequencies are very large and go to infinity like ε^{-1} as ε goes to 0.

To gain some intuition about these waves, it is interesting to make a comparison with the incompressible limit of compressible Euler or Navier–Stokes equations, in the isentropic case, to simplify. The compressible isentropic Euler equations are as follows

$$(E_{\text{comp}}) \qquad \begin{cases} \partial_t \rho + \operatorname{div}(\rho u) = 0 \\ \partial_t(\rho u) + \operatorname{div}(\rho u \otimes u) + \dfrac{\nabla \rho^\gamma}{\gamma \varepsilon^2} = 0, \end{cases}$$

where γ is a real number greater than 1. In this case, ε denotes the Mach number, i.e. the ratio of the typical velocity of the fluid to the speed of sound. Here the typical velocity is $O(1)$ and the sound speed is of order ε^{-1}. The Mach number plays the role of the Rossby number. As ε goes to 0, formally, $\nabla \rho = 0$ therefore ρ is a constant, say 1, and $\operatorname{div} u = 0$. This gives the incompressible Euler system (E), hence the limit flow has a constant density and is incompressible. This is the analog of the two-dimensional property of the limit flow in the case of rotating fluids. However, in general the initial data have a varying density and the problem is not divergence-free. If we write $\rho = 1 + \tilde{\rho}$, forget all the non-linear terms of (E_{comp}) and set $\varepsilon = 1$ we get

$$\begin{cases} \partial_t \tilde{\rho} + \operatorname{div} u = 0 \\ \partial_t u + \gamma \nabla \tilde{\rho} = 0, \end{cases}$$

which are exactly the equations of acoustic waves. Therefore, general initial data split as the Mach number goes to 0 into an incompressible part (that is the projection of u on divergence-free fields and of ρ on constants) evolving according to the incompressible Euler equations, and a compressible part which creates acoustic waves (with bounded spatial frequencies) with very high time frequency (of order ε^{-1}). The time average of acoustic waves is zero, but they carry a non-vanishing part of the total energy in the general case. The incompressible limit and the high rotation limit are therefore two highly linked problems, with similar features, the first one being maybe more familiar to the reader. We refer to [92],[93],[50], [41] and [42] for a mathematical justification of this limit.

Let us go back to the high rotation limit and to Poincaré waves. Those waves propagate very fast in the domain. Therefore if the domain is unbounded they go rapidly to infinity and in fact disappear. Only the x_3 independent part remains and solutions converge to solutions of two-dimensional Euler or Navier–Stokes equations. At the mathematical level, this leads to the setup and use of Strichartz type estimates on Poincaré waves, which are dispersive waves. This is detailed in Part III, Chapter 5.

If the domain is bounded, however, or if the flow is periodic, Poincaré waves persist for long times, and interact not only with the limit two-dimensional flow, but also with themselves. The first striking fact is that the interaction between two Poincaré waves does not create x_3 independent fields: the interaction of waves does not affect the limit x_3 independent field. In particular if we forget the waves (by projection on x_3 independent vector fields, or by time averaging, or by approaching a weak limit), the limit x_3 independent field satisfies Euler or Navier–Stokes equations and is completely decoupled from the waves (which may not be the case in the framework of low Mach number limit in non-homogeneous fluids [19]). Next, interaction between the limit flow and Poincaré waves always takes place, alters the waves and can be seen as a kind of "diffraction" by the medium. Interaction between two Poincaré waves to create another Poincaré wave, however, is less frequent: let us consider a periodic box of lengths a_1, a_2 and a_3. A wave ξ interacts with a wave ξ' to give birth to a wave ξ'' provided

$$\xi + \xi' = \xi'', \quad \text{and} \quad \frac{\xi_3}{|\widetilde{\xi}|} + \frac{\xi_3'}{|\widetilde{\xi}'|} = \frac{\xi_3''}{|\widetilde{\xi}''|} \text{ where } \widetilde{\xi} = \left(\frac{\xi_1}{a_1}, \frac{\xi_2}{a_2}, \frac{\xi_3}{a_3}\right),$$

which are the usual resonance conditions in the three-wave interaction problem. In the periodic case, all the components of ξ, ξ' and ξ'' are integers. Thus the above conditions turn out to be diophantine equations which in general have no solutions. More precisely, if we consider a_1, a_2 and a_3 as parameters. for almost every (a_1, a_2, a_3) there is no integer solution to the above equations except the trivial solutions given by symmetries. Therefore generically in the sizes of the periodic box, the waves do not interact with themselves and only interact with the two-dimensional underlying flow.

Mathematically to handle waves we introduce the Poincaré group as follows. Let $\mathcal{L}(t)v_0$ be the solution of the Coriolis system (C_ε) with $\varepsilon = 1$ and initial data v_0. We describe the solution of the $(\mathrm{NSC}_\varepsilon)$ system by

$$u(t, x_h, x_3) = \overline{u}(t, x_h) + \mathcal{L}\left(\frac{t}{\varepsilon}\right) u_{\mathrm{osc}}(t, x_h, x_3)$$

where $\overline{u}$ is a two-dimensional divergence-free vector field and u_{osc} is a divergence-free, zero x_3 average vector field. The main point is the weak convergence of $\mathcal{L}(t/\varepsilon)u_{\mathrm{osc}}$ to 0 as ε goes to 0, leading to the weak convergence of u to $\overline{u}$ which satisfies a two-dimensional Euler or Navier–Stokes equation (depending on whether ν_h vanishes or not). The oscillatory profile u_{osc} satisfies a three-dimensional Navier–Stokes type system, with special properties because of the rare occurrence of wave interactions. Even in the general case, the non-linearity of this system contains few terms, and behaves like a two-dimensional non-linearity. It therefore turns out to be possible to prove global well-posedness of this equation, and hence global well-posedness of the limit system on $(\overline{u}, u_{\mathrm{osc}})$, which was quite unexpected before the work of [2] since the limit system is a priori three dimensional. This analysis will be detailed in Part III, Chapter 6.

In particular if at $t = 0$, u_{osc} is identically 0 then u_{osc} remains identically 0 for any positive time. Such initial data are called "well prepared" in contrast with general initial data ($u_{\text{osc}} \neq 0$ at $t = 0$) which are called "ill prepared".

The existence of these fast waves is a first difficulty for the study of the high rotation limit of (NSC_ε), since two time-scales have to be considered. A second difficulty arises in the presence of boundaries and will be detailed in the next paragraph.

Ekman layers

Let us now study in more detail the case of a fluid between two parallel infinite planes $\Omega = \{0 \leq x_3 \leq 1\}$. In the limit of high rotation, the fluid velocity is independent of x_3 (in the first step we forget the propagation of Poincaré waves studied in the preceding paragraph). However it is classical to enforce Dirichlet boundary conditions on $\partial\Omega$, namely $u = 0$ on $\partial\Omega$ (the fluid stops on the boundary), which is incompatible with the Taylor–Proudman theorem (except if $u = 0$ in the whole domain!). This incompatibility leads to boundary layers which appear near $\partial\Omega$. Boundary layers are located in very thin parts of the domain, usually near walls where the velocity of the fluid has very large gradients: the velocity changes are of $O(1)$ within lengths of order λ with λ tending to 0 with ε. Let us derive the size, the equations and the main features of horizontal boundary layers in a physical way (a rigorous approach is given in Part III, Chapter 7). In the boundary layers, the viscosity is of order $\nu\lambda^{-2}$, the pressure of order ε^{-1} and the Coriolis force of order ε^{-1}. Hence to get equilibrium we must take λ of order $\sqrt{\varepsilon\nu}$. In particular, if ν is of order ε, λ is also of order ε. The equation of the boundary layer expresses the balance between the Coriolis force, the vertical viscosity and the pressure. After rescaling we get, with $u = u\,(x_h, x_3/\varepsilon)$:

$$\begin{aligned} -\partial_\zeta^2 u^1 - u^2 + \partial_1 p &= 0 \\ -\partial_\zeta^2 u^2 + u^1 + \partial_2 p &= 0 \\ \partial_\zeta p &= 0. \end{aligned}$$

In particular the pressure does not change, to first order, in the boundary layer and is given by the pressure in the interior of the domain. We can therefore set the pressure to 0 in the layer. We thus obtain

$$-\partial_\zeta^2 u^1 - u^2 = 0 \quad \text{and} \quad -\partial_\zeta^2 u^2 + u^1 = 0,$$

which is a fourth-order ordinary differential equation in u^1 and u^2. It is completed by the boundary conditions $u \to u_\infty(t, x_h)$ as $x_3 \to \infty$ for the tangential components, where u_∞ is the velocity in the interior of the fluid (outer limit of the boundary layer), and by $u = 0$ at $x_3 = 0$ (Dirichlet boundary condition). Solving that differential equation is then straightforward and provides the classical expression for the tangential velocity,

$$u_{\tan}(t, x_h, x_3) = \left(\mathrm{Id} - \exp\left(-\frac{x_3}{\sqrt{2\varepsilon\nu}}\right) R\left(-\frac{x_3}{\sqrt{2\varepsilon\nu}}\right)\right) u_\infty(t, x_h)$$

where $R(\zeta)$ denotes the rotation of angle ζ. This boundary layer is called the Ekman layer, or the Ekman spiral taking into account the shape of $u_{\tan}(t, x_h, x_3)$.

Now to enforce incompressibility we must have $\partial_3 u^3 = -\partial_1 u^1 - \partial_2 u^2$, which leads, together with $u^3 = 0$ for $x_3 = 0$, to

$$u^3(t, x_h, x_3) = \sqrt{\frac{\varepsilon\nu}{2}}\, \mathrm{curl}_h u_\infty(t, x_h) f\left(-\frac{x_3}{\sqrt{2\varepsilon\nu}}\right)$$

with $f(\zeta) = -1/2e^{-\zeta}(\sin\zeta + \cos\zeta)$. As a consequence, u^3 does not go to 0 as x_3 tends to infinity: if the velocity in the interior of the domain is not constant, a small amount of the fluid, of order $\sqrt{\varepsilon\nu}$, enters the domain (or the boundary layer, depending on the sign of the two-dimensional curl). This phenomenon is called Ekman suction, and $u^3(t, x_h, \infty)$ is called the Ekman suction velocity or Ekman transpiration velocity. This velocity is responsible for global circulation in the whole domain, of order $\sqrt{\varepsilon\nu}$, but not limited to the boundary layer, a first three-dimensional effect in the interior of the domain.

This small velocity has a very important effect in the energy balance. Namely, let us compute the energy dissipation in the Ekman layer, that is let us evaluate the order of magnitude of $\nu \int |\nabla u|^2$ in the layer. The gradient $|\nabla u|$ is of order $\lambda^{-1} \sim \sqrt{\varepsilon\nu}^{-1}$, therefore $\nu \int |\nabla u|^2$ scales as

$$\nu \sqrt{\varepsilon\nu}^{-2} \sqrt{\varepsilon\nu} \sim \sqrt{\frac{\nu}{\varepsilon}},$$

the last term $\sqrt{\varepsilon\nu}$ corresponding to the volume of integration. In the most interesting case $\nu \sim \varepsilon$, the dissipation of energy per unit of time is of order one: despite its smallness, the Ekman layer dissipates a significant amount of energy, and cannot be neglected in the energy balance. In other words, the Ekman layer damps the interior motion, like an order-one friction term. This phenomenon is called Ekman pumping. Note that if ν remains constant while ε goes to 0, then the dissipation is infinite! In fact the fluid is immediately stopped by its boundary layers. This remains true if $\nu \gg \varepsilon$ when ε tends to 0. On the contrary, if $\nu \ll \varepsilon$ then Ekman pumping is negligible.

Let us compute the Ekman pumping precisely . The dissipation in one layer equals

$$\nu \int \left(|\partial_3 u^1|^2 + |\partial_3 u^2|^2\right) dx \quad \text{where } |\partial_3 u| = \frac{|u_\infty|}{\sqrt{\varepsilon\nu}} \exp\left(-\frac{x_3}{\sqrt{2\varepsilon\nu}}\right).$$

Hence the dissipation equals

$$\nu \frac{|u_\infty|^2}{\varepsilon\nu} \frac{\sqrt{\varepsilon\nu}}{\sqrt{2}} = |u_\infty|^2 \sqrt{\frac{\nu}{2\varepsilon}}.$$

As there are two layers (one for each boundary), the total dissipation equals $\sqrt{2\beta}|u_\infty|^2$ where $\beta = \nu/\varepsilon$. Hence Ekman pumping is a linear dissipation, and adds the linear term $\sqrt{2\beta}u$ in the Euler or Navier–Stokes equations. The limit equation on the x_3 independent field is

$$\begin{cases} \partial_t u + u \cdot \nabla u + \sqrt{2\beta} u + \nabla p = 0 \\ \operatorname{div} u = 0. \end{cases}$$

As will be clear in the rigorous derivation of Ekman layers, this friction term is deeply connected to the Ekman suction velocity. The role of this vertical velocity is to move the fluid from the interior to the boundary layers, where it slows down because of the large vertical viscosity. The slow fluid then goes back to the interior of the domain again with the help of the Ekman suction velocity. This can be observed very simply, by making a cup of tea turn rapidly with a spoon [67]. Everyday experience shows that once one stops turning the spoon, the tea stops moving within a dozen seconds. The first idea is that the dissipation is due to viscosity. Let us make a short computation of the associated order of magnitudes: the typical size of the cup is $L = 5$ cm, the kinematic viscosity of water is of the order of $10^{-2}\text{cm}^2\cdot\text{s}^{-1}$, and the rotation is, say $\Omega = 4\pi\,\text{s}^{-1}$. The diffusion time-scale is therefore $T_{\text{diff}} \stackrel{\text{def}}{=} L^2/\nu \sim 40$ minutes, whereas the suction time scale is $T_{\text{suct}} \stackrel{\text{def}}{=} L/\sqrt{\nu\Omega} = 14$ seconds. Tea can be considered as a rotating fluid and the main effect is in fact Ekman suction which brings tea from the interior of the cup to boundaries where dissipation is high. The time-scale is then the time needed for a particle of fluid to cross the cup and reach a boundary. This is the right time-scale.

Now in the general case Poincaré waves propagate in the medium. They also violate in general the Dirichlet boundary condition, and have their own boundary layers, which are very similar to Ekman boundary layers, except for their slightly different size and their dependence on the frequency. Again those boundary layers dissipate energy and damp the waves like a linear friction term (which depends on the frequency and the wavenumber). The detailed computations of the boundary layers of Poincaré waves is the object of Part III, Chapter 7.

To summarize, rotating fluids in $\mathbf{T}^2 \times [0,1]$ or $\mathbf{R}^2 \times]0,1[$ consist of:

- a limit two-dimensional, divergence-free flow;
- horizontal boundary layers at $x_3 = 0$ and $x_3 = 1$ to match the interior flow with Dirichlet boundary condition (Ekman layers);
- Poincaré waves, with high time frequency;
- horizontal boundary layers at $x_3 = 0$ and $x_3 = 1$ to match Poincaré waves with Dirichlet boundary conditions.

Moreover, the limit flow is damped by Ekman pumping and satisfies a damped Euler (or Navier–Stokes) equation. Poincaré waves are also damped by Ekman pumping, and in the periodic case satisfy a quadratic damped equation. In the case $\mathbf{R}^2 \times [0,1]$, Poincaré waves go to infinity very fast (with speed ε^{-1}) and go to 0 locally in time and space (for $t > 0$). All this will be detailed in Part III, Chapter 7.

When, instead of e^3, the direction of rotation r is not perpendicular to the boundaries, the boundary layer is still an Ekman layer of size $\sqrt{\varepsilon\nu/|r \cdot e^3|}$ provided $r \cdot e^3$ does not vanish. The situation is, however, different if $r \cdot e^3 = 0$: the vertical layers are very different and more difficult to analyze. We shall discuss the layers in a general three-dimensional domain in the last chapter of this book.

Stability of Ekman layers

A general problem in boundary layer theory is to know whether the layer remains laminar or becomes turbulent, that is, to know whether the characteristics of the flow vary over lengths of size $\sqrt{\varepsilon\nu}$ in x_3 and of size 1 in x_h in times of order 1, or whether small scales also appear in the horizontal directions, or whether the flow evolves in small time, of order say $\sqrt{\varepsilon\nu}$. The answer is not straightforward, since the velocity, being of order 1 in the layer, a particle may cross it in time scales of order $\sqrt{\varepsilon\nu}$ if by chance its velocity is not parallel to the boundary. Hence naturally, the system could evolve in a significant manner in times of order $\sqrt{\varepsilon\nu}$. This would in fact create tangential structures of typical size $\sqrt{\varepsilon\nu}$. In other words there is a priori no reason why the flow in the boundary layer would remain so anisotropic (the x_3-direction being the only direction of high variation), as the transport term has a natural destabilizing effect.

On the other hand, we can think that in the boundary layer the viscosity is so important that it suffices to stabilize everything and to cancel motions in the vertical direction.

The answer lies in between these two limit cases, and depends on the ratio between inertial forces and viscous forces. Let us define the Reynolds number Re of the boundary layer as the typical ratio between inertial forces and viscous forces in the boundary layer. The inertial forces are of order $U^2\lambda^{-1}$ where U is the typical velocity in the layer ($|u_\infty|$ in our case), and the viscous forces are of order $\nu U \lambda^{-2}$. Therefore we can define Reynolds number by

$$Re = |u_\infty|\frac{\lambda}{\nu} = |u_\infty|\sqrt{\frac{\varepsilon}{\nu}}\,\cdot$$

If the Reynolds is small, then viscous forces prevail and the flow is expected to be stable. On the contrary if the Reynolds is large, inertial forces are important and the flow may be unstable. This is a classical phenomenon in fluid mechanics, a feature common to many different physical cases (boundary layers of viscous flows, rotating flows, magnetohydrodynamics (MHD)): there exists a critical Reynolds number Re_c such that the flow is stable and the boundary layer remains stable provided $Re < Re_c$, and such that the flow is unstable and the boundary becomes more complicated or even turbulent if $Re > Re_c$. In the case of Ekman layers. the critical Reynolds number can be computed, and equals approximately 54.

Therefore if $Re < Re_c \sim 54$, the flow of a highly rotating fluid between two plates splits into an interior two-dimensional flow and a boundary layer flow, completely laminar.

If $Re > Re_c$, on the contrary, the boundary layer is no longer laminar. If the Reynolds is not too high (say $Re < 120$) rolls appear, of size $\sqrt{\nu\varepsilon} \times \sqrt{\nu\varepsilon} \times 1$ and move with a velocity of order 1. At high Reynolds number (say $Re > 120$), the rolls themselves become unstable and are destroyed. The flow near the boundary is fully chaotic and turbulent. An interesting question is to know whether this turbulent layer remains near the boundary or goes into the interior of the domain and destabilizes it. In this latter case, the whole flow would be turbulent (and of course the Taylor–Proudman theorem would no longer be true).

This leads to a restriction of the size of the limit solution to get convergence. Note that if the viscosity is anisotropic and if ν_h/ν_v tends to $+\infty$, all those problems disappear since the critical Reynolds number is infinite: Ekman layers are always stable. This is not surprising since anisotropic viscosity accounts for turbulent behavior. This only says that the model is strong enough and that no other phenomenon (instability of the boundary layer) appears in this turbulent model.

Let us end this introduction with the proof that at small Reynolds numbers $Re < Re_1$ for some $Re_1 < Re_c$, Ekman layers are linearly stable, by using energy estimates. Let u_{E} be a pure Ekman layer given by

$$u_{\mathrm{E}} = \begin{cases} \left(\mathrm{Id} - \exp\left(-\dfrac{x_3}{\sqrt{2\varepsilon\nu}}\right) R\left(-\dfrac{x_3}{\sqrt{2\varepsilon\nu}}\right)\right) u_\infty(t, x_h) \\ \sqrt{\dfrac{\varepsilon\nu}{2}}\mathrm{curl}_h u_\infty(t, x_h) f\left(\dfrac{x_3}{\sqrt{2\varepsilon\nu}}\right) \end{cases}$$

and let us linearize the Navier–Stokes Coriolis equations (NSC_ε) around u_{E}. This yields, denoting the perturbation w,

$$(\mathrm{LNSC}) \quad \begin{cases} \partial_t w + u_{\mathrm{E}} \cdot \nabla w + w \cdot \nabla u_{\mathrm{E}} + \dfrac{e_3 \wedge w}{\varepsilon} - \nu\Delta w + \nabla p = 0 \\ \mathrm{div}\, w = 0 \\ w_{|\partial\Omega} = 0. \end{cases}$$

Let us estimate the L^2 norm of the perturbation w. We have, using the divergence-free condition and integrating over Ω,

$$\frac{1}{2}\frac{d}{dt}\int |w|^2\, dx + \nu\int |\nabla w|^2 dx + \int w \cdot (w \cdot \nabla u_{\mathrm{E}})\, dx = 0.$$

The main term of $\int w \cdot (w \cdot \nabla u_{\mathrm{E}})\, dx$ turns out to be $\int w_3 w \cdot \partial_3 u_{\mathrm{E}}\, dx$. As

$$\|\partial_3 u_{\mathrm{E}}(\cdot, x_3)\|_{L^\infty(\Omega_h)} \leq \|u_\infty\|_{L^\infty(\Omega_h)} \frac{1}{\sqrt{\varepsilon\nu}} \exp\left(-\frac{x_3}{\sqrt{2\varepsilon\nu}}\right).$$

the following estimate holds:

$$\begin{aligned} N &\stackrel{\text{def}}{=} \left| \int w_3 w \cdot \partial_3 u_{\mathrm{E}}\, dx \right| \\ &\leq \frac{\|u_\infty\|_{L^\infty(\Omega_h)}}{\sqrt{\varepsilon\nu}} \int \left| \left(\int_0^{x_3} \partial_3 w(t, x_h, z) dz \right) \left(\int_0^{x_3} \partial_3 w_3(t, x_h, z) dz \right) e^{-x_3/\sqrt{2\varepsilon\nu}} \right| dx \\ &\leq \frac{\|u_\infty\|_{L^\infty(\Omega_h)}}{\sqrt{\varepsilon\nu}} \|\partial_3 w\|_{L^2} \|\partial_3 w_3\|_{L^2} \int_{\mathbf{R}} z e^{-z/\sqrt{2\varepsilon\nu}} dz \\ &\leq C_0 \sqrt{\varepsilon\nu} \|u_\infty\|_{L^\infty(\Omega_h)} \|\partial_3 u\|_{L^2} \|\partial_3 u^3\|_{L^2}, \end{aligned}$$

where C_0 is a numerical constant. This last term can be absorbed by the viscous term $\nu \int |\nabla u|^2\, dx$ provided $C_0 \|u_\infty\|_{L^\infty(\Omega_h)} \sqrt{\varepsilon\nu} \leq \nu$, namely provided

$$Re = \|u_\infty\|_{L^\infty(\Omega_h)} \sqrt{\frac{\varepsilon}{\nu}} \leq Re_1 = \frac{1}{C_0}.$$

Hence (LNSC) is stable provided $Re_c < Re_1$. Of course $Re_1 < Re_c$ and the computation of C_0 leads to $Re_1 \sim 4$. There is therefore a large gap between Re_1 and Re_c. Note that this proof uses little of the precise description of u_{E} since it only uses the exponential decay of $\partial_3 u_{\mathrm{E}}$ with respect to $x_3/\sqrt{2\varepsilon\nu}$. It is therefore not surprising that the result is not optimal at all. The method, however, can be used in a wide range of situations, and we refer to [52] for other applications.

References

After each part of this book we will give specific references related to the topics treated in that part. Here we wish to present to the reader a (far from exhaustive) selection of classical physical monographs on geophysical fluids.

- G.K. Batchelor: *An introduction to fluid dynamics* [9]
- P. Bougeault and R. Sadourny: *Dynamique de l'atmosphère et de l'océan* [11]
- B. Cushman-Roisin: *Introduction to geophysical fluid dynamics* [40]
- P.G. Drazin, W.H. Reid: *Hydrodynamic stability* [52]
- M. Ghil and S. Childress: *Topics in Geophysical Fluid Dynamics: Atmospheric Dynamics, Dynamo Theory, and Climate Dynamics* [63]
- A.E. Gill: *Atmosphere–ocean dynamics* [65]
- H.P. Greenspan: *The theory of rotating fluids* [67]

- H. Lamb: *Hydrodynamics* [85]
- R. Lewandowski. *Analyse Mathématique et Océanographie* [88]
- C.C. Lin: *The Theory of Hydrodynamic Stability* [90]
- A. Majda: *Introduction to PDEs and Waves for the Atmosphere and Ocean* [94]
- J. Pedlosky: *Geophysical Fluid dynamics* [103]
- H. Schlichting: *Boundary layer theory* [111].

PART II

On the Navier–Stokes equations

This part is devoted to the mathematical study of the Navier–Stokes equations for incompressible homogeneous fluids evolving in a domain (i.e. an open connected subset) Ω of $\mathbf{R}^d$, where $d = 2$ or 3 denotes the space dimension. For instance, the space domain may be either the whole space $\mathbf{R}^d$, or a domain of $\mathbf{R}^d$, or a three-dimensional unbounded domain such as $\mathbf{R}^2 \times]0, 1[$, which will be of particular interest in the framework of geophysical flows. Denoting by $u \in \mathbf{R}^d$ the velocity field, $p \in \mathbf{R}$ the pressure field, and $\nu > 0$ the kinematic viscosity, the Cauchy problem can be written as follows:

$$(\mathrm{NS}_\nu) \qquad \begin{cases} \partial_t u + u \cdot \nabla u - \nu \Delta u + \nabla p = f \\ \operatorname{div} u = 0 \\ u_{|t=0} = u_0, \end{cases}$$

where f is a given bulk force. Note that the density ρ has been chosen equal to 1 for simplicity. It corresponds to the case of the non-dimensionalized Navier–Stokes equations, in which the viscosity is expressed as the inverse Reynolds number. The system (NS_ν) is supplemented with the no-slip boundary conditions $u_{|\partial\Omega} = 0$. Multiplying (NS_ν) by u and integrating over Ω formally yields the well-known energy equality

$$\frac{1}{2} \int_\Omega |u(t,x)|^2 \, dx + \nu \int_0^t \int_\Omega |\nabla u(t',x)|^2 \, dx dt'$$
$$= \frac{1}{2} \int_\Omega |u_0(x)|^2 \, dx + \int_0^t \int_\Omega f \cdot u(t',x) \, dx dt'.$$

It follows that the natural regularity assumptions for the data are u_0 square integrable and divergence-free, while f is a vector field the components of which should be in $L^2_{\mathrm{loc}}(\mathbf{R}^+; H^{-1}(\Omega))$ where $H^{-1}(\Omega)$ denotes the set of functions which are the sum of an L^2 function and the divergence of an L^2 vector field. This implies the introduction of a little bit of functional analysis. In particular, the basic properties of the Stokes operator and Sobolev embeddings are recalled in Chapter 1, followed by the proof of Leray's global existence theorem in Chapter 2. The question of uniqueness and stability is addressed in Chapter 3: the stability of Leray solutions is proved in two dimensions, and the existence of stable solutions is proved in three dimensions for bounded domains and for domains without boundary, namely $\mathbf{R}^3$ or $\mathbf{T}^3$.

1

Some elements of functional analysis

Before introducing the concept of Leray's weak solutions to the incompressible Navier–Stokes equations, classical definitions of Sobolev spaces are required. In particular, when it comes to the analysis of the Stokes operator, suitable functional spaces of incompressible vector fields have to be defined. Several issues regarding the associated dual spaces, embedding properties, and the mathematical way of considering the pressure field are also discussed.

1.1 Function spaces

Let us first recall the definition of some functional spaces that we shall use throughout this book. In the framework of weak solutions of the Navier–Stokes equations, incompressible vector fields with finite viscous dissipation and the no-slip property on the boundary are considered. Such H^1-type spaces of incompressible vector fields, and the corresponding dual spaces, are important ingredients in the analysis of the Stokes operator.

Definition 1.1 *Let Ω be a domain of $\mathbf{R}^d$ ($d = 2$ or 3). The space $\mathcal{D}(\Omega)$ is defined as the space of smooth functions compactly supported in the domain Ω, and $\mathcal{D}'(\Omega)$ as the space of distributions on Ω.*

The space $H^1(\Omega)$ denotes the space of L^2 functions f on Ω such that ∇f also belongs to $L^2(\Omega)$. The Hilbertian norm is defined by

$$\|f\|^2_{H^1(\Omega)} \stackrel{def}{=} \|f\|^2_{L^2(\Omega)} + \|\nabla f\|^2_{L^2(\Omega)}.$$

The space $H^1_0(\Omega)$ is defined as the closure of $\mathcal{D}(\Omega)$ for the $H^1(\Omega)$ norm, and the space $H^{-1}(\Omega)$ as the dual space of $H^1_0(\Omega)$ for the $\mathcal{D}' \times \mathcal{D}$ duality and we state

$$\|f\|_{H^{-1}(\Omega)} \stackrel{def}{=} \sup_{\substack{\varphi \in H^1_0(\Omega) \\ \|\varphi\|_{H^1(\Omega)} \leq 1}} \langle f, \varphi \rangle.$$

Let us recall the Poincaré inequality. When Ω is a bounded domain and f is in $\mathcal{D}(\Omega)$, writing

$$f(x_1, x') = \int_{-\infty}^{x_1} \partial_{y_1} f(y_1, x')\, dx_1,$$

and using the Cauchy–Schwarz inequality, we get

$$|f(x_1, x')|^2 \leq C_\Omega \int_{-\infty}^{\infty} |\nabla f(y_1, x')|^2\, dy_1.$$

By integration, it follows that

$$\|f\|^2_{L^2(\Omega)} \leq C_\Omega^2 \|\nabla f\|^2_{L^2(\Omega)},$$

and by density this also holds for all functions $f \in H_0^1(\Omega)$. This means that the norm $\|\cdot\|_{H_0^1}$ defined by

$$\|f\|_{H_0^1} \stackrel{\text{def}}{=} \|\nabla f\|_{L^2}$$

is equivalent to the $\|\cdot\|_{H^1}$ norm if the domain is bounded.

As we shall be working with divergence-free vector fields, we have to adapt the above definition to that setting.

Definition 1.2 *Let Ω be a domain of $\mathbf{R}^d$ ($d = 2$ or 3). We shall denote by $\mathcal{V}(\Omega)$ the space of vector fields, the components of which belong to $H_0^1(\Omega)$, and by $\mathcal{V}_\sigma(\Omega)$ the space of divergence-free vector fields in $\mathcal{V}(\Omega)$. The closure of $\mathcal{V}_\sigma(\Omega)$ in $L^2(\Omega)$ will be denoted $\mathcal{H}(\Omega)$. Finally, we shall denote by $\mathcal{V}'(\Omega)$ the space of vector fields with $H^{-1}(\Omega)$ components.*

If E is a subspace of $\mathcal{V}(\Omega)$, the polar space E° is the space of vector fields f in $\mathcal{V}'(\Omega)$ such that for all $v \in E$.

$$\langle f, v\rangle \stackrel{\text{def}}{=} \sum_{j=1}^{d} \langle f_j, v_j\rangle_{H^{-1}\times H_0^1} = 0.$$

If F is a subspace of $\mathcal{V}'(\Omega)$, the polar space $F^\star$ is the space of vector fields v in $\mathcal{V}(\Omega)$ such that for all $f \in F$, $\langle f, v\rangle = 0$.

To make the notation lighter, we shall omit mention of Ω when no confusion is likely.

Remark The space $\mathcal{V}'(\Omega)$ is a Hilbert space, endowed with the following norm

$$\begin{aligned}\|f\|_{\mathcal{V}'(\Omega)} &= \sup_{v\in\mathcal{V}(\Omega)} \langle f, v\rangle \\ &= \|u\|_{H_0^1(\Omega)}.\end{aligned}$$

where u is the unique solution of the Dirichlet problem

$$(\mathrm{Id} - \Delta)u = f.$$

We recall, indeed, that $\mathrm{Id} - \Delta$ is a one-to-one isometry from H^1_0 to H^{-1}.

When the boundary of the domain is regular, namely the boundary is a C^1-hypersurface, it can be checked that the map $f \mapsto f \cdot n$ is well defined on the space of L^2 vector fields with L^2 divergence (with values in the space $H^{-\frac{1}{2}}$ on the boundary). In that case, the space $\mathcal{H}$ is exactly the space of L^2 divergence-free vector fields such that $f \cdot n_{|\partial\Omega} = 0$.

Now we are ready to define the well-known Leray projector.

Definition 1.3 *We denote by* $\mathbf{P}$ *the orthogonal projection of* $(L^2(\Omega))^d$ *on* $\mathcal{H}$.

1.2 The Stokes problem

The Stokes system is defined as follows. Let $f \in \mathcal{V}'$. We shall say that u in $\mathcal{V}_\sigma$ solves the inhomogeneous Stokes problem $Au = f$ if, for all $v \in \mathcal{V}_\sigma$,

$$\langle u - \Delta u, v \rangle = \langle f, v \rangle. \tag{1.2.1}$$

In other words, $u - \Delta u - f \in \mathcal{V}_\sigma^\circ$.

Again assuming that $f \in \mathcal{V}'$, we shall say that u in $\mathcal{V}_\sigma$ solves the homogeneous Stokes problem $\mathcal{A}u = f$ if, for all $v \in \mathcal{V}_\sigma$,

$$\langle -\Delta u, v \rangle = \langle f, v \rangle. \tag{1.2.2}$$

In other words, $-\Delta u - f \in \mathcal{V}_\sigma^\circ$.

When considering the Navier–Stokes equations (NS_ν), the homogeneous version of the Stokes system arises in a very natural way. However, the inhomogeneous Stokes problem allows us to control the L^2 norm of the velocity, which is particularly convenient in the case of unbounded domains. As a matter of fact, the following existence and uniqueness result holds.

Theorem 1.1 *Given* $f \in \mathcal{V}'$, *there exists a unique solution* u *in* $\mathcal{V}_\sigma$ *of the inhomogeneous Stokes problem (1.2.1). When the domain* Ω *is bounded, there is a unique solution* u *in* $\mathcal{V}_\sigma$ *of the homogeneous Stokes problem (1.2.2).*

Proof The proof is nothing but the Lax–Milgram theorem. For the reader's convenience, we recall it. Given f in $\mathcal{V}'$, a linear map $\mathcal{V}_\sigma \to \mathbf{R}$ on the Hilbert space $\mathcal{V}_\sigma$ (a closed subspace of $\mathcal{V}$ endowed with the H^1 scalar product) can be defined as $v \mapsto \langle f, v \rangle$. Thanks to the Riesz theorem, a unique u exists in $\mathcal{V}_\sigma$ such that

$$\forall v \in \mathcal{V}_\sigma\,,\ (u|v)_{H^1} = \langle f, v \rangle.$$

By definition of the H^1 scalar product, we get

$$\forall v \in \mathcal{V}_\sigma\,,\ (u|v)_{L^2} + (\nabla u|\nabla v)_{L^2} = \langle f, v\rangle.$$

As u and v belong to $\mathcal{V}$, we get

$$\forall v \in \mathcal{V}_\sigma\,,\ \langle u - \Delta u, v\rangle = \langle f, v\rangle,$$

and the theorem is proved (the case of the homogeneous Stokes problem in a bounded domain is strictly analogous thanks to Poincaré's inequality). □

Let us point out that relations (1.2.1) and (1.2.2) are equalities on linear forms on $\mathcal{V}_\sigma$. It is therefore natural to introduce the following definition.

Definition 1.4 *We denote by $\mathcal{V}'_\sigma(\Omega)$ the space of continuous linear forms on the space $\mathcal{V}_\sigma(\Omega)$. The norm in $\mathcal{V}'_\sigma(\Omega)$ is given by*

$$\|f\|_{\mathcal{V}_\sigma(\Omega)} \stackrel{def}{=} \sup_{\substack{v\in\mathcal{V}_\sigma\\ \|v\|_{\mathcal{V}}\leq 1}} \langle f, v\rangle.$$

The following proposition will be very useful in the following.

Proposition 1.1 *For any function $f \in \mathcal{V}'(\Omega)$, a sequence of functions $(f_n)_{n\in\mathbf{N}}$ in $\mathcal{H}(\Omega)$ exists such that*

$$\lim_{n\to\infty} \|f_n - f\|_{\mathcal{V}'_\sigma(\Omega)} = 0.$$

Proof By the density of $(L^2)^d$ in $\mathcal{V}'$ there exists a sequence $(g_n)_{n\in\mathbf{N}}$ in $L^2(\Omega)$ (we drop the exponent d to simplify the notation) such that

$$\lim_{n\to\infty} \|g_n - g\|_{\mathcal{V}'(\Omega)} = 0.$$

We have in particular, of course,

$$\lim_{n\to\infty} \|g_n - g\|_{\mathcal{V}'_\sigma(\Omega)} = 0.$$

Then we just need to write

$$\|f - \mathbf{P}g_n\|_{\mathcal{V}'_\sigma(\Omega)} \leq \|g_n - \mathbf{P}g_n\|_{\mathcal{V}'_\sigma(\Omega)} + \|g_n - f\|_{\mathcal{V}'(\Omega)},$$

and to notice that

$$\begin{aligned}\|g_n - \mathbf{P}g_n\|_{\mathcal{V}'_\sigma(\Omega)} &= \sup_{\substack{v\in\mathcal{V}_\sigma \\ \|v\|_{\mathcal{V}}\leq 1}} \langle g_n - \mathbf{P}g_n, v\rangle \\ &= \sup_{\substack{v\in\mathcal{V}_\sigma \\ \|v\|_{\mathcal{V}}\leq 1}} (g_n - \mathbf{P}g_n | v)_{L^2} \\ &= \sup_{\substack{v\in\mathcal{V}_\sigma \\ \|v\|_{\mathcal{V}}\leq 1}} (g_n | v - \mathbf{P}v)_{L^2} \\ &= 0.\end{aligned}$$

The proposition is proved, on choosing $f_n = \mathbf{P}y_n$. ⊓

Remarks

- The fact that a vector field $g \in \mathcal{V}'$ belongs to the polar space $\mathcal{V}_\sigma^\circ$ of $\mathcal{V}_\sigma$ implies in particular that for all $\varphi \in \mathcal{D}$, one has for all $1 \leq i,j \leq d$

$$\langle g, \psi_{ij}\rangle = 0, \quad \text{where } \psi_{ij} = \varepsilon_i \partial_j \varphi - \varepsilon_j \partial_i \varphi \in \mathcal{V}_\sigma,$$

  where $\varepsilon_i \in \mathbf{R}^d$ has zero components except 1 on the ith component, so that

$$\langle g^i, -\partial_j\varphi\rangle + \langle g^j, \partial_i\varphi\rangle = 0,$$

  hence $\partial_j g^i - \partial_i g^j = 0$ in $\mathcal{D}'$, i.e. $\operatorname{curl} g = 0$.
- There exist simple examples of smooth domains Ω and vector fields g such that $\operatorname{curl} g = 0$ but which do not appear as a gradient. For instance, consider the two-dimensional ring

$$\Omega = \{x \in \mathbf{R}^2 \,/\, 0 < R_1 < |x| < R_2\}$$

  and $g = (-\partial_2 \log|x|, \partial_1 \log|x|)$. Then g is clearly divergence-free, and irrotational since $x \mapsto \log|x|$ is harmonic on Ω. Assume that g can be written as a gradient of a function p. Since g is smooth on Ω, so is p. Let now $x_0 \in \Omega$ such that $g(x_0) \neq 0$, and let $t \mapsto \gamma(t)$ be the unique solution of

$$\frac{d\gamma}{dt}(t) = g(\gamma(t)), \quad \gamma(0) = x_0.$$

  Note that the fact that the level lines of $x \mapsto \log|x|$ are circles implies that γ is periodic. Then, we have

$$\frac{d}{dt} p\circ\gamma(t) = \frac{d\gamma}{dt}\cdot\nabla p(\gamma(t)) = |\nabla p(\gamma(t))|^2 \geq 0,$$

  which yields a contradiction to the periodicity of γ, since the derivative at time $t = 0$ does not vanish.
- However, the condition for a vector field $g \in \mathcal{V}'$ to belong to $\mathcal{V}_\sigma^\circ$ is stronger than the assumption that g is irrotational. As a matter of fact,

when g belongs to $\mathcal{V}_\sigma^\circ$, one can prove that a locally L^2 function p exists such that $g = \nabla p$. Moreover, this function p is $L^2_{\rm loc}(\overline{\Omega})$ when Ω has a C^1 boundary. In order to give the flavor of this kind of result, let us state the following proposition, which will be proved in Section 1.4.

Proposition 1.2 *Let Ω be a C^1 bounded domain of $\mathbf{R}^d$ and $g \in \mathcal{V}_\sigma^\circ$. Then there exists $p \in L^2(\Omega)$ such that $g = \nabla p$.*

- In all that follows, we shall denote by ∇p any element of $\mathcal{V}_\sigma^\circ$.

1.3 A brief overview of Sobolev spaces

The aim of this section is to recall the basic features of Sobolev spaces in the whole space $\mathbf{R}^d$ and useful compactness results in the case of bounded domains.

1.3.1 *Definition in the case of the whole space $\mathbf{R}^d$*

In the whole space $\mathbf{R}^d$ ($d = 2$ or 3), Sobolev spaces are defined in terms of integrability properties in frequency space, using the Fourier transform. Let us recall that for all $u \in \mathcal{D}'$, the Fourier transform $\mathcal{F}u$, also denoted by $\widehat{u}$, is defined by

$$\forall \xi \in \mathbf{R}^d, \quad \mathcal{F}u(\xi) = \widehat{u}(\xi) \stackrel{\text{def}}{=} \int_{\mathbf{R}^d} e^{-ix\cdot\xi} u(x)\, dx.$$

The inverse Fourier transform allows us to recover u from $\widehat{u}$:

$$u(x) = \mathcal{F}^{-1}\widehat{u}(x) = (2\pi)^{-d} \int_{\mathbf{R}^d} e^{ix\cdot\xi}\widehat{u}(\xi)\, d\xi.$$

For all $s \in \mathbf{R}$ we introduce the inhomogeneous Sobolev spaces

$$H^s \stackrel{\text{def}}{=} \left\{ u \in \mathcal{S}'/\|u\|^2_{H^s} \stackrel{\text{def}}{=} \int_{\mathbf{R}^d} (1+|\xi|^2)^s |\widehat{u}(\xi)|^2\, d\xi < +\infty \right\}$$

and similarly the homogeneous Sobolev spaces

$$\dot{H}^s \stackrel{\text{def}}{=} \left\{ u \in \mathcal{S}'/\widehat{u} \in L^1_{\rm loc} \quad \text{and} \quad \|u\|^2_{\dot{H}^s} \stackrel{\text{def}}{=} \int_{\mathbf{R}^d} |\xi|^{2s} |\widehat{u}(\xi)|^2\, d\xi < +\infty \right\}.$$

Let us notice that when $s \geq d/2$, $\dot{H}^s$ is not a Hilbert space. Let us define the following sequence. Let $\mathcal{C}$ be a ring included in the unit ball $B(0,1)$ such that $\mathcal{C} \cap 2\mathcal{C} = \emptyset$ and let us define

$$f_n \stackrel{\text{def}}{=} \mathcal{F}^{-1} \sum_{q=1}^{n} \frac{2^{q(s+\frac{d}{2})}}{q} \mathbf{1}_{2^{-q}\mathcal{C}}(\xi).$$

It is left as an exercise to the reader to check that the sequence $(f_n)_{n\in\mathbf{N}}$ is a Cauchy sequence in $\dot{H}^s$ which does not converge in $\dot{H}^s$ if $s \geq d/2$.

1.3.2 Sobolev embeddings

The purpose of this section is to recall Sobolev embeddings, which will be extensively used in this book. In the whole space $\mathbf{R}^d$, the statement is the following.

Theorem 1.2 *If s is positive and smaller than $d/2$ then the space $\dot{H}^s(\mathbf{R}^d)$ is continuously embedded in $L^{\frac{2d}{d-2s}}(\mathbf{R}^d)$.*

Proof First of all, let us show how a scaling argument allows us to find the critical index $p = 2d/(d-2s)$. Let f be a function on $\mathbf{R}^d$, and let us denote by f_ℓ the function $f_\ell(x) \stackrel{\text{def}}{=} f(\ell x)$ $(\ell > 0)$. We have, for all $p \in [1, \infty[$,

$$\|f_\ell\|_{L^p(\mathbf{R}^d)} = \ell^{-\frac{d}{p}} \|f\|_{L^p(\mathbf{R}^d)}$$

and

$$\begin{aligned}\|f_\ell\|^2_{\dot{H}^s(\mathbf{R}^d)} &= \int_{\mathbf{R}^d} |\xi|^{2s} |\widehat{f}_\ell(\xi)|^2 \, d\xi \\ &= \ell^{-2d} \int_{\mathbf{R}^d} |\xi|^{2s} |\widehat{f}(\ell^{-1}\xi)|^2 \, d\xi \\ &= \ell^{-d+2s} \|f\|^2_{\dot{H}^s(\mathbf{R}^d)}.\end{aligned}$$

As soon as an inequality of the type $\|f\|_{L^p(\mathbf{R}^d)} \leq C\|f\|_{\dot{H}^s(\mathbf{R}^d)}$ holds for any smooth function f, it holds also for f_ℓ for any positive ℓ. This leads to the relation $p = 2d/(d-2s)$.

Let us now prove the theorem. In order to simplify the computations, we may assume without loss of generality that $\|f\|_{\dot{H}^s(\mathbf{R}^d)} = 1$.

Let us start by observing that for any $p \in [1, +\infty[$, Fubini's theorem allows us to write for any measurable function f,

$$\begin{aligned}\|f\|^p_{L^p(\mathbf{R}^d)} &\stackrel{\text{def}}{=} \int_{\mathbf{R}^d} |f(x)|^p \, dx \\ &= p \int_{\mathbf{R}^d} \int_0^{|f(x)|} \Lambda^{p-1} \, d\Lambda \, dx \\ &= p \int_0^\infty \Lambda^{p-1} \text{meas}\left(\left\{x \in \mathbf{R}^d \,/\, |f(x)| > \Lambda\right\}\right) d\Lambda.\end{aligned}$$

As quite often in this book, let us decompose the function into low and high frequencies. More precisely, we shall write $f = f_{1,A} + f_{2,A}$, with

$$f_{1,A} = \mathcal{F}^{-1}(\mathbf{1}_{B(0,A)} \widehat{f}) \quad \text{and} \quad f_{2,A} = \mathcal{F}^{-1}(\mathbf{1}_{B^c(0,A)} \widehat{f}), \tag{1.3.1}$$

where $A > 0$ will be determined later. As the support of the Fourier transform of $f_{1,A}$ is compact, the function $f_{1,A}$ is bounded. More precisely we have the following lemma.

Lemma 1.1 *Let s be in $]-\infty, d/2[$ and let K be a compact subset of $\mathbf{R}^d$. If f belongs to $\dot{H}^s$ and if* Supp $\widehat{f}$ *is included in K, then we have*

$$\|f\|_{L^\infty} \leq (2\pi)^{-d} \left(\int_K \frac{d\xi}{|\xi|^{2s}} \right)^{\frac{1}{2}} \|f\|_{\dot{H}^s}.$$

Proof The inverse Fourier formula together with the Cauchy–Schwarz inequality allows us to write

$$\begin{aligned} \|f\|_{L^\infty(\mathbf{R}^d)} &\leq (2\pi)^{-d} \|\widehat{f}\|_{L^1(\mathbf{R}^d)} \\ &\leq (2\pi)^{-d} \int_K |\xi|^{-s} |\xi|^s |\widehat{f}(\xi)| \, d\xi \\ &\leq (2\pi)^{-d} \left(\int_K \frac{d\xi}{|\xi|^{2s}} \right)^{\frac{1}{2}} \|f\|_{\dot{H}^s}. \end{aligned}$$

The lemma is proved. □

Applying this lemma, we get

$$\|f_{1,A}\|_{L^\infty} \leq C_s A^{\frac{d}{2}-s}. \tag{1.3.2}$$

The triangle inequality implies that for any positive A we have

$$\left\{ x \in \mathbf{R}^d \, / |f(x)| > \Lambda \right\} \subset \left\{ x \in \mathbf{R}^d \, / 2|f_{1,A}(x)| > \Lambda \right\} \cup \left\{ x \in \mathbf{R}^d \, / 2|f_{2,A}(x)| > \Lambda \right\}.$$

From the above inequality (1.3.2) we infer that

$$A = A_\Lambda \stackrel{\text{def}}{=} \left(\frac{\Lambda}{4C_s} \right)^{\frac{p}{d}} \Longrightarrow \text{meas} \left(\left\{ x \in \mathbf{R}^d \, / |f_{1,A}(x)| > \Lambda/2 \right\} \right) = 0.$$

From this we deduce that

$$\|f\|^p_{L^p(\mathbf{R}^d)} \leq p \int_0^\infty \Lambda^{p-1} \text{meas} \left(\left\{ x \in \mathbf{R}^d \, / |f_{2,A_\Lambda}(x)| > \Lambda/2 \right\} \right) d\Lambda.$$

It is well known (this is the so-called Bienaimé–Tchebychev inequality) that

$$\begin{aligned} \text{meas} \left(\left\{ x \in \mathbf{R}^d \, / |f_{2,A_\Lambda}(x)| > \Lambda/2 \right\} \right) &= \int_{\left\{ x \in \mathbf{R}^d \, / |f_{2,A_\Lambda}(x)| > \Lambda/2 \right\}} dx \\ &\leq \int_{\left\{ x \in \mathbf{R}^d \, / |f_{2,A_\Lambda}(x)| > \Lambda/2 \right\}} \frac{4|f_{2,A_\Lambda}(x)|^2}{\Lambda^2} \, dx \\ &\leq \frac{4}{\Lambda^2} \|f_{2,A_\Lambda}\|^2_{L^2(\mathbf{R}^d)}. \end{aligned}$$

Thus we infer

$$\|f\|^p_{L^p(\mathbf{R}^d)} \leq 4p \int_0^\infty \Lambda^{p-3} \|f_{2,A_\Lambda}\|^2_{L^2(\mathbf{R}^d)} \, d\Lambda.$$

But we know that the Fourier transform is (up to a constant) a unitary transform of L^2. Thus we have

$$\|f\|^p_{L^p(\mathbf{R}^d)} \leq 4p(2\pi)^{-d}\int_0^\infty \Lambda^{p-3}\int_{(|\xi|\geq A_\Lambda)} |\widehat{f}(\xi)|^2\, d\xi d\Lambda. \tag{1.3.3}$$

Then by the definition of A_Λ we have

$$|\xi| \geq A_\Lambda \Longleftrightarrow \Lambda \leq 4C_s|\xi|^{\frac{d}{p}}.$$

Then, Fubini's theorem implies that

$$\begin{aligned}\|f\|^p_{L^p(\mathbf{R}^d)} &\leq 4p(2\pi)^{-d}\int_{\mathbf{R}^d}\left(\int_0^{4C_s|\xi|^{\frac{d}{p}}} \Lambda^{p-3}\, d\Lambda\right)|\widehat{f}(\xi)|^2\, d\xi\\ &\leq \frac{4p}{p-2}(2\pi)^{-d}\,(4C_s)^{p-2}\int_{\mathbf{R}^d}|\xi|^{\frac{d(p-2)}{p}}|\widehat{f}(\xi)|^2\, d\xi.\end{aligned}$$

As $2s = d(p-2)/p$ the theorem is proved. □

The following corollary will be useful in the future.

Corollary 1.1 *If p belongs to $]1,2]$, then*

$$L^p(\mathbf{R}^d) \subset \dot{H}^s(\mathbf{R}^d) \quad \textit{with} \quad s = -d\left(\frac{1}{p}-\frac{1}{2}\right)\cdot$$

Proof This corollary is proved by duality. Let us write

$$\|a\|_{\dot{H}^s} = \sup_{\|\varphi\|_{\dot{H}^{-s}}\leq 1} \langle a,\varphi\rangle.$$

As $-s = d\left(\frac{1}{p}-\frac{1}{2}\right) = d\left(\frac{1}{2}-\left(1-\frac{1}{p}\right)\right)$, we have that $\|\varphi\|_{\dot{H}^{-s}} \geq C\|\varphi\|_{L^{p'}}$, and thus

$$\begin{aligned}\|a\|_{\dot{H}^s} &\leq C \sup_{\|\varphi\|_{L^{p'}}\leq 1} \langle a,\varphi\rangle\\ &\leq C\|a\|_{L^p}.\end{aligned}$$

This concludes the proof. □

Other useful estimates are Gagliardo–Nirenberg inequalities.

Corollary 1.2 *If $p \in [2,\infty[$ such that $1/p > 1/2 - 1/d$, then a constant C exists such that for any domain Ω in $\mathbf{R}^d$, we have for any $u \in H^1_0(\Omega)$,*

$$\|u\|_{L^p(\Omega)} \leq C\|u\|^{1-\sigma}_{L^2(\Omega)}\|\nabla u\|^\sigma_{L^2(\Omega)} \quad \textit{where} \quad \sigma = \frac{d(p-2)}{2p}\cdot \tag{1.3.4}$$

Proof By density arguments, we may suppose that u is in $\mathcal{D}(\Omega)$. Then, Sobolev embeddings yield

$$\|u\|_{L^p(\Omega)} \leq C\|u\|_{\dot{H}^{\frac{d(p-2)}{2p}}(\mathbf{R}^d)}.$$

The convexity of the Sobolev norms

$$\|u\|_{\dot{H}^\sigma(\mathbf{R}^d)} \leq \|u\|_{L^2(\mathbf{R}^d)}^{1-\sigma}\|u\|_{\dot{H}^1(\mathbf{R}^d)}^{\sigma} \quad \text{for all} \quad \sigma \in [0,1],$$

allows us to conclude. □

The following lemma, known as Bernstein inequalities, will be useful in the future, especially in Part III.

Lemma 1.2 *Let B be a ball of $\mathbf{R}^d$. A constant C exists so that, for any non-negative integer k, any couple of real numbers (p,q) such that $1 \leq p \leq q \leq +\infty$ and any function u of L^p, we have*

$$\operatorname{Supp} \widehat{u} \subset \lambda B \Rightarrow \sup_{|\alpha|=k} \|\partial^\alpha u\|_{L^q} \leq C^{k+1}\lambda^{k+d(\frac{1}{p}-\frac{1}{q})}\|u\|_{L^p}.$$

Proof Using a dilation of size λ, we can assume throughout the proof that $\lambda = 1$. Let ϕ be a function of $\mathcal{D}(\mathbf{R}^d)$ the value of which is 1 near B. As $\widehat{u}(\xi) = \phi(\xi)\widehat{u}(\xi)$, we can write, if g denotes the inverse Fourier transform of ϕ,

$$\partial^\alpha u = \partial^\alpha g \star u$$

where $\star$ denotes the convolution operator

$$g \star u(x) = \int_{\mathbf{R}^d} g(x-y)u(y)\,dy.$$

By Young's inequality, we get

$$\|\partial^\alpha u\|_{L^q} \leq \|\partial^\alpha g\|_{L^r}\|u\|_{L^p} \quad \text{with} \quad \frac{1}{q}+1 = \frac{1}{r}+\frac{1}{p}\cdot$$

By a Hölder estimate

$$\|\partial^\alpha g\|_{L^r} \leq \|\partial^\alpha g\|_{L^\infty}^{1-\frac{1}{r}}\|\partial^\alpha g\|_{L^1}^{\frac{1}{r}}.$$

Then using the general convexity inequality

$$\forall (a,b) \in \mathbf{R}^+ \times \mathbf{R}^+, \forall \theta \in]0,1[, \quad ab \leq \theta a^{\frac{1}{\theta}} + (1-\theta)b^{\frac{1}{1-\theta}}, \tag{1.3.5}$$

we get

$$\begin{aligned}\|\partial^\alpha g\|_{L^r} &\leq \|\partial^\alpha g\|_{L^\infty} + \|\partial^\alpha g\|_{L^1}\\ &\leq 2\|(1+|\cdot|^2)^d\partial^\alpha g\|_{L^\infty}\\ &\leq 2\|(\mathrm{Id}-\Delta)^d((\cdot)^\alpha\phi)\|_{L^1}.\end{aligned}$$

The lemma is proved. □

1.3.3 *A compactness result*

In the next chapter, we shall use the following compactness theorem, known as Rellich's theorem.

Theorem 1.3 *Let Ω be a bounded domain of $\mathbf{R}^d$. The embedding of $H_0^1(\Omega)$ into $L^2(\Omega)$ is compact and so is that of $L^2(\Omega)$ into $H^{-1}(\Omega)$.*

Proof For the reader's convenience, we present here a proof based on the Fourier transform of periodic functions. Without any loss of generality, we may assume that $\overline{\Omega}$ is included in $Q \stackrel{\text{def}}{=}]0, 2\pi[^d$. Let us define, for any function u in $H_0^1(\Omega)$, the $2\pi\,\mathbf{Z}$-periodic function

$$\widetilde{u}(x) = \sum_{j \in \mathbf{Z}^d} u(x - 2\pi j).$$

Let us define its Fourier coefficients for $k \in \mathbf{Z}^d$ by

$$\widetilde{u}_k \stackrel{\text{def}}{=} \int_Q e^{-ik\cdot x} u(x) \frac{dx}{(2\pi)^d} \cdot$$

Using integration by parts and the Fourier–Plancherel theorem, we obtain

$$\sum_{k \in \mathbf{Z}^d} (1 + |k|^2)|\widetilde{u}_k|^2 \leq C\|\nabla u\|_{L^2}^2.$$

Let us define the sequence $(T_N)_{N\in\mathbf{N}}$ of linear maps by

$$T_N \begin{cases} L^2(\Omega) & \to L^2(Q) \\ u & \mapsto \displaystyle\sum_{|k| \leq N} \widetilde{u}_k e^{ik\cdot x} \end{cases}$$

where $L^2(Q)$ is identified with $2\pi\,\mathbf{Z}^d$-periodic L^2 functions. Obviously, the range of T_N is a finite-dimensional vector space. Moreover, the following property holds

$$\begin{aligned} \|u - (T_N u)_{|\Omega}\|_{L^2(\Omega)}^2 &\leq \|\widetilde{u} - T_N u\|_{L^2(Q)}^2 \\ &\leq \left\| \sum_{|k|>N} \widetilde{u}_k e^{ik\cdot x} \right\|_{L^2(Q)}^2 \\ &\leq \frac{C^2}{(N+1)^2} \|\nabla u\|_{L^2(\Omega)}^2. \end{aligned} \tag{1.3.6}$$

Thus we end up with the inequality

$$\|\operatorname{Id} - T_{N|\Omega}\|_{\mathcal{L}(H_0^1(\Omega); L^2(\Omega))} \leq \frac{C}{N+1} \cdot$$

The operator Id appears as a limit in the norm of $\mathcal{L}(H_0^1(\Omega); L^2(\Omega))$ of operators of finite rank. This proves the compactness of the identity as a map from $H_0^1(\Omega)$ into $L^2(\Omega)$.

Now let us prove the compactness of the identity as a map from $L^2(\Omega)$ into $H^{-1}(\Omega)$. For any u in $L^2(\Omega)$, we have

$$\begin{aligned}
\|u - (T_N u)_{|\Omega}\|_{H^{-1}(\Omega)} &= \sup_{\substack{\varphi \in H_0^1(\Omega) \\ \|\nabla \varphi\|_{L^2}=1}} \langle u - (T_N u)_{|\Omega}, \varphi \rangle \\
&= \sup_{\substack{\varphi \in H_0^1(\Omega) \\ \|\nabla \varphi\|_{L^2}=1}} \int_\Omega (u - T_N u)(x)\varphi(x)\, dx \\
&\leq \sup_{\substack{\varphi \in H^1(Q) \\ \|\varphi\|_{H^1}=1}} \int_Q (u - T_N u)(x)\varphi(x)\, dx.
\end{aligned}$$

The Fourier–Plancherel theorem implies, by definition of T_N, that

$$\begin{aligned}
\|u - (T_N u)_{|\Omega}\|_{H^{-1}(\Omega)} &\leq (2\pi)^{-d} \sup_{\substack{\varphi \in H^1(Q) \\ \|\varphi\|_{H^1}=1}} \sum_{|k|>N} \widetilde{u}_k \widetilde{\varphi}_k \\
&\leq N^{-1}(2\pi)^{-d} \sup_{\substack{\varphi \in H^1(Q) \\ \|\varphi\|_{H^1}=1}} \sum_{|k|>N} \widetilde{u}_k (1+|k|^2)^{\frac{1}{2}} \widetilde{\varphi}_k.
\end{aligned}$$

The Cauchy–Schwarz inequality implies that

$$\|u - (T_N u)_{|\Omega}\|_{H^{-1}(\Omega)} \leq N^{-1} \|u\|_{L^2}.$$

The theorem is proved. □

1.4 Proof of the regularity of the pressure

The aim of this section is the proof of Proposition 1.2. Let Ω be a C^1 bounded domain of $\mathbf{R}^d$ and $\nabla L^2(\Omega)$ be the set of vector fields $f \in \mathcal{V}'(\Omega)$ be such that there exists a function $p \in L^2(\Omega)$ satisfying $f = \nabla p$. Let $v \in \mathcal{V}(\Omega)$ such that v belongs to $(\nabla L^2(\Omega))^\star$. Then, for all functions $p \in L^2(\Omega)$,

$$\begin{aligned}
\langle \operatorname{div} v, p \rangle &= -\langle \nabla p, v \rangle \\
&= -\langle f, v \rangle \\
&= 0.
\end{aligned}$$

It follows that $\operatorname{div} v = 0$, which means that $(\nabla L^2(\Omega))^\star \subset \mathcal{V}_\sigma$. Thus $\mathcal{V}_\sigma^\circ$ is a subset of $((\nabla L^2)^\star)^\circ$. Thanks to the Hahn–Banach theorem, one has

$$\mathcal{V}_\sigma^\circ \subset ((\nabla L^2)^\star)^\circ = \overline{\nabla L^2}^{H^{-1}}.$$

In other words, if $f \in \mathcal{V}_\sigma^\circ$, there exists a sequence $(p_n)_{n \in \mathbf{N}}$ of $L^2(\Omega)$ functions such that ∇p_n converges to f in $\mathcal{V}'$.

In order to prove Proposition 1.2, it remains to prove that the range of

$$\nabla \quad \begin{cases} L^2 \to \mathcal{V}' \\ p \;\; \mapsto \nabla p \end{cases}$$

is closed. The regularity of the boundary $\partial\Omega$ turns out to be essential at this point of the proof.

Let us first state the following lemma.

Lemma 1.3 *Let Ω be a C^1 bounded domain of $\mathbf{R}^d$ and $p \in H^{-1}(\Omega)$. Then p belongs to $L^2(\Omega)$ if and only if ∇p belongs to $\mathcal{V}'(\Omega)$. Moreover, there exists a positive constant C such that*

$$\|p\|_{L^2}^2 \leq C\left(\|p\|_{H^{-1}}^2 + \|\nabla p\|_{\mathcal{V}'}^2\right). \tag{1.4.1}$$

Postponing the proof of Lemma 1.3, we deduce the following corollary, which implies that $\nabla L^2(\Omega)$ is closed and this ends the proof of Proposition 1.2.

Corollary 1.3 *Under the assumptions of Lemma 1.3, there exists a positive constant C such that for all functions p in*

$$L_0^2(\Omega) \stackrel{def}{=} \left\{p \in L^2(\Omega) \,/\, \int_\Omega p(x)\,dx = 0\right\},$$

one has

$$\|p\|_{L^2(\Omega)} \leq C\|\nabla p\|_{\mathcal{V}'(\Omega)}.$$

Proof Let us assume that there exists a sequence $(p_n)_{n\in\mathbf{N}}$ in $L_0^2(\Omega)$ such that $\|p_n\|_{L^2(\Omega)} = 1$ and $\|\nabla p_n\|_{\mathcal{V}'(\Omega)}$ converges to 0.

The embedding of $L^2(\Omega)$ into $H^{-1}(\Omega)$ is compact due to Theorem 1.3. Thus, up to an extraction of a subsequence, we may assume that a function p in $L^2(\Omega)$ exists such that

$$p_n \to p \quad \text{weakly in } L^2(\Omega) \quad \text{and} \quad \lim_{n\to\infty} \|p_n - p\|_{H^{-1}(\Omega)} = 0.$$

As the domain Ω is bounded, the constant functions belong to $L^2(\Omega)$. Thus the function p is in $L_0^2(\Omega)$. Moreover, the assertion above implies in particular that $(p_n)_{n\in\mathbf{N}}$ converges to p in the distribution sense, and so does $(\nabla p_n)_{n\in\mathbf{N}}$ to ∇p. As $\|\nabla p_n\|_{\mathcal{V}'(\Omega)}$ converges to 0, this implies that $p = 0$, hence $\|p_n\|_{H^{-1}(\Omega)}$ converges to 0. Passing to the limit in the inequality of Lemma 1.3 gives a contradiction. □

Proof of Lemma 1.3 Since Ω is a C^1 bounded domain of $\mathbf{R}^d$, there exists a finite family $(U_j)_{1\leq j\leq N}$ of open subsets of $\mathbf{R}^d$ such that

$$\partial\Omega \subset \bigcup_{1\leq j\leq N} U_j \tag{1.4.2}$$

and, for all j, there exists a C^1 diffeomorphism χ_j from U_j to W_j such that

$$\chi_j(U_j \cap \overline{\Omega}) = W_j \cap \mathbf{R}^d_+ \quad \text{with} \quad \mathbf{R}^d_+ = \Big\{x = (x', x_d) \in \mathbf{R}^{d-1} \times \mathbf{R}^+\Big\}. \tag{1.4.3}$$

Finally, there exists an open subset U_0 of $\mathbf{R}^d$ such that $\overline{U}_0 \subset \Omega$ and

$$\overline{\Omega} \subset \bigcup_{0 \le j \le N} U_j. \tag{1.4.4}$$

Now let us consider a partition of unity $(\varphi_j)_{0 \le j \le N}$ associated with the family $(U_j)_{0 \le j \le N}$ and $p \in H^{-1}(\Omega)$. Then, we write

$$p = \sum_{j=0}^{N} p_j \quad \text{with} \quad p_j = \varphi_j p.$$

By definition of $\|\cdot\|_{H^{-1}(\Omega)}$, we have, if $\mathcal{B}$ denotes the unit ball of $H^1_0(\Omega)$,

$$\begin{aligned}
\|ap\|_{H^{-1}(\Omega)} &= \sup_{u \in \mathcal{B}} \langle ap, u\rangle \\
&= \sup_{u \in \mathcal{B}} \langle p, au\rangle \\
&\le \|p\|_{H^{-1}(\Omega)} \sup_{u \in \mathcal{B}} \|au\|_{H^1_0(\Omega)}.
\end{aligned}$$

The Leibnitz formula (and Poincaré's inequality) implies that

$$\|\nabla(au)\|_{L^2} \le C_a \|u\|_{H^1_0(\Omega)}.$$

This gives that $\|ap\|_{H^{-1}(\Omega)} \le C_a \|p\|_{H^{-1}(\Omega)}$ and in particular that

$$\|p_j\|_{H^{-1}(\Omega)} \le C\|p\|_{H^{-1}(\Omega)}.$$

Moreover, if ∇p belongs to $\mathcal{V}'(\Omega)$, then using the fact that $\nabla p_j = p\nabla\varphi_j + \varphi_j \nabla p$, we deduce that, for all $j \in \{0, \ldots, N\}$,

$$\|\nabla p_j\|_{\mathcal{V}'} \le C\|p\|_{H^{-1}(\Omega)} + C\|\nabla p\|_{\mathcal{V}'}.$$

Then it suffices to prove that estimate (1.4.1) holds for each p_j.

Since p_0 is compactly supported in Ω, estimate (1.4.1) is obtained easily for p_0. As a matter of fact, p_0 and its derivatives belong to $H^{-1}(\mathbf{R}^d)$ and the Fourier transform can be used: let us write that, for any function p in $L^2(\mathbf{R}^d)$,

$$\begin{aligned}
|\widehat{p}(\xi)| &\le |\mathbf{1}_{B(0,1)}\widehat{p}(\xi)| + |\mathbf{1}_{B^c(0,1)}\widehat{p}(\xi)| \\
&\le (1+|\xi|^2)^{\frac{1}{2}} \left|\mathbf{1}_{B(0,1)}(1+|\xi|^2)^{-\frac{1}{2}}|\widehat{p}(\xi)|\right| \\
&\quad + \left|\sum_{k=1}^{d} (1+|\xi|^2)^{\frac{1}{2}} \mathbf{1}_{B^c(0,1)} \frac{i\xi_k}{|\xi|^2}(1+|\xi|^2)^{-\frac{1}{2}} \mathcal{F}(\partial_k p)(\xi)\right| \\
&\le \sqrt{2}(1+|\xi|^2)^{-\frac{1}{2}}|\widehat{p}(\xi)| + C\sum_{k=1}^{d}(1+|\xi|^2)^{-\frac{1}{2}}|\mathcal{F}(\partial_k p)(\xi)|.
\end{aligned}$$

The Fourier–Plancherel theorem implies that

$$\|p\|_{L^2} \leq C\|p\|_{H^{-1}(\mathbf{R}^2)} + \|\nabla p\|_{H^{-1}(\mathbf{R}^2)}, \tag{1.4.5}$$

which proves the result for p_0. For each $j \geq 1$, χ_j being a C^1 diffeomorphism, one has

$$\chi_j^{-1} \begin{cases} H_0^1(W_j \cap \mathbf{R}_+^d) \to H_0^1(U_j \cap \Omega) \\ v \qquad\qquad\quad\; \mapsto v \circ \chi_j^{-1}. \end{cases}$$

A change of variables allows us to deduce that

$$\chi_j^* \begin{cases} H^{-1}(U_j \cap \Omega) \to H^{-1}(W_j \cap \mathbf{R}_+^d) \\ f \qquad\qquad\quad\;\; \mapsto \left(v \mapsto \langle f, J\chi_j \circ \chi_j^{-1} \times v \circ \chi_j^{-1}\rangle\right). \end{cases}$$

As a result, the proof reduces to the case when $\Omega = \mathbf{R}_+^d$. Let us introduce the mappings Q and R from $H^1(\mathbf{R}^d)$ to $H_0^1(\mathbf{R}_+^d)$ given for $u \in H^1(\mathbf{R}^d)$ by

$$Qu(x) = \begin{cases} 0 \quad \text{if} \quad x_d < 0, \\ u(x) + 3u(x', -x_d) - 4u(x', -2x_d) \quad \text{otherwise,} \end{cases}$$

and

$$Ru(x) = \begin{cases} 0 \quad \text{if} \quad x_d < 0, \\ u(x) - 3u(x', -x_d) + 2u(x', -2x_d) \quad \text{otherwise.} \end{cases}$$

It can easily be checked that Q and R map continuously $H^1(\mathbf{R}^d)$ to $H_0^1(\mathbf{R}_+^d)$ and $L^2(\mathbf{R}^d)$ to $L^2(\mathbf{R}_+^d)$. Moreover, one has

$$\frac{\partial}{\partial x_j} \circ Q = Q \circ \frac{\partial}{\partial x_j} \quad \text{if} \quad j \neq d \quad \text{and} \quad \frac{\partial}{\partial x_d} \circ R = Q \circ \frac{\partial}{\partial x_d} \cdot \tag{1.4.6}$$

Next, we consider the transposed maps on H^{-1}, namely

$${}^tQ \quad \begin{cases} H^{-1}(\mathbf{R}_+^d) \to H^{-1}(\mathbf{R}^d) \\ f \qquad\qquad\; \mapsto {}^tQf / \langle {}^tQf, v\rangle = \langle f, Qv\rangle \end{cases}$$

and similarly for R. Taking the transpose of (1.4.6) leads to

$$\frac{\partial}{\partial x_j} \circ {}^tQ = {}^tQ \circ \frac{\partial}{\partial x_j} \quad \text{if} \quad j \neq d \quad \text{and} \quad \frac{\partial}{\partial x_d} \circ {}^tQ = {}^tR \circ \frac{\partial}{\partial x_d} \cdot \tag{1.4.7}$$

Applying (1.4.7) to $\widetilde{p}_j = \chi_j^* p_j \in H^{-1}(\mathbf{R}_+^d)$ yields

$${}^tQ\widetilde{p}_j \in H^{-1}(\mathbf{R}^d) \quad \text{and} \quad \frac{\partial}{\partial x_j}({}^tQ\widetilde{p}_j) \in H^{-1}(\mathbf{R}^d).$$

Using (1.4.5), we deduce that ${}^tQ\widetilde{p}_j \in L^2(\mathbf{R}^d)$ and

$$\|{}^tQ\widetilde{p}_j\|_{L^2(\mathbf{R}^d)}^2 \leq C\left(\|{}^tQ\widetilde{p}_j\|_{H^{-1}(\mathbf{R}^d)}^2 + \|\nabla {}^tQ\widetilde{p}_j\|_{\mathcal{V}'(\mathbf{R}^d)}^2\right).$$

Since the restriction of Q on compactly supported functions in $\mathbf{R}^d_+$ is the identity map, the restriction of ${}^tQ\widetilde{p}_j$ to $\mathbf{R}^d_+$ is $\widetilde{p}_j$. Therefore, we have

$$\|\widetilde{p}_j\|^2_{L^2(\mathbf{R}^d_+)} \leq C\left(\|\widetilde{p}_j\|^2_{H^{-1}(\mathbf{R}^d_+)} + \|\nabla\widetilde{p}_j\|^2_{\mathcal{V}'(\mathbf{R}^d_+)}\right),$$

so that p_j belongs to $L^2(\Omega)$ and

$$\|p_j\|^2_{L^2(\Omega)} \leq C\left(\|p\|^2_{H^{-1}(\mathbf{R}^d_+)} + \|\nabla p\|^2_{\mathcal{V}'(\mathbf{R}^d_+)}\right),$$

which proves Lemma 1.3, hence Proposition 1.2. □

Remark For a general open subset Ω, it can be proved that there exists p in $L^2_{\text{loc}}(\Omega)$ such that if $f \in \mathcal{V}^\circ_\sigma(\Omega)$, then $f = \nabla p$.

2

Weak solutions of the Navier–Stokes equations

The mathematical analysis of the incompressible Stokes and Navier–Stokes equations in a possibly unbounded domain Ω of $\mathbf{R}^d$ ($d = 2$ or 3) is the purpose of this chapter. Notice that no regularity assumptions will be required on the domain Ω.

2.1 Spectral properties of the Stokes operator

Because of the compactness result stated in Theorem 1.3, page 27, the case of bounded domains will be different (in fact slightly simpler) than the case of general domains.

2.1.1 The case of bounded domains

The study of the spectral properties of the Stokes operator previously defined relies on the study of its inverse, which is in fact much easier. We shall restrict ourselves here to the case of the homogeneous Stokes operator which is adapted to the case of a bounded domain.

Definition 2.1 *Let us denote by $\mathcal{B}$ the following operator*

$$\mathcal{B}\begin{cases}\mathcal{H} \to \mathcal{V}_\sigma \subset \mathcal{H}\\ f \mapsto u \,/ - \Delta u - f \in \mathcal{V}_\sigma^\circ.\end{cases}$$

This operator has the following properties.

Proposition 2.1 *The operator $\mathcal{B}$ is continuous, self-adjoint, positive, one-to-one and thus the range of $\mathcal{B}$ is dense in $\mathcal{H}$.*

Proof Let us show that $\mathcal{B}$ is symmetric: for all vector fields f and g in $\mathcal{H}$, denoting $u = \mathcal{B}f$ and $v = \mathcal{B}g$, one has

$$(\mathcal{B}f|g)_{L^2} = \langle -\Delta v, u\rangle = (\nabla u|\nabla v)_{L^2} = \langle -\Delta u, v\rangle = (f|\mathcal{B}g)_{L^2}.$$

The fact that $\mathcal{B}$ is bounded is due to the following computation. Let f be a vector field in $\mathcal{H}(\Omega)$ and denote $u = \mathcal{B}f$. Then, using the Poincaré inequality, we have

$$(\mathcal{B}f|f)_{L^2} = \langle -\Delta u, u\rangle \geq c\|u\|_{L^2}^2 = c\|\mathcal{B}f\|_{L^2}^2.$$

This proves that $\mathcal{B}$ is continuous and positive. Finally to show that $\mathcal{B}$ is one-to-one, we observe that the kernel of $\mathcal{B}$ is equal to $\mathcal{V}_\sigma^\circ \cap \mathcal{H}$, which is exactly $\mathcal{V}_\sigma^\perp$. By definition of $\mathcal{H}$ as the closure of $\mathcal{V}_\sigma$ in L^2, this space reduces to $\{0\}$ and the proposition follows. As the closure in $\mathcal{H}$ of the range of $\mathcal{B}$ (denoted by $R(\mathcal{B})$) is equal to the orthogonal space of $\ker \mathcal{B}$, we infer that $R(\mathcal{B})$ is dense in $\mathcal{H}$. □

The basic theorem is the following.

Theorem 2.1 *Let Ω be a bounded domain of $\mathbf{R}^d$. A Hilbertian basis $(e_k)_{k\in\mathbf{N}}$ of $\mathcal{H}$ and a non-decreasing sequence of positive eigenvalues converging to infinity $(\mu_k^2)_{k\in\mathbf{N}}$ exist such that*

$$-\Delta e_k + \nabla \pi_k = \mu_k^2 e_k, \tag{2.1.1}$$

where $(\pi_k)_{k\in\mathbf{N}}$ is a sequence of $L^2_{\text{loc}}(\Omega)$ functions, the gradient of which belongs to $\mathcal{V}_\sigma^\circ$. Moreover, $(\mu_k^{-1} e_k)_{k\in\mathbf{N}}$ is an orthonormal basis of $\mathcal{V}_\sigma$ endowed with the H_0^1 scalar product.

Moreover for any $f \in \mathcal{V}'$ we have

$$\|f\|^2_{\mathcal{V}'_\sigma} = \sum_{j\in\mathbf{N}} \mu_j^{-2} \langle f, e_j\rangle^2.$$

Remark As claimed by Proposition 1.2, the functions π_k are in $L^2(\Omega)$ when the boundary of Ω is C^1.

Proof of Theorem 2.1 Let us consider the operator $\mathcal{B}$ defined in Definition 2.1. Proposition 2.1 claims that the operator $\mathcal{B}$ is positive, self-adjoint and one-to-one. As the range of $\mathcal{B}$ is included in $\mathcal{V}_\sigma$ which is included in H_0^1, the operator is compact as inferred by Theorem 1.3. Applying the spectral theorem for self-adjoint compact operators in a Hilbert space, we get the existence of a Hilbertian basis $(e_k)_{k\in\mathbf{N}}$ of $\mathcal{H}$ and a non-decreasing sequence of positive eigenvalues converging to infinity $(\mu_k^2)_{k\in\mathbf{N}}$ such that $\mathcal{B}e_k = \mu_k^{-2} e_k$. By definition of $\mathcal{B}$, this implies that there exists $\widetilde{\pi}_k \in L^2_{\text{loc}}(\Omega)$ such that

$$\begin{aligned} e_k &= -\Delta \mathcal{B} e_k + \nabla \widetilde{\pi}_k \\ &= -\mu_k^{-2} \Delta e_k + \nabla \widetilde{\pi}_k. \end{aligned}$$

Moreover, denoting $\pi_k = \mu_k^2 \widetilde{\pi}_k$, it is obvious that $(e_k)_{k\in\mathbf{N}}$ satisfies (2.1.1) and that

$$\begin{aligned} (e_k | e_{k'})_{\mathcal{V}} &= \langle -\Delta e_k, e_{k'} \rangle \\ &= \langle -\nabla \pi_k + \mu_k^2 e_k, e_{k'} \rangle \\ &= \mu_k^2 (e_k | e_{k'})_{\mathcal{H}} \\ &= \mu_k^2 \delta_{k,k'}. \end{aligned}$$

This proves $(\mu_k^{-1} e_k)_{k\in\mathbf{N}}$ is an orthonormal family of $\mathcal{V}_\sigma$ endowed with the H_0^1 scalar product. Let us prove that the vector space generated by $(e_k)_{k\in\mathbf{N}}$ is dense in $\mathcal{V}_\sigma$. Let us consider any u in $\mathcal{V}_\sigma$ such that

$$\forall k \in \mathbf{N}, \quad (u|e_k)_{H_0^1} = 0.$$

We have, by definition of the H_0^1 scalar product,

$$(u|e_k)_{H_0^1} = (\nabla u|\nabla e_k)_{L^2} = \langle -\Delta e_k, u\rangle.$$

Using (2.1.1), we infer that

$$\begin{aligned}(u|e_k)_{H_0^1} &= \langle \mu_k^2 e_k + \nabla \pi_k, u\rangle\\ &= \mu_k^2 \langle e_k, u\rangle\\ &= \mu_k^2 (e_k|u)_{L^2}.\end{aligned}$$

As the vector space generated in $\mathcal{H}$ by the sequence $(e_k)_{k\in\mathbf{N}}$ is dense in $\mathcal{H}$, we deduce that $u = 0$ and thus that $(e_k)_{k\in\mathbf{N}}$ is an orthonormal basis of $\mathcal{V}_\sigma$.

Now let us consider a vector field $f \in \mathcal{V}'$ and let us compute its norm in $\mathcal{V}'_\sigma$. By definition of $\|f\|_{\mathcal{V}'_\sigma}$ we have

$$\begin{aligned}\|f\|^2_{\mathcal{V}'_\sigma} &= \sup_{\substack{v\in\mathcal{V}_\sigma\\ \|v\|\leq 1}} \langle f, v\rangle^2\\ &= \sup_{\substack{\|(\alpha_j)\|_{\ell^2}\leq 1\\ k}} \left\langle f, \sum_{j\leq k} \alpha_j \mu_j^{-1} e_j \right\rangle^2.\end{aligned}$$

Using the characterization of $\ell^2(\mathbf{N})$, we infer that

$$\begin{aligned}\|f\|^2_{\mathcal{V}'_\sigma} &= \sup_{\|(\alpha_j)\|_{\ell^2}\leq 1} \sup_k \left(\sum_{j\leq k} \alpha_j \mu_j^{-1} \langle f, e_j\rangle \right)^2\\ &= \left\| (\mu_j^{-1}\langle f, e_j\rangle)_j \right\|^2_{\ell^2}.\end{aligned}$$

This concludes the proof of Theorem 2.1. □

Definition 2.2 *Let us define*

$$P_k \left\{ \begin{array}{l} \mathcal{V}' \to \mathcal{V}_\sigma \\ f \mapsto \displaystyle\sum_{j\leq k} \langle f, e_j\rangle\, e_j. \end{array}\right.$$

Remark

- The operators P_k are the spectral projectors of the Stokes problem (see the forthcoming Theorem 2.2 on page 38 for the general definition of spectral projectors).
- As $(e_k)_{k\in\mathbf{N}}$ is a Hilbert basis on $\mathcal{H}$, we have, for any vector field in $L^2(\Omega)^d$,

$$\mathbf{P}v = \sum_j (v|e_j)e_j = \sum_j \langle v, e_j\rangle e_j$$

where $\mathbf{P}$ is the Leray projector on L^2 divergence-free vector fields defined in Definition 1.3.

Corollary 2.1 *The operators P_k satisfy*

$$\|P_k f\|_{\mathcal{V}'_\sigma} \leq \|f\|_{\mathcal{V}'_\sigma} \quad \textit{and} \quad \forall f \in \mathcal{V}', \lim_{k\to\infty} \|P_k f - f\|_{\mathcal{V}'_\sigma} = 0.$$

Proof The first point of the corollary is clear. As to the second one, according to Theorem 2.1 we have

$$\|P_k f - f\|^2_{\mathcal{V}'_\sigma} = \sum_{j\geq k} \mu_j^{-2} \langle f, e_j\rangle^2.$$

This goes to zero as k goes to infinity, as the remainder of a convergent series. □

2.1.2 The general case

Let us first define the inverse of the inhomogeneous Stokes operator.

Definition 2.3 *Let us denote by B the following operator*

$$B \begin{cases} \mathcal{H} \to \mathcal{V}_\sigma \subset \mathcal{H} \\ f \mapsto u \,/ u - \Delta u - f \in \mathcal{V}^\circ_\sigma. \end{cases}$$

This operator has the following properties.

Proposition 2.2 *The operator B is continuous, self-adjoint and one-to-one. Moreover it satisfies $\|B\|_{\mathcal{L}(\mathcal{H})} \leq 1$.*

Proof The fact that B is bounded is due to the following computation. Let f be a vector field in $\mathcal{H}(\Omega)$ and denote $u = Bf$. Then

$$(Bf|f)_{L^2} = \langle u|u - \Delta u\rangle \geq \|u\|^2_{L^2} = \|Bf\|^2_{L^2}.$$

This proves that B is continuous and that $\|B\|_{\mathcal{L}(\mathcal{H})} \leq 1$.

Now let us show that B is symmetric. For all vector fields f and g in $\mathcal{H}(\Omega)$, denoting $u = Bf$ and $v = Bg$, one has

$$(Bf|g)_{L^2} = \langle u, v - \Delta v\rangle = (u|v)_{L^2} + (\nabla u|\nabla v)_{L^2} = \langle u - \Delta u, v\rangle = (f|Bg)_{L^2}.$$

Finally to show that B is one-to-one, we observe that the kernel of B is equal to $\mathcal{V}_\sigma^\circ \cap \mathcal{H}$, which is exactly $\mathcal{V}_\sigma^\perp$. By the definition of $\mathcal{H}$ as the closure of $\mathcal{V}_\sigma$ in L^2, this space reduces to $\{0\}$ and the proposition follows. □

As the closure in $\mathcal{H}$ of the range of B (denoted by $R(B)$) is equal to the orthogonal space of $\ker B$, we infer from Proposition 2.2 that $R(B)$ is dense in $\mathcal{H}$. This allows us to define the inverse of B as an unbounded operator A with a dense domain of definition.

Definition 2.4 *Let us denote by A the following operator*

$$A \begin{cases} R(B) \to \mathcal{H} \\ \quad u \mapsto f \;/\; Bf = u. \end{cases}$$

This is exactly the operator A defined in (1.2.1) on page 19.

Lemma 2.1 *The operator A is self-adjoint with domain $R(B)$.*

This lemma is a classical result of operator theory. For the reader's convenience, we give a proof of it.

Proof of Lemma 2.1 Of course A is symmetric, so the only point to check is that the domain $\mathcal{D}(A^*)$ of A^* is $R(B)$. By definition,

$$\mathcal{D}(A^*) = \{v \in \mathcal{H} \mid \exists C > 0,\ \forall u \in R(B),\ (Au|v)_\mathcal{H} \leq C\|u\|_\mathcal{H}\}\,.$$

Since A is symmetric, we have of course $R(B) \subset \mathcal{D}(A^*)$. Now let us prove that $\mathcal{D}(A^*) \subset R(B)$. Let us define the graph norm

$$\|v\|_{A^*}^2 \stackrel{\text{def}}{=} \|v\|_\mathcal{H}^2 + \|A^*v\|_\mathcal{H}^2.$$

The fact that $(\mathcal{D}(A^*), \|\cdot\|_{A^*})$ is a Hilbert space is left as an exercise to the reader. The equality $\mathcal{D}(A^*) = R(B)$ will result from the fact that $R(B)$ is closed and dense for the $\|\cdot\|_{A^*}$ norm. For any $f \in \mathcal{H}$, we have $A^*Bf = ABf = f$ hence

$$\|Bf\|_{A^*}^2 = \|Bf\|_\mathcal{H}^2 + \|f\|_\mathcal{H}^2$$

which immediately gives that $\|f\|_\mathcal{H}^2 \leq \|Bf\|_{A^*}^2$, hence that the space $R(B)$ is closed in $(\mathcal{D}(A^*), \|\cdot\|_{A^*})$. Now let v be in the orthogonal space of $R(B)$ in the sense of the $(\cdot\,|\,\cdot)_{A^*}$ scalar product. By definition, we have, for any f in $\mathcal{H}$,

$$(Bf|v)_\mathcal{H} + (A^*Bf|A^*v)_\mathcal{H} = 0.$$

As B is self-adjoint and $A^*B = \mathrm{Id}$, we get for all f in $\mathcal{H}$,

$$(f|Bv)_\mathcal{H} + (f|A^*v)_\mathcal{H} = 0.$$

This implies that $Bv + A^*v = 0$. In particular A^*v belongs to $R(B) = \mathcal{D}(A)$ and by application of the operator A we get $v + AA^*v = 0$. By definition of A^*, we have

$$(AA^*v|v)_{\mathcal{H}} = \|A^*v\|_{\mathcal{H}}^2,$$

thus $\|v\|_{A^*} = 0$ and the lemma is proved. □

As the Stokes operator $\mathcal{A} = A - \mathrm{Id}$ is also self-adjoint, we are now ready to apply the spectral theorem to the operator $\mathcal{A}$ (see, for instance, [108], Chapter VII for a proof).

Theorem 2.2 *There exists a family of orthogonal projections on $\mathcal{H}$, denoted by $(\mathbf{P}_\lambda)_{\lambda\in\mathbf{R}}$, which commutes with $\mathcal{A}$ and satisfies the following properties.*
The family $(\mathbf{P}_\lambda)_{\lambda\in\mathbf{R}}$ is increasing, in the following sense:

$$\mathbf{P}_\lambda \mathbf{P}_{\lambda'} = \mathbf{P}_{\inf(\lambda,\lambda')}, \qquad \textit{for any } (\lambda, \lambda') \in \mathbf{R}^2. \tag{2.1.2}$$

For $\lambda < 0$, $\mathbf{P}_\lambda = 0$ and for any $u \in \mathcal{H}$,

$$\lim_{\lambda\to\infty} \|\mathbf{P}_\lambda u - u\|_{\mathcal{H}} = 0. \tag{2.1.3}$$

The family $(\mathbf{P}_\lambda)_{\lambda\in\mathbf{R}}$ is continuous on the right, which means that

$$\forall u \in \mathcal{H}, \quad \lim_{\lambda'\to\lambda,\, \lambda'>\lambda} \|\mathbf{P}_{\lambda'} u - \mathbf{P}_\lambda u\|_{\mathcal{H}} = 0. \tag{2.1.4}$$

For any $u \in \mathcal{H}$, the function $\lambda \mapsto (\mathbf{P}_\lambda u|u)_{\mathcal{H}} = \|\mathbf{P}_\lambda u\|_{\mathcal{H}}^2$ is increasing, and

$$\|u\|_{L^2}^2 = \int_{\mathbf{R}} d(\mathbf{P}_\lambda u|u) \quad \textit{and} \quad \|\nabla u\|_{L^2}^2 = (\mathcal{A}u|u)_{\mathcal{H}} = \int_{\mathbf{R}} \lambda\, d(\mathbf{P}_\lambda u|u). \tag{2.1.5}$$

Let us state a corollary which shows that the operators $\mathbf{P}_\lambda$ should be understood as smoothing operators; they are an extension of frequency cut-off operators.

Corollary 2.2 *For any $u \in \mathcal{H}$, the vector field $\mathbf{P}_\lambda u$ is an element of $\mathcal{V}_\sigma$ and*

$$\|\nabla \mathbf{P}_\lambda u\|_{L^2} \le \lambda^{\frac{1}{2}} \|u\|_{L^2}.$$

For any $u \in \mathcal{V}_\sigma$, we have

$$\|(\mathrm{Id} - \mathbf{P}_\lambda)u\|_{L^2} \le \frac{1}{\lambda^{\frac{1}{2}}} \|u\|_{\mathcal{V}_\sigma}.$$

Proof Using (2.1.5) we get

$$\|\nabla \mathbf{P}_\lambda u\|_{L^2}^2 = \int_{\mathbf{R}} \lambda' d(\mathbf{P}_\lambda \mathbf{P}_{\lambda'} u|u).$$

Using (2.1.2) this means that

$$\begin{aligned}\|\nabla \mathbf{P}_\lambda u\|_{L^2}^2 &= \int_0^\lambda \lambda' d(\mathbf{P}_{\lambda'} u|u)\\ &\leq \lambda \int_{\mathbf{R}} d(\mathbf{P}_{\lambda'} u|u)\\ &\leq \lambda \|u\|_{L^2}^2,\end{aligned}$$

which proves the first part of the corollary. For the second part we notice that if u is in $\mathcal{V}_\sigma$ then

$$\|(\mathrm{Id} - \mathbf{P}_\lambda)u\|_{L^2}^2 = \int_{\mathbf{R}} d((\mathrm{Id} - \mathbf{P}_\lambda)u|u).$$

Using (2.1.2) again, we get

$$\begin{aligned}\|(\mathrm{Id} - \mathbf{P}_\lambda)u\|_{L^2}^2 &= \int_\lambda^\infty d(\mathbf{P}_{\lambda'} u|u)\\ &\leq \int_\lambda^\infty \frac{\lambda'}{\lambda} d(\mathbf{P}_{\lambda'} u|u)\\ &\leq \frac{1}{\lambda}\|\nabla u\|_{L^2}^2\end{aligned}$$

which ends the proof of Corollary 2.2. □

Let us give a few examples in which the operators $\mathbf{P}_\lambda$ can (and will) be explicitly computed.

In the case of bounded domains, the operators $\mathbf{P}_\lambda$ are given by $\mathbf{P}_\lambda \stackrel{\text{def}}{=} P_{n(\lambda)}$ with $n(\lambda) \stackrel{\text{def}}{=} \max\{k \;/\; \mu_k^2 < \lambda\}$ where the sequences $(\mu_k)_{k\in\mathbf{N}}$ and $(P_k)_{k\in\mathbf{N}}$ are given by Theorem 2.1 and Definition 2.2.

The case when $\Omega = \mathbf{T}^d$, i.e. the periodic box $(0, 2\pi)^d$ with periodic boundary conditions, is very similar to the bounded case. Using the Galilean invariance of the Navier–Stokes system, we shall only consider mean free vector fields. Everything that will be done in that framework is based on the discrete Fourier transform. Let us recall that for all distributions u defined on $\mathbf{T}^d$,

$$\widehat{u}_k \stackrel{\text{def}}{=} (2\pi)^{-d} \int_{\mathbf{T}^d} e^{-ik\cdot x} u(x)\, dx. \quad \text{for} \quad k \in \mathbf{Z}^d.$$

Now let us define Sobolev spaces for any real number s:

$$H^s \stackrel{\text{def}}{=} \left\{ u \in \mathcal{D}'(\mathbf{T}^d) \;/\; \|u\|_{H^s}^2 \stackrel{\text{def}}{=} \sum_{k\in\mathbf{Z}^d} |k|^{2s} |\widehat{u}_k|^2 < +\infty \right\}.$$

Note that $\mathcal{V}_\sigma$ is given by

$$\mathcal{V}_\sigma = \left\{ u \in H^1 \;/\; \forall k \in \mathbf{Z}^d,\; k \cdot \widehat{u}_k = 0 \right\}.$$

The spectral projection $\mathbf{P}_\lambda$ can be easily expressed in terms of the Fourier transform. Let us note that, as u is assumed to be real valued, we have $\overline{\widehat{u}_k} = \widehat{u}_{-k}$. If $k \neq 0$, let us denote by $(f_k^m)_{1\leq m\leq d-1}$ an orthogonal basis of $\{k\}^\perp$ in $\mathbf{R}^d$. It is easily checked that the family

$$e_k^{m,n}(x) \stackrel{\text{def}}{=} \frac{1}{2(2\pi)^{\frac{d}{2}}}\left((f_k^m + if_k^n)e^{ik\cdot x} + (f_k^m - if_k^n)e^{-ik\cdot x}\right)$$

is an orthonormal basis of $\mathcal{H}$ which satisfies $-\Delta e_k^{m,n} = |k|^2 e_k^{m,n}$. Of course we have

$$\mathbf{P}_\lambda(u)(x) = \sum_{\substack{|k|^2<\lambda\\ 1\leq m,n\leq d-1}} (u|e_k^{m,n})_{L^2}\, e_k^{m,n}(x).$$

In the case of the whole space, the Fourier transform plays, of course, again a crucial role: the operator $\mathbf{P}_\lambda$ reduces to the Fourier cut-off operator

$$\mathbf{P}_\lambda(u) = \mathcal{F}^{-1}(\mathbf{1}_{|\xi|^2\leq\lambda}\widehat{u}(\xi)). \tag{2.1.6}$$

Let us return to the general aspect of the spectral theorem. We would like to define, as in the bounded case, the spectral projections on $\mathcal{V}'$ and also to prove an approximation result analogous to Proposition 2.1. The extension to $\mathcal{V}'$ is achieved simply by transposition, as defined in the following proposition.

Proposition 2.3 *The map*

$$\widetilde{\mathbf{P}}_\lambda \begin{cases} \mathcal{V}' \to \mathcal{V}'_\sigma \\ f \mapsto v \mapsto \langle f, \mathbf{P}_\lambda v\rangle \end{cases}$$

satisfies the following properties:

$$\|\widetilde{\mathbf{P}}_\lambda f\|_{\mathcal{V}'_\sigma} \leq \|f\|_{\mathcal{V}'_\sigma}; \tag{2.1.7}$$

$$\forall f \in \mathcal{V}'_\sigma,\ \widetilde{\mathbf{P}}_\lambda f \in \mathcal{V}_\sigma \quad \text{and} \quad \lim_{\lambda\to+\infty} \|\widetilde{\mathbf{P}}_\lambda f - f\|_{\mathcal{V}'_\sigma} = 0. \tag{2.1.8}$$

Proof Let us first observe that, as $\mathbf{P}_\lambda$ is an orthogonal projection on $\mathcal{H}$, for any u and v in $\mathcal{H}$, we have

$$\langle \mathbf{P}_\lambda u, v\rangle = (\mathbf{P}_\lambda u|v)_{\mathcal{H}} = (u|\mathbf{P}_\lambda v)_{\mathcal{H}} = \langle u, \mathbf{P}_\lambda v\rangle.$$

Thus the operator $\widetilde{\mathbf{P}}_\lambda$ is an extension of $\widetilde{\mathbf{P}}_\lambda$ on $\mathcal{V}'$. By definition of $\|f\|_{\mathcal{V}'_\sigma}$ we have

$$\begin{aligned}\|\widetilde{\mathbf{P}}_\lambda f\|_{\mathcal{V}'_\sigma} &= \sup_{v\in\mathcal{V}_\sigma} \langle \widetilde{\mathbf{P}}_\lambda f, v\rangle \\ &= \sup_{v\in\mathcal{V}_\sigma} \langle f, \mathbf{P}_\lambda v\rangle \\ &\leq \|f\|_{\mathcal{V}'_\sigma},\end{aligned}$$

which proves (2.1.7). Let us now prove (2.1.8). We shall first show that $\widetilde{\mathbf{P}}_\lambda f$ is in $\mathcal{H}$. For any $v \in \mathcal{V}_\sigma$ we have

$$\begin{aligned}\langle \widetilde{\mathbf{P}}_\lambda f, v\rangle &= \langle f, \mathbf{P}_\lambda v\rangle\\ &\leq \|f\|_{\mathcal{V}'}\|\mathbf{P}_\lambda v\|_{\mathcal{V}_\sigma}.\end{aligned}$$

Using Corollary 2.2 as well as (2.1.5) we infer that for any $v \in \mathcal{V}_\sigma$,

$$\langle \widetilde{\mathbf{P}}_\lambda f, v\rangle \leq \|f\|_{\mathcal{V}'}(1+\lambda^{\frac{1}{2}})\|v\|_{\mathcal{H}}.$$

As $\mathcal{V}_\sigma$ is dense in $\mathcal{H}$ we find that $\widetilde{\mathbf{P}}_\lambda f$ is in $\mathcal{H}$. Then as $\widetilde{\mathbf{P}}_\lambda^2 f = \widetilde{\mathbf{P}}_\lambda f$ we deduce that $\mathbf{P}_\lambda f$ is in $\mathcal{V}_\sigma$ thanks again to Corollary 2.2.

Then using Proposition 1.1, page 20 we know that for any positive ε there is a vector field f_ε in $\mathcal{H}$ such that

$$\|f - f_\varepsilon\|_{\mathcal{V}'_\sigma} \leq \frac{\varepsilon}{4}.$$

As f_ε belongs to $\mathcal{H}$, we have $\widetilde{\mathbf{P}}_\lambda f_\varepsilon = \mathbf{P}_\lambda f_\varepsilon$. Thus we can write

$$\|\widetilde{\mathbf{P}}_\lambda f - f\|_{\mathcal{V}'_\sigma} \leq \|\widetilde{\mathbf{P}}_\lambda(f - f_\varepsilon)\|_{\mathcal{V}'_\sigma} + \|\mathbf{P}_\lambda f_\varepsilon - f_\varepsilon\|_{\mathcal{V}'_\sigma} + \|f - f_\varepsilon\|_{\mathcal{V}'_\sigma}.$$

Then (2.1.7) implies that

$$\begin{aligned}\|\mathbf{P}_\lambda f - f\|_{\mathcal{V}'_\sigma} &\leq 2\|f - f_\varepsilon\|_{\mathcal{V}'_\sigma} + \|\mathbf{P}_\lambda f_\varepsilon - f_\varepsilon\|_{\mathcal{V}'_\sigma}\\ &\leq \frac{\varepsilon}{2} + C\|\mathbf{P}_\lambda f_\varepsilon - f_\varepsilon\|_{\mathcal{H}}.\end{aligned}$$

Identity (2.1.3) allows us to conclude the proof. □

Notation In all that follows, we shall denote $\widetilde{P}_\lambda$ by $\mathbf{P}_\lambda$.

The following proposition makes the link between the family $(\mathbf{P}\lambda)_{\lambda>0}$ and the Leray projector $\mathbf{P}$ defined in Definition 1.3 page 19.

Proposition 2.4 *For any f in $(L^2)^d$, we have*

$$\lim_{\lambda\to\infty} \mathbf{P}_\lambda f = \mathbf{P}f$$

where $\mathbf{P}$ denotes the Leray projector on $\mathcal{H}$ of Definition 1.3.

Proof By the definition of $\mathbf{P}_\lambda$, we have, for any v in $\mathcal{V}_\sigma$,

$$\langle \mathbf{P}_\lambda \mathbf{P}f, v\rangle = \langle \mathbf{P}f, \mathbf{P}_\lambda v\rangle = (\mathbf{P}f|\mathbf{P}_\lambda v)_{\mathcal{H}}.$$

As the Leray projector $\mathbf{P}$ is the projector of $(L^2)^d$ on $\mathcal{H}$, we have

$$(\mathbf{P}f|\mathbf{P}_\lambda v)_{\mathcal{H}} = (f|\mathbf{P}\mathbf{P}_\lambda v)_{\mathcal{H}} = (f|\mathbf{P}_\lambda v)_{\mathcal{H}}.$$

By the definition of $\mathbf{P}_\lambda$ on $\mathcal{V}'_\sigma$, which contains $(L^2)^d$, we infer that

$$\langle P_\lambda \mathbf{P}f, v\rangle = \langle f, \mathbf{P}_\lambda v\rangle$$

and thus that $\mathbf{P}_\lambda \mathbf{P}f = \mathbf{P}_\lambda f$. Then the assertion (2.1.4) ensures the proposition. □

2.2 The Leray theorem

In this section we present a general approach to proving the global existence of weak solutions to the Navier–Stokes equations in a general domain Ω of $\mathbf{R}^d$ with $d = 2$ or 3. Let us notice that no assumption about the regularity of the boundary is made. Let us now state the weak formulation of the incompressible Navier–Stokes system (NS_ν).

Definition 2.5 *Given a domain Ω in $\mathbf{R}^d$, we shall say that u is a weak solution of the Navier–Stokes equations on $\mathbf{R}^+ \times\Omega$ with initial data u_0 in $\mathcal{H}$ and an external force f in $L^2_{\mathrm{loc}}(\mathbf{R}^+;\mathcal{V}')$ if and only if u belongs to the space*

$$C(\mathbf{R}^+;\mathcal{V}'_\sigma)\cap L^\infty_{\mathrm{loc}}(\mathbf{R}^+;\mathcal{H})\cap L^2_{\mathrm{loc}}(\mathbf{R}^+;\mathcal{V}_\sigma)$$

and for any function Ψ in $C^1(\mathbf{R}^+;\mathcal{V}_\sigma)$, the vector field u satisfies the following condition (S_Ψ):

$$\int_\Omega (u\cdot\Psi)(t,x)\,dx + \int_0^t\int_\Omega (\nu\nabla u:\nabla\Psi - u\otimes u:\nabla\Psi - u\cdot\partial_t\Psi)\,(t',x)\,dxdt'$$
$$= \int_\Omega u_0(x)\cdot\Psi(0,x)\,dx + \int_0^t \langle f(t'),\Psi(t')\rangle\,dt'$$

with

$$\nabla u:\nabla\Psi = \sum_{j,k=1}^d \partial_j u^k\partial_j\Psi^k \quad \textit{and} \quad u\otimes u:\nabla\Psi = \sum_{j,k=1}^d u^j u^k\partial_j\Psi^k.$$

Let us remark that the above relation means that the equality in (NS_ν) must be understood as an equality in the sense of $\mathcal{V}'_\sigma$.

Now let us state the Leray theorem.

Theorem 2.3 *Let Ω be a domain of $\mathbf{R}^d$ and u_0 a vector field in $\mathcal{H}$. Then, there exists a global weak solution u to (NS_ν) in the sense of Definition 2.5. Moreover, this solution satisfies the energy inequality for all $t\geq 0$,*

$$\frac{1}{2}\int_\Omega |u(t,x)|^2\,dx + \nu\int_0^t\int_\Omega |\nabla u(t',x)|^2\,dxdt'$$
$$\leq \frac{1}{2}\int_\Omega |u_0(x)|^2\,dx + \int_0^t \langle f(t',\cdot),u(t',\cdot)\rangle\,dt'. \tag{2.2.1}$$

It is convenient to state the following definition.

Definition 2.6 *A solution of* (NS_ν) *in the sense of the above Definition 2.5 which moreover satisfies the energy inequality (2.2.1) is called a Leray solution of* (NS_ν).

Let us remark that the energy inequality implies a control on the energy. This control depends on the domain Ω.

Proposition 2.5 *Any Leray solution* u *of* (NS_ν) *satisfies*

$$\|u(t)\|_{L^2}^2 + \nu \int_0^t \|\nabla u(t')\|_{L^2}^2 \, dt' \leq e^{\nu t} \left(\|u_0\|_{L^2} + \frac{1}{\nu} \int_0^t \|f(t')\|_{\mathcal{V}'_\sigma}^2 \, dt' \right).$$

If the domain Ω *satisfies the Poincaré inequality, a constant* C *exists such that any Leray solution* u *of* (NS_ν) *satisfies*

$$\|u(t)\|_{L^2}^2 + \nu \int_0^t \|\nabla u(t')\|_{L^2}^2 dt' \leq \|u_0\|_{L^2}^2 + \frac{C}{\nu} \int_0^t \|f(t')\|_{\mathcal{V}'_\sigma}^2 dt'.$$

Proof By definition of the norm $\|\cdot\|_{\mathcal{V}'_\sigma}$, we have

$$\langle f(t,\cdot), u(t,\cdot)\rangle \leq \|f(t,\cdot)\|_{\mathcal{V}'_\sigma} \|u(t,\cdot)\|_{\mathcal{V}_\sigma}.$$

Inequality (2.2.1) becomes

$$\|u(t)\|_{L^2}^2 + 2\nu \int_0^t \|\nabla u(t')\|_{L^2}^2 \, dt' \leq \|u_0\|_{L^2}^2 + \int_0^t \|f(t')\|_{\mathcal{V}'_\sigma} \|u(t',\cdot)\|_{\mathcal{V}}^2 \, dt'.$$

As $\|u(t',\cdot)\|_{\mathcal{V}}^2 = \|u(t',\cdot)\|_{L^2}^2 + \|\nabla u(t',\cdot)\|_{L^2}^2$, we get, using the fact that $2ab \leq a^2 + b^2$,

$$\begin{aligned} \|u(t)\|_{L^2}^2 + 2\nu \int_0^t \|\nabla u(t')\|_{L^2}^2 \, dt' \leq \|u_0\|_{L^2} &+ \frac{C}{\nu} \int_0^t \|f(t')\|_{\mathcal{V}'_\sigma}^2 \, dt' \\ &+ \nu \int_0^t \|f(t')\|_{\mathcal{V}'_\sigma}^2 \|u(t',\cdot)\|_{L^2}^2 \, dt'. \end{aligned}$$

Then Gronwall's lemma gives the result for any domain Ω. When the domain Ω satisfies the Poincaré inequality, then the last term on the right disappears in the above inequality and then the second inequality of the proposition is obvious. □

The outline of this section is now the following.

- First approximate solutions are built in spaces with finite frequencies by using simple ordinary differential equation results in L^2-type spaces.
- Next, a compactness result is derived.
- Finally the conclusion is obtained by passing to the limit in the weak formulation, taking especial care of the non-linear terms.

2.2.1 Construction of approximate solutions

In this section, we intend to build approximate solutions of the Navier–Stokes equations.

Let $(\mathbf{P}_\lambda)_\lambda$ be the family of spectral projections given by Theorem 2.2; we shall consider in the present proof only integer values of λ. Thus let us denote by $\mathcal{H}_k$ the space $\mathbf{P}_k\mathcal{H}$.

Lemma 2.2 *For any bulk force f in $L^2_{\text{loc}}(\mathbf{R}^+;\mathcal{V}')$, a sequence $(f_k)_{k\in\mathbf{N}}$ exists in the space $C^1(\mathbf{R}^+;\mathcal{V}_\sigma)$ such that for any $k\in\mathbf{N}$ and for any $t>0$, the vector field $f_k(t)$ belongs to $\mathcal{H}_k$, and*

$$\lim_{k\to\infty}\|f_k-f\|_{L^2([0,T];\mathcal{V}'_\sigma)}=0.$$

Proof Thanks to Proposition 2.3 and to the Lebesgue theorem, a sequence $(\widetilde{f}_k)_{k\in\mathbf{N}}$ exists in $L^2_{\text{loc}}(\mathbf{R}^+;\mathcal{V}_\sigma)$ such that for any positive integer k and for almost all positive t, the vector field $\widetilde{f}_k(t)$ belongs to $\mathcal{H}_k$ and

$$\forall T>0,\quad \lim_{k\to\infty}\|\widetilde{f}_k-f\|_{L^2([0,T];\mathcal{V}'_\sigma)}=0.$$

A standard (and omitted) time regularization procedure concludes the proof of the lemma. □

In order to construct the approximate solution, let us establish some properties of the non-linear term.

Definition 2.7 *Let us define the bilinear map*

$$Q\begin{cases}\mathcal{V}\times\mathcal{V}\to\mathcal{V}'\\ (u,v)\mapsto -\operatorname{div}(u\otimes v).\end{cases}$$

Sobolev embeddings, stated in Theorem 1.2, ensure that Q is continuous. In the sequel, the following lemma will be useful.

Lemma 2.3 *For any u and v in $\mathcal{V}$, the following estimates hold. For d in $\{2,3\}$, a constant C exists such that, for any $\varphi\in\mathcal{V}$,*

$$\langle Q(u,v),\varphi\rangle\le C\|\nabla u\|_{L^2}^{\frac{d}{4}}\|\nabla v\|_{L^2}^{\frac{d}{4}}\|u\|_{L^2}^{1-\frac{d}{4}}\|v\|_{L^2}^{1-\frac{d}{4}}\|\nabla\varphi\|_{L^2}.$$

Moreover for any u in $\mathcal{V}_\sigma$ and any v in $\mathcal{V}$,

$$\langle Q(u,v),v\rangle=0.$$

Proof The first two inequalities follow directly from the Gagliardo–Nirenberg inequality stated in Corollary 1.2, once it is noticed that

$$\begin{aligned}\langle Q(u,v),\varphi\rangle&\le\|u\otimes v\|_{L^2}\|\nabla\varphi\|_{L^2}\\&\le\|u\|_{L^4}\|v\|_{L^4}\|\nabla\varphi\|_{L^2}.\end{aligned}$$

In order to prove the third assertion, let us assume that u and v are two vector fields the components of which belong to $\mathcal{D}(\Omega)$. Then we deduce from integration by parts that

$$\begin{aligned}
\langle Q(u,v),v\rangle &= -\int_\Omega (\operatorname{div}(u\otimes v)\cdot v)(x)\,dx \\
&= -\sum_{\ell,m=1}^{d}\int_\Omega \partial_m(u^m(x)v^\ell(x))v^\ell(x)\,dx \\
&= \sum_{\ell,m=1}^{d}\int_\Omega u^m(x)v^\ell(x)\partial_m v^\ell(x)\,dx \\
&= -\int_\Omega |v(x)|^2 \operatorname{div} u(x)\,dx - \langle Q(u,v),v\rangle.
\end{aligned}$$

Thus, we have

$$\langle Q(u,v),v\rangle = -\frac{1}{2}\int_\Omega |v(x)|^2 \operatorname{div} u(x)\,dx.$$

The two expressions are continuous on $\mathcal{V}$ and, by definition, $\mathcal{D}$ is dense in $\mathcal{V}$. Thus the formula is true for any $(u,v)\in\mathcal{V}_\sigma\times\mathcal{V}$, which completes the proof. □

Thanks to Proposition 2.3 and the above lemma, we can state

$$F_k(u) \stackrel{\text{def}}{=} \mathbf{P}_k Q(u,u).$$

Now let us introduce the following ordinary differential equation

$$(\mathrm{NS}_{\nu,k})\qquad \begin{cases} \dot u_k(t) = \nu\mathbf{P}_k\Delta u_k(t) + F_k(u_k(t)) + f_k(t)\\ u_k(0) = \mathbf{P}_k u_0.\end{cases}$$

Theorem 2.2 implies that $\mathbf{P}_k\Delta$ is a linear map from $\mathcal{H}_k$ into itself. Thus the continuity properties on Q and $\mathbf{P}_k$ allow us to apply the Cauchy–Lipschitz theorem. This gives the existence of $T_k\in\,]0,+\infty]$ and a unique maximal solution u_k of $(\mathrm{NS}_{\nu,k})$ in $C^\infty([0,T_k[;\mathcal{H}_k)$. In order to prove that $T_k=+\infty$, let us observe that, thanks to Lemma 2.3 and Theorem 2.2,

$$\|\dot u_k(t)\|_{L^2} \le \nu k\|u_k(t)\|_{L^2} + Ck^{\frac{d}{4}}\|u_k(t)\|_{L^2}^2 + \|f_k(t)\|_{L^2}.$$

If $\|u_k(t)\|_{L^2}$ remains bounded on some interval $[0,T[$, so does $\|\dot u_k(t)\|_{L^2}$. Thus, for any k, the function u_k satisfies the Cauchy criteria when t tends to T. Thus the solution can be extended beyond T. It follows that a uniform bound on $\|u_k(t)\|_{L^2}$ will imply that $T_k=+\infty$.

2.2.2 A priori bounds

The purpose of this section is the proof of the following proposition.

Proposition 2.6 *The sequence $(u_k)_{k\in\mathbf{N}}$ is bounded in the space*

$$L^\infty_{\rm loc}(\mathbf{R}^+;\mathcal{H})\cap L^2_{\rm loc}(\mathbf{R}^+;\mathcal{V}_\sigma)\cap L^{\frac{8}{d}}_{\rm loc}(L^4(\Omega)).$$

Moreover, the sequence $(\Delta u_k)_{k\in\mathbf{N}}$ is bounded in the space $L^2_{\rm loc}(\mathbf{R}^+;\mathcal{V}'_\sigma)$.

Proof Let us now estimate the L^2 norm of $u_k(t)$. Taking the L^2 scalar product of equation $(\mathrm{NS}_{\nu,k})$ with $u_k(t)$, we get

$$\frac{1}{2}\frac{d}{dt}\|u_k(t)\|^2_{L^2} = \nu(\Delta u_k(t)|u_k(t))_{L^2} + (F_k(u_k(t))|u_k(t))_{L^2} + (f_k(t)|u_k(t))_{L^2}.$$

By the definition of F_k, Lemma 2.3 implies that

$$(F_k(u_k(t))|u_k(t))_{L^2} = \langle Q(u_k(t),u_k(t)),u_k\rangle = 0.$$

Thus we infer that

$$\frac{1}{2}\frac{d}{dt}\|u_k(t)\|^2_{L^2} + \nu(\nabla u_k(t)|\nabla u_k(t))_{L^2} = (f_k(t)|u_k(t))_{L^2}. \tag{2.2.2}$$

By time integration, we get the fundamental energy estimate for the approximate Navier–Stokes system: for all $t\in[0,T_k)$

$$\frac{1}{2}\|u_k(t)\|^2_{L^2} + \nu\int_0^t \|\nabla u_k(t')\|^2_{L^2}\,dt' = \frac{1}{2}\|u_k(0)\|^2_{L^2} + \int_0^t (f_k(t')|u_k(t'))_{L^2}\,dt'. \tag{2.2.3}$$

Using the (well known) fact that $2ab\le a^2+b^2$, we get

$$\|u_k(t)\|^2_{L^2} + \nu\int_0^t \|\nabla u_k(t')\|^2_{L^2}\,dt' \le \|u_k(0)\|^2_{L^2} + \frac{C}{\nu}\int_0^t \|f_k(t')\|^2_{\mathcal{V}'_\sigma}\,dt'$$
$$+\nu\int_0^t \|u_k(t')\|^2_{L^2}\,dt'. \tag{2.2.4}$$

Gronwall's lemma implies that $(u_k)_{k\in\mathbf{N}}$ remains uniformly bounded in $\mathcal{H}$ for all time, hence that $T_k=+\infty$. In addition, we obtain that the sequence $(u_k)_{k\in\mathbf{N}}$ is bounded in the space $L^\infty_{\rm loc}(\mathbf{R}^+;\mathcal{H})\cap L^2_{\rm loc}(\mathbf{R}^+;\mathcal{V}_\sigma)$. Using the Gagliardo–Nirenberg inequalities (see Corollary 1.2, page 25), we deduce that the sequence $(u_k)_{k\in\mathbf{N}}$ is bounded in the space

$$L^\infty_{\rm loc}(\mathbf{R}^+;\mathcal{H})\cap L^2_{\rm loc}(\mathbf{R}^+;\mathcal{V}_\sigma)\cap L^{\frac{8}{d}}_{\rm loc}(L^4(\Omega)).$$

Moreover, we have, for any $v\in\mathcal{V}_\sigma$,

$$\begin{aligned}\langle -\Delta u_k, v\rangle &= (\nabla u_k|\nabla v)_{L^2}\\ &\le \|u_k\|_{H^1_0}\|v\|_{\mathcal{V}}.\end{aligned}$$

By definition of the norm $\|\cdot\|_{\mathcal{V}'_\sigma}$, we infer that the sequence $(\Delta u_k)_{k\in\mathbf{N}}$ is bounded in $L^2_{\rm loc}(\mathbf{R}^+;\mathcal{V}'_\sigma)$. The whole proposition is proved. □

2.2.3 Compactness properties

Let us now prove the following fundamental result.

Proposition 2.7 *A vector field u exists in $L^2_{\rm loc}(\mathbf{R}^+;\mathcal{V}_\sigma)$ such that up to an extraction (which we omit to note) we have for any positive real number T and for any compact subset K of Ω*

$$\lim_{k\to\infty}\int_{[0,T]\times K}|u_k(t,x)-u(t,x)|^2\,dtdx=0. \tag{2.2.5}$$

Moreover for all vector fields $\Psi\in L^2([0,T];\mathcal{V})$ and $\Phi\in L^2([0,T]\times\Omega)$ we can write

$$\begin{aligned}\lim_{k\to\infty}\int_{[0,T]\times\Omega}(\nabla u_k(t,x)-\nabla u(t,x)):\nabla\Psi(t,x)\,dtdx=0;\\ \lim_{k\to\infty}\int_{[0,T]\times\Omega}(u_k(t,x)-u(t,x))\,\Phi(t,x)\,dtdx=0.\end{aligned} \tag{2.2.6}$$

Finally for any $\psi\in C^1(\mathbf{R}^+;\mathcal{V}_\sigma)$ the following limit holds:

$$\lim_{k\to\infty}\sup_{t\in[0,T]}|\langle u_k(t),\psi(t)\rangle-\langle u(t),\psi(t)\rangle|=0. \tag{2.2.7}$$

Proof A standard diagonal process with an increasing sequence of positive real numbers T_n and an exhaustive sequence of compact subsets K_n of Ω reduces the proof of (2.2.5) to the proof of the relative compactness of the sequence $(u_k)_{k\in\mathbf{N}}$ in $L^2([0,T]\times K)$. So let us consider a positive real number ε. As the sequence $(u_k)_{k\in\mathbf{N}}$ is bounded in $L^2([0,T];\mathcal{V}_\sigma)$, Corollary 2.2 (together with Lebesgue's theorem) implies that an integer k_0 exists such that

$$\forall k\in\mathbf{N},\quad \|u_k-\mathbf{P}_{k_0}u_k\|_{L^2([0,T]\times\Omega)}\leq\frac{\varepsilon}{2}\cdot \tag{2.2.8}$$

Now we claim that there is an $L^{\frac{4}{d}}([0,T])$ function f_{k_0}, independent of k, such that

$$\|\partial_t\mathbf{P}_{k_0}u_k(t)\|_{L^2}\leq f_{k_0}(t). \tag{2.2.9}$$

Let us prove the claim. Proposition 2.6 tells us, in particular first, using the fact that the sequence $(u_k)_{k\in\mathbf{N}}$ is bounded in the space $L^2_{\rm loc}(\mathbf{R}^+;\mathcal{V}_\sigma)$, we infer that the sequence $(-\Delta u_k)_{k\in\mathbf{N}}$ is bounded in $L^2_{\rm loc}(\mathbf{R}^+;\mathcal{V}'_\sigma)$. The fact that the sequence $(f_k)_{k\in\mathbf{N}}$ is bounded in $L^2_{\rm loc}(\mathbf{R}^+;\mathcal{V}'_\sigma)$ is clear by definition of f_k, and finally using Lemma 2.3, we have

$$\|F_k(u_k(t))\|_{\mathcal{V}'_\sigma}\leq C\|\nabla u_k\|^{\frac{d}{2}}_{L^2}\|u_k\|^{2-\frac{d}{2}}_{L^2}. \tag{2.2.10}$$

Then using the energy estimate (2.2.4) with (2.2.10), we deduce that

$$\forall k\in\mathbf{N},\quad \|\partial_t u_k\|_{L^{\frac{4}{d}}(\mathbf{R}^+;\mathcal{V}'_\sigma)}\leq C. \tag{2.2.11}$$

The claim (2.2.9) is obtained.

But for any $t \in [0,T]$, the family $(\mathbf{P}_{k_0} u_k)_{k\in\mathbf{N}}$ is bounded in H^1_0. Let us denote by L^2_K the space of vector fields in L^2 supported in K. As the embedding of H^1_0 into L^2_K is compact, Ascoli's theorem implies that the family $(\mathbf{P}_{k_0} u_k)_{k\in\mathbf{N}}$ is relatively compact in the space $L^\infty([0,T];L^2_K)$, thus in the space $L^2([0,T]\times K)$. It can therefore be covered by a finite number of balls of $L^2([0,T]\times K)$ of radius $\varepsilon/2$, and the existence of u follows, such that (2.2.5) is satisfied.

In particular this implies of course that u_k converges towards u in the space $\mathcal{D}'(]0,T[\times\Omega)$. The sequence $(u_k)_{k\in\mathbf{N}}$ is bounded in the space $L^2([0,T];H^1_0)$. So as $\mathcal{D}(]0,T[\times\Omega)$ is dense both in $L^2([0,T]\times\Omega)$ and in $L^2([0,T];H^1_0)$, the weak convergence holds in the spaces $L^2([0,T]\times\Omega)$ and $L^2([0,T];H^1_0)$. This proves (2.2.6). Moreover since $\operatorname{div} u_k = 0$ we also infer that $\operatorname{div} u = 0$.

In order to prove (2.2.7), we consider a vector field ψ in $C^1([0,T];\mathcal{V}_\sigma)$ and the function

$$g_k \begin{cases} [0,T] \to \mathbf{R} \\ t \mapsto \langle u_k(t), \psi(t)\rangle. \end{cases}$$

The sequence $(g_k)_{k\in\mathbf{N}}$ is bounded in $L^\infty([0,T];\mathbf{R})$. Moreover

$$\dot g_k(t) = \langle \dot u_k(t), \psi(t)\rangle + \langle u_k(t), \partial_t \psi(t)\rangle.$$

We therefore have

$$|\dot g_k(t)| \leq \|\dot u_k(t)\|_{\mathcal{V}'_\sigma} \sup_{t\in[0,T]} \|\psi(t)\|_{\mathcal{V}_\sigma} + \|u_k(t)\|_{L^2} \sup_{t\in[0,T]} \|\partial_t\psi(t)\|_{L^2}$$

and using (2.2.11) this implies that $(\dot g_k)_{k\in\mathbf{N}}$ is bounded in $L^{\frac{4}{d}}([0,T];\mathbf{R})$. The sequence $(g_k)_{k\in\mathbf{N}}$ is therefore a bounded sequence of $C^{1-\frac{d}{4}}([0,T];\mathbf{R})$ which by Ascoli's theorem again implies that, up to an extraction, $g_k(t)$ converges strongly towards $g(t)$ in $L^\infty([0,T];\mathbf{R})$.

But using (2.2.6) we know that $g_k(t)$ converges strongly in $L^2([0,T])$ towards $\int_\Omega u(t,x)\psi(t,x)\,dx$. Thus we have (2.2.7). □

2.2.4 End of the proof of the Leray theorem

The local strong convergence of $(u_k)_{k\in\mathbf{N}}$ will be crucial in order to pass to the limit in $(\mathrm{NS}_{\nu,k})$ to obtain solutions of (NS_ν).

According to the definition of a weak solution of (NS_ν), let us consider a test function Ψ in $C^1(\mathbf{R}^+;\mathcal{V}_\sigma)$. Because u_k is a solution of $(\mathrm{NS}_{\nu,k})$, we have

$$\begin{aligned}\frac{d}{dt}\langle u_k(t), \Psi(t)\rangle &= \langle \dot u_k(t), \Psi(t)\rangle + \langle u_k(t), \dot\Psi(t)\rangle \\ &= \nu\langle \mathbf{P}_k \Delta u_k(t), \Psi(t)\rangle + \langle \mathbf{P}_k Q(u_k(t), u_k(t)), \Psi(t)\rangle \\ &\quad + \langle f_k(t), \Psi(t)\rangle + \langle u_k(t), \dot\Psi(t)\rangle.\end{aligned}$$

We have, after integration by parts,

$$\langle \mathbf{P}_k \Delta u_k(t), \Psi(t)\rangle = -\nu \int_\Omega \nabla u_k(t,x) : \nabla \mathbf{P}_k \Psi(t,x)\, dx,$$

$$\langle \mathbf{P}_k Q(u_k(t), u_k(t)), \Psi(t)\rangle = \int_\Omega u_k(t,x) \otimes u_k(t,x) : \nabla \mathbf{P}_k \Psi(t,x)\, dx$$

and

$$\langle u_k(t), \dot{\Psi}(t)\rangle = \int_\Omega u_k(t,x) \cdot \partial_t \Psi(t,x)\, dx.$$

By time integration between 0 and t, we infer that

$$\begin{aligned}
&\int_\Omega u_k(t,x)\cdot \Psi(t,x)\, dx \\
&\quad + \int_0^t \int_\Omega (\nu \nabla u_k : \nabla \mathbf{P}_k \Psi - u_k \otimes u_k : \nabla \mathbf{P}_k \Psi - u_k \cdot \partial_t \Psi)(t',x)\, dx dt' \\
&\quad = \int_\Omega u_k(0,x)\cdot \Psi(0,x)\, dx + \int_0^t \langle f_k(t'), \Psi(t')\rangle\, dt'.
\end{aligned}$$

We now have to pass to the limit.

Let us start by proving the following preliminary lemma.

Lemma 2.4 *Let* H *be a Hilbert space, and let* $(A_n)_{n\in\mathbf{N}}$ *be a bounded sequence of linear operators on* H *such that*

$$\forall h \in \mathrm{H}, \quad \lim_{n\to\infty} \|A_n h - h\|_{\mathrm{H}} = 0.$$

Then if $\psi \in C([0,T];\mathrm{H})$ *we have*

$$\lim_{n\to\infty} \sup_{t\in[0,T]} \|A_n\psi(t) - \psi(t)\|_{\mathrm{H}} = 0.$$

Proof The function ψ is continuous in time with values in H, so for all positive ε, the compact $\psi([0,T])$ can be covered by a finite number of balls of radius

$$\frac{\varepsilon}{2(\mathcal{A}+1)} \quad \text{with } \mathcal{A} \stackrel{\text{def}}{=} \sup_n \|A_n\|_{\mathcal{L}(\mathrm{H})}$$

and center $(\psi(t_\ell))_{0\le \ell\le N}$. Then we have, for all t in $[0,T]$ and ℓ in $\{0,\dots,N\}$,

$$\begin{aligned}
\|A_n\psi(t) - \psi(t)\|_{\mathrm{H}} &\le \|A_n\psi(t) - A_n\psi(t_\ell)\|_{\mathrm{H}} \\
&\quad + \|A_n\psi(t_\ell) - \psi(t_\ell)\|_{\mathrm{H}} + \|\psi(t_\ell) - \psi(t)\|_{\mathrm{H}}.
\end{aligned}$$

The assumption on A_n implies that for any ℓ, the sequence $(A_n\psi(t_\ell))_{n\in\mathbf{N}}$ tends to $\psi(t_\ell)$. Thus, an integer n_N exists such that, if $n \ge n_N$,

$$\forall \ell \in \{0,\dots,N\}, \quad \|A_n\psi(t_\ell) - \psi(t_\ell)\|_{\mathrm{H}} < \frac{\varepsilon}{2}.$$

We infer that, if $n \geq n_N$, for all $t \in [0,T]$ and all $\ell \in \{0,\ldots,N\}$,

$$\|A_n\psi(t)-\psi(t)\|_{\mathrm{H}} \leq \|A_n\psi(t)-A_n\psi(t_\ell)\|_{\mathrm{H}} + \|\psi(t_\ell)-\psi(t)\|_{\mathrm{H}} + \frac{\varepsilon}{2}\cdot$$

For any t, let us choose ℓ such that

$$\|\psi(t)-\psi(t_\ell)\|_{\mathrm{H}} \leq \frac{\varepsilon}{2(\mathcal{A}+1)}\cdot$$

The lemma is proved. □

We can apply the lemma to $\mathrm{H}=\mathcal{V}$, and to $\Psi \in C([0,T];\mathcal{V}_\sigma)$. We find that

$$\lim_{k\to\infty} \sup_{t\in[0,T]} \|\mathbf{P}_k\Psi(t)-\Psi(t)\|_{\mathcal{V}} = 0.$$

This implies that

$$\lim_{k\to\infty} \nu\int_0^t\int_\Omega \nabla u_k : \nabla\mathbf{P}_k\Psi\,dxdt' = \lim_{k\to\infty} \nu\int_0^t\int_\Omega \nabla u_k : \nabla\Psi\,dxdt',$$

and the weak convergence of $(u_k)_{k\in\mathbf{N}}$ to u in $L^2_{\mathrm{loc}}(\mathbf{R}^+;\mathcal{V}_\sigma)$ ensures that

$$\lim_{k\to\infty} \nu\int_0^t\int_\Omega \nabla u_k : \nabla\Psi\,dxdt' = \nu\int_0^t\int_\Omega \nabla u : \nabla\Psi\,dxdt'.$$

By (2.2.7) we have, for any non-negative t,

$$\lim_{k\to\infty}\int_\Omega u_k(t,x)\cdot\Psi(t,x)\,dx = \lim_{k\to\infty}\langle u_k(t),\Psi(t)\rangle = \langle u(t),\Psi(t)\rangle$$

and by (2.2.6)

$$\lim_{k\to\infty}\sup_{t\in[0,T]}\left|\int_\Omega u_k(t,x)\cdot\partial_t\Psi(t,x)\,dx - \langle u(t),\partial_t\Psi(t)\rangle\right| = 0.$$

The two terms associated with the initial data and the bulk forces are convergent by construction of the sequence $(f_k)_{k\in\mathbf{N}}$ and because of the fact that, thanks to Theorem 2.2, $(\mathbf{P}_k u_0)_{k\in\mathbf{N}}$ tends to u_0 in L^2.

Now let us pass to the limit in the non-linear term. As above we have in fact

$$\lim_{k\to\infty}\int_0^t\int_\Omega (u_k\otimes u_k : \nabla\mathbf{P}_k\Psi)(t',x)\,dxdt'$$
$$= \lim_{k\to\infty}\int_0^t\int_\Omega (u_k\otimes u_k : \nabla\Psi)(t',x)\,dxdt',$$

so it is enough to prove that

$$\lim_{k\to\infty}\int_0^t\int_\Omega (u_k\otimes u_k : \nabla\Psi)(t',x)\,dxdt' = \int_0^t\int_\Omega (u\otimes u : \nabla\Psi)(t',x)\,dxdt'.$$

Let $(K_n)_{n\in\mathbf{N}}$ be an exhaustive sequence of compact subsets of Ω. Applying Lemma 2.4 with $\mathrm{H} = (L^2)^d$ and $A_n f \stackrel{\text{def}}{=} \mathbf{1}_{K_n} f$, we have

$$\lim_{n\infty} \|\mathbf{1}_{K_n}\nabla\Psi - \nabla\Psi\|_{L^2([0,T];L^2)} = 0.$$

Using the fact that the sequence $(u_k \otimes u_k)_{k\in\mathbf{N}}$ is bounded in $L^1([0,T];L^2(\Omega))$, it is enough to prove that, for any compact subset K of Ω, we have

$$\lim_{k\to\infty} \|u_k \otimes u_k - u\otimes u\|_{L^1([0,T];L^2_K)} = 0 \tag{2.2.12}$$

which will be implied by

$$\lim_{k\to\infty} \|u_k - u\|_{L^2([0,T];L^4_K)} = 0. \tag{2.2.13}$$

Recall that L^p_K denotes the space of L^p functions supported in K. Using Corollary 1.2, we have, for any vector field $a\in\mathcal{V}$,

$$\|a(t)\|_{L^4_K} \leq C\|a(t)\|_{L^2_K}^{1-\frac{d}{4}} \|\nabla a(t)\|_{L^2}^{\frac{d}{4}}.$$

Hölder's inequality implies that

$$\|a\|_{L^2([0,T];L^4_K)} \leq C\|a\|_{L^2([0,T]\times K)}^{1-\frac{d}{4}} \|\nabla a\|_{L^2([0,T]\times\Omega)}^{\frac{d}{4}}.$$

We therefore have

$$\|u_k - u\|_{L^2([0,T];L^4_K)} \leq C\|u_k - u\|_{L^2([0,T]\times K)}^{1-\frac{d}{4}} \|\nabla(u_k - u)\|_{L^2([0,T]\times\Omega)}^{\frac{d}{4}}.$$

Proposition 2.7 allows us to conclude the proof of the fact that u is a solution of (NS_ν) in the sense of Definition 2.5.

It remains to prove the energy inequality (2.2.1). Assertion (2.2.7) of Proposition 2.7 implies in particular that for any time $t\geq 0$ and any $v\in\mathcal{V}_\sigma$,

$$\liminf_{k\to\infty} (u_k(t)|v)_{\mathcal{H}} = (u(t)|v)_{\mathcal{H}}.$$

As $\mathcal{V}_\sigma$ is dense in $\mathcal{H}$, we get that for any $t\geq 0$, the sequence $(u_k(t))_{k\in\mathbf{N}}$ converges weakly towards $u(t)$ in the Hilbert space $\mathcal{H}$. Hence

$$\|u(t)\|_{L^2}^2 \leq \liminf_{k\to\infty} \|u_k(t)\|_{L^2}^2 \quad \text{for all } t\geq 0.$$

On the other hand, $(u_k)_{k\in\mathbf{N}}$ converges weakly to u in $L^2_{\mathrm{loc}}(\mathbf{R}^+;\mathcal{V})$, so that for all non-negative t, we have

$$\int_0^t \|\nabla u(t')\|_{L^2}^2\,dt' \leq \liminf_{k\to\infty} \int_0^t \|\nabla u_k(t')\|_{L^2}^2\,dt'.$$

Taking the $\liminf_{k\to\infty}$ in the energy equality for approximate solutions (2.2.3) yields the energy inequality (2.2.1).

To conclude the proof of Theorem 2.3 we just need to prove the time continuity of u with values in $\mathcal{V}'_\sigma$. That result is obtained by using the fact that u satisfies (S_Ψ) in particular with a function Ψ independent of time.

Choosing such a function yields that for any $\Psi \in \mathcal{V}_\sigma$,

$$\langle u(t), \Psi\rangle - \langle u(t'), \Psi\rangle = \int_t^{t'} \langle f(t''), \Psi\rangle \, dt''$$

$$+ \int_t^{t'}\int_\Omega (\nu \nabla u(t'') : \nabla\Psi - u \otimes u(t'') : \nabla\Psi) \, dx dt''.$$

Using the inequality

$$\|u(t)\|_{L^4} \leq \|u(t)\|_{L^2}^{1-\frac{d}{4}} \|\nabla u(t)\|_{L^2}^{\frac{d}{4}},$$

we infer, by energy inequality, that u belongs to $L^{\frac{8}{d}}([0,T]; L^4(\Omega))$. Then we deduce that

$$|\langle u(t), \Psi\rangle - \langle u(t'), \Psi\rangle| \leq |t-t'|^{1-\frac{d}{4}} \|u\|^2_{L^{\frac{8}{d}}([0,T];L^4)} \|\Psi\|_{\mathcal{V}_\sigma}$$

$$+ |t-t'|^{\frac{1}{2}} (\|f\|_{L^2([0,T];\mathcal{V}'_\sigma)} + \nu \|\nabla u\|_{L^2([0,T];L^2)}) \|\Psi\|_{\mathcal{V}_\sigma}$$

which concludes the proof of Theorem 2.3.

3

Stability of Navier–Stokes equations

In this chapter we intend to investigate the stability of the Leray solutions constructed in the previous chapter. It is useful to start by analyzing the linearized version of the Navier–Stokes equations, so the first section of the chapter is devoted to the proof of the well-posedness of the time-dependent Stokes system. The study will be applied in Section 3.2 to the two-dimensional Navier–Stokes equations, and the more delicate case of three space dimensions will be dealt with in Sections 3.3–3.5.

3.1 The time-dependent Stokes problem

Given a positive viscosity ν, the time-dependent Stokes problem reads as follows:

$$(\mathrm{ES}_\nu)\begin{cases}\partial_t u - \nu\Delta u = f - \nabla p\\ \operatorname{div} u = 0\\ u_{|\partial\Omega} = 0\\ u_{|t=0} = u_0 \in \mathcal{H}.\end{cases}$$

Let us define what a solution of this problem is.

Definition 3.1 *Let u_0 be in $\mathcal{H}$ and f in $L^2_{\mathrm{loc}}(\mathbf{R}^+;\mathcal{V}')$. We shall say that u is a solution of (ES_ν) with initial data u_0 and external force f if and only if u belongs to the space*

$$C(\mathbf{R}^+;\mathcal{V}'_\sigma)\cap L^\infty_{\mathrm{loc}}(\mathbf{R}^+;\mathcal{H})\cap L^2_{\mathrm{loc}}(\mathbf{R}^+;\mathcal{V}_\sigma)$$

and satisfies, for any Ψ in $C^1(\mathbf{R}^+;\mathcal{V}_\sigma)$,

$$\langle u(t),\Psi(t)\rangle + \int_{[0,t]\times\Omega}(\nu\nabla u:\nabla\Psi - u\cdot\partial_t\Psi)(t',x)\,dxdt'$$
$$= \int_\Omega u_0(x)\cdot\Psi(0,x)\,dx + \int_0^t\langle f(t'),\Psi(t')\rangle\,dt'.$$

The following theorem holds.

Theorem 3.1 *The problem (ES_ν) has a unique solution in the sense of the above definition. Moreover this solution belongs to $C(\mathbf{R}^+;\mathcal{H})$ and satisfies*

$$\frac{1}{2}\|u(t)\|^2_{L^2} + \nu\int_0^t\|\nabla u(t')\|^2_{L^2}\,dt' = \frac{1}{2}\|u_0\|^2_{L^2} + \int_0^t\langle f(t'),u(t')\rangle\,dt'.$$

Proof In order to prove uniqueness, let us consider some function u in $C(\mathbf{R}^+;\mathcal{V}'_\sigma)\cap L^2_{\rm loc}(\mathbf{R}^+;\mathcal{V}_\sigma)$ such that, for all Ψ in $C^1(\mathbf{R}^+;\mathcal{V}_\sigma)$,

$$\langle u(t),\Psi(t)\rangle+\int_0^t\int_\Omega(\nu\nabla u:\nabla\Psi-u\cdot\partial_t\Psi)(t',x)\,dxdt'=0.$$

This is valid in particular for a time-independent function $\mathbf{P}_k\Psi$ where Ψ is any given vector field in $\mathcal{V}_\sigma$. We may write

$$\begin{aligned}\langle\mathbf{P}_k u(t),\Psi\rangle&=\langle u(t),\mathbf{P}_k\Psi\rangle\\&=-\nu\int_0^t\int_\Omega\nabla u(t',x):\nabla\mathbf{P}_k\Psi(x)\,dxdt'\\&=\nu\int_0^t\langle u(t'),\Delta\mathbf{P}_k\Psi\rangle\,dt'.\end{aligned}$$

We have, thanks to the spectral Theorem 2.2,

$$\begin{aligned}\langle\mathbf{P}_k u(t),\Psi\rangle&=\nu\int_0^t\langle\mathbf{P}_k u(t'),\Delta\mathbf{P}_k\Psi\rangle\,dt'\\&\leq\nu\|\Delta\mathbf{P}_k\Psi\|_{\mathcal{H}}\int_0^t\|\mathbf{P}_k u(t')\|_{\mathcal{H}}\,dt'\\&\leq\nu k\|\mathbf{P}_k\Psi\|_{\mathcal{H}}\int_0^t\|\mathbf{P}_k u(t')\|_{\mathcal{H}}\,dt'.\end{aligned}$$

By the definition of $\mathcal{H}$, the space $\mathcal{V}_\sigma$ is dense in $\mathcal{H}$, and we have

$$\begin{aligned}\|\mathbf{P}_k u(t)\|_{\mathcal{H}}&=\sup_{\substack{\|\Psi\|_{L^2}=1\\\Psi\in\mathcal{V}_\sigma}}\langle\mathbf{P}_k(t)u,\Psi\rangle\\&\leq\nu k\int_0^t\|\mathbf{P}_k u(t')\|_{\mathcal{H}}\,dt'.\end{aligned}$$

The Gronwall lemma ensures that $\mathbf{P}_k u(t)=0$ for any t and k. This implies uniqueness.

In order to prove existence, let us consider a sequence $(f_k)_{k\in\mathbf{N}}$ associated with f by Lemma 2.2, page 44, and then the approximated problem

$$(\mathrm{ES}_{\nu,k})\begin{cases}\partial_t u_k-\nu\mathbf{P}_k\Delta u_k=f_k\\u_{k|t=0}=\mathbf{P}_k u_0.\end{cases}$$

Again thanks to the spectral Theorem 2.2, page 38, it is a linear ordinary differential equation on $\mathcal{H}_k$ which has a global solution u_k which is $C^1(\mathbf{R}^+;\mathcal{H}_k)$. By the energy estimate in $(\mathrm{ES}_{\nu,k})$ we get that

$$\frac{1}{2}\frac{d}{dt}\|u_k(t)\|^2_{L^2}+\nu\|\nabla u_k(t)\|^2_{L^2}=\langle f_k(t),u_k(t)\rangle.$$

A time integration gives

$$\frac{1}{2}\|u_k(t)\|_{L^2}^2 + \nu \int_0^t \|\nabla u_k(t')\|_{L^2}^2 \, dt' = \frac{1}{2}\|\mathbf{P}_k u(0)\|_{L^2}^2 + \int_0^t \langle f_k(t'), u_k(t')\rangle \, dt'. \tag{3.1.1}$$

In order to pass to the limit, we write the energy estimate for $u_k - u_{k+\ell}$, which gives

$$\begin{aligned}
\delta_{k,\ell}(t) &\stackrel{\text{def}}{=} \frac{1}{2}\|(u_k - u_{k+\ell})(t)\|_{L^2}^2 + \nu \int_0^t \|\nabla(u_k - u_{k+\ell})(t')\|_{L^2}^2 \, dt' \\
&= \frac{1}{2}\|(\mathbf{P}_k - \mathbf{P}_{k+\ell})u(0)\|_{L^2}^2 + \int_0^t \langle (f_k - f_{k+\ell})(t'), u_k(t')\rangle \, dt' \\
&\leq \frac{1}{2}\|(\mathbf{P}_k - \mathbf{P}_{k+\ell})u(0)\|_{L^2}^2 + \frac{C}{\nu}\int_0^t \|(f_k - f_{k+\ell})(t')\|_{\mathcal{V}'_\sigma}^2 \\
&\quad + \frac{\nu}{2}\int_0^t \|\nabla(u_k - u_{k+\ell})(t')\|_{L^2}^2 \, dt' + \frac{\nu}{2}\int_0^t \|(u_k - u_{k+\ell})(t')\|_{L^2}^2 \, dt'.
\end{aligned}$$

This implies that

$$\begin{aligned}
&e^{-\nu t}\|(u_k - u_{k+\ell})(t)\|_{L^2}^2 + \nu e^{-\nu t}\int_0^t \|\nabla(u_k - u_{k+\ell})(t')\|_{L^2}^2 \, dt' \\
&\quad \leq \left(\|(\mathbf{P}_k - \mathbf{P}_{k+\ell})u(0)\|_{L^2}^2 + \frac{C}{\nu}\int_0^t \|(f_k - f_{k+\ell})(t')\|_{\mathcal{V}'_\sigma}^2 \, dt' \right).
\end{aligned}$$

This implies immediately that the sequence $(u_k)_{k\in\mathbf{N}}$ is a Cauchy sequence in the space $C(\mathbf{R}^+;\mathcal{H}) \cap L^2_{\text{loc}}(\mathbf{R}^+;\mathcal{V}_\sigma)$. Let us denote by u the limit and prove that u is a solution in the sense of Definition 3.1. As u_k is a C^1 solution of the ordinary differential equation $(\text{ES}_{\nu,k})$, we have, for a Ψ in $C^1(\mathbf{R}^+;\mathcal{V}_\sigma)$.

$$\frac{d}{dt}\langle u_k(t), \Psi(t)\rangle = \nu\langle \Delta u_k(t), \Psi(t)\rangle + \langle f_k(t), \Psi(t)\rangle + \langle u_k(t), \partial_t \Psi(t)\rangle.$$

By time integration, we get

$$\begin{aligned}
\langle u_k(t), \Psi(t)\rangle &= -\nu \int_0^t \int_\Omega \nabla u_k(t', x) : \nabla \Psi(t', x) \, dx dt' \\
&\quad + \int_0^t \langle f_k(t'), \Psi(t')\rangle \, dt' + \langle \mathbf{P}_k u(0), \Psi(0)\rangle + \int_0^t \langle u_k(t'), \partial_t \Psi(t')\rangle \, dt'.
\end{aligned}$$

Passing to the limit in the above equality and in (3.1.1) gives the theorem. □

Remark This proof works independently of the nature of the domain Ω. In the case when the domain Ω is bounded, the solution is given by the explicit formula

$$u(t)=\sum_{j\in\mathbf{N}}U_j(t)e_j \quad \text{with}$$

$$U_j(t)\stackrel{\text{def}}{=}e^{-\nu\mu_j^2t}(u_0|e_j)_{L^2}+\int_0^t e^{-\nu\mu_j^2(t-t')}\langle f(t'),e_j\rangle\,dt'. \tag{3.1.2}$$

In the case of the whole space $\mathbf{R}^d$, we have the following analogous formula

$$u(t,x)=(2\pi)^{-d}\int_{\mathbf{R}^d}e^{ix\cdot\xi}\widehat{u}(t,\xi)\,d\xi \quad \text{with}$$

$$\widehat{u}(t,\xi)\stackrel{\text{def}}{=}e^{-\nu|\xi|^2t}\widehat{u}_0(\xi)+\int_0^t e^{-\nu|\xi|^2(t-t')}\widehat{f}(t',\xi)\,dt'. \tag{3.1.3}$$

3.2 Stability in two dimensions

In a two-dimensional domain, the Leray weak solutions are unique and even stable. More precisely, we have the following theorem.

Theorem 3.2 *For any data u_0 in $\mathcal{H}$ and f in $L^2_{\text{loc}}(\mathbf{R}^+;\mathcal{V}')$, the Leray weak solution is unique. Moreover, it belongs to $C(\mathbf{R}^+;\mathcal{H})$ and satisfies, for any (s,t) such that $0\le s\le t$,*

$$\frac{1}{2}\|u(t)\|^2_{L^2}+\nu\int_s^t\|\nabla u(t')\|^2_{L^2}\,dt'=\frac{1}{2}\|u(s)\|^2_{L^2}+\int_s^t\langle f(t'),u(t')\rangle\,dt'. \tag{3.2.1}$$

Furthermore, the Leray solutions are stable in the following sense. Let u (resp. v) be the Leray solution associated with u_0 (resp. v_0) in $\mathcal{H}$ and f (resp. g) in the space $L^2_{\text{loc}}(\mathbf{R}^+;\mathcal{V}')$. Then

$$\begin{aligned}&e^{-\nu t}\|(u-v)(t)\|^2_{L^2}+\nu e^{-\nu t}\int_0^t\|\nabla(u-v)(t')\|^2_{L^2}\,dt'\\&\quad\le\left(\|u_0-v_0\|^2_{L^2}+\frac{1}{\nu}\int_0^t\|(f-g)(t')\|^2_{\mathcal{V}'_\sigma}\,dt'\right)\exp\left(\frac{CE^2(t)}{\nu^4}\right)\end{aligned}$$

with

$$E(t)\stackrel{\text{def}}{=}e^{\nu t}\min\left\{\|u_0\|^2_{L^2}+\frac{1}{\nu}\int_0^t\|f(t')\|^2_{\mathcal{V}'_\sigma}\,dt'\,,\,\|v_0\|^2_{L^2}+\frac{1}{\nu}\int_0^t\|g(t')\|^2_{\mathcal{V}'_\sigma}\,dt'\right\}.$$

Remark When the domain Ω satisfies the Poincaré inequality, the estimate becomes

$$\begin{aligned}&\|(u-v)(t)\|^2_{L^2}+\nu\int_0^t\|\nabla(u-v)(t')\|^2_{L^2}\,dt'\\&\quad\le\left(\|u_0-v_0\|^2_{L^2}+\frac{1}{\nu}\int_0^t\|(f-g)(t')\|^2_{\mathcal{V}'_\sigma}\,dt'\right)\exp\left(\frac{CE^2_{\mathcal{P}}(t)}{\nu^4}\right)\end{aligned}$$

with $E_{\mathcal{P}}(t)\stackrel{\text{def}}{=}e^{-\nu t}E(t)$.

Proof of Theorem 3.2 As u belongs to $L^\infty_{\rm loc}(\mathbf{R}^+;\mathcal{H})\cap L^2_{\rm loc}(\mathbf{R}^+;\mathcal{V}_\sigma)$, thanks to Lemma 2.3, page 44, the non-linear term $Q(u,u)$ belongs to $L^2_{\rm loc}(\mathbf{R}^+;\mathcal{V}')$. Thus u is the solution of (ES_ν) with initial data u_0 and external force $f+Q(u,u)$. Theorem 3.1 immediately implies that u belongs to $C(\mathbf{R}^+;\mathcal{H})$ and satisfies, for any (s,t) such that $0\leq s\leq t$,

$$\frac{1}{2}\|u(t)\|^2_{L^2}+\nu\int_s^t\|\nabla u(t')\|^2_{L^2}\,dt'$$
$$=\frac{1}{2}\|u(s)\|^2_{L^2}+\int_s^t\langle f(t'),u(t')\rangle\,dt'+\int_s^t\langle Q(u(t'),u(t')),u(t')\rangle\,dt'.$$

Using Lemma 2.3, we get the energy equality (3.2.1).

To prove the stability, let us observe that the difference $w\stackrel{\rm def}{=}u-v$ is the solution of (ES_ν) with data u_0-v_0 and external force

$$f-g+Q(u,u)-Q(v,v).$$

Theorem 3.1 implies that

$$\|w(t)\|^2_{L^2}+2\nu\int_0^t\|\nabla w(t')\|^2_{L^2}\,dt'=\|w(0)\|^2_{L^2}$$
$$+2\int_0^t\langle(f-g)(t'),w(t')\rangle\,dt'+2\int_0^t\langle(Q(u,u)-Q(v,v))(t'),w(t')\rangle\,dt'.$$

The non-linear term is estimated thanks to the following lemma.

Lemma 3.1 *In two-dimensional domains, if a and b belong to $\mathcal{V}_\sigma$, we have*

$$|\langle(Q(a,a)-Q(b,b)),a-b\rangle|\leq C\|\nabla(a-b)\|^{\frac{3}{2}}_{L^2}\|a-b\|^{\frac{1}{2}}_{L^2}\|\nabla a\|^{\frac{1}{2}}_{L^2}\|a\|^{\frac{1}{2}}_{L^2}.$$

Proof It is a nice exercise in elementary algebra to deduce from Lemma 2.3 that

$$\langle Q(a,a)-Q(b,b),a-b\rangle=\langle Q(a-b,a),a-b\rangle. \tag{3.2.2}$$

Using Lemma 2.3 again, we get the result. □

Let us go back to the proof of Theorem 3.2. Using the well-known fact that $2ab\leq a^2+b^2$, we get

$$\|w(t)\|^2_{L^2}+\frac{3}{2}\nu\int_0^t\|\nabla w(t')\|^2_{L^2}\,dt'\leq\|w(0)\|^2_{L^2}+\frac{2}{\nu}\int_0^t\|(f-g)(t')\|^2_{\mathcal{V}'_\sigma}\,dt'$$
$$+C\int_0^t\|\nabla w(t')\|^{\frac{3}{2}}_{L^2}\|w(t')\|^{\frac{1}{2}}_{L^2}\|\nabla u(t')\|^{\frac{1}{2}}_{L^2}\|u(t')\|^{\frac{1}{2}}_{L^2}\,dt'+C\nu\int_0^t\|w(t')\|_{L^2}\,dt'.$$

Note that if the domain satisfies the Poincaré inequality, then the last term on the right-hand side can be omitted.

Using (with $\theta = 1/4$) the convexity inequality (1.3.5), page 26 we infer that

$$\begin{aligned}\|w(t)\|_{L^2}^2 + \nu \int_0^t \|\nabla w(t')\|_{L^2}^2\,dt' \\ \leq \|w(0)\|_{L^2}^2 + \frac{2}{\nu}\int_0^t \|(f-g)(t')\|_{\mathcal{V}'_\sigma}^2\,dt' \\ + \frac{C}{\nu^3}\int_0^t \|w(t')\|_{L^2}^2 \left(\|\nabla u(t')\|_{L^2}^2\|u(t')\|_{L^2}^2 + \nu\right)dt'.\end{aligned}$$

Gronwall's lemma implies that

$$\begin{aligned}\|w(t)\|_{L^2}^2 + \nu \int_0^t \|\nabla w(t')\|_{L^2}^2\,dt' \\ \leq \left(\|w(0)\|_{L^2}^2 + \frac{2}{\nu}\int_0^t \|(f-g)(t')\|_{\mathcal{V}'_\sigma}^2\,dt'\right) \\ \times \exp\left(\nu t + \frac{C}{\nu^3}\sup_{\tau\in[0,t]}\|u(\tau)\|_{L^2}^2\int_0^t \|\nabla u(t')\|_{L^2}^2\,dt'\right).\end{aligned}$$

The energy estimate tells us that

$$\sup_{\tau\in[0,t]}\|u(\tau)\|_{L^2}^2\int_0^t \|\nabla u(t')\|_{L^2}^2\,dt' \leq \frac{1}{\nu}\left(\|u_0\|_{L^2}^2 + \frac{2}{\nu}\int_0^t \|f(t')\|_{\mathcal{V}'_\sigma}^2\,dt'\right)^2 e^{2\nu t}.$$

As u and v play the same role, the theorem is proved. □

3.3 Stability in three dimensions

In order to obtain stability, we need to enforce the time regularity of the Leray solution. The precise stability theorem is the following.

Theorem 3.3 *Let u be a Leray solution associated with initial velocity u_0 in $\mathcal{H}$ and bulk force f in $L^2([0,T];\mathcal{V}')$. We assume that u belongs to the space $L^4([0,T];\mathcal{V}_\sigma)$ for some positive T. Then u is unique, belongs to $C([0,T];\mathcal{H})$ and satisfies, for any (s,t) such that $0 \leq s \leq t \leq T$,*

$$\frac{1}{2}\|u(t)\|_{L^2}^2 + \nu\int_s^t \|\nabla u(t')\|_{L^2}^2\,dt' = \frac{1}{2}\|u(s)\|_{L^2}^2 + \int_s^t \langle f(t'), u(t')\rangle\,dt'. \quad (3.3.1)$$

Let v be any solution associated with v_0 in $\mathcal{H}$ and g in $L^2_{\mathrm{loc}}([0,T];\mathcal{V}')$. Then, for all t in $[0,T]$,

$$\begin{aligned}e^{-\nu t}\|(u-v)(t)\|_{L^2}^2 + \nu e^{-\nu t}\int_0^t \|\nabla(u-v)(t')\|_{L^2}^2\,dt' \\ \leq \left(\|u_0 - v_0\|_{L^2}^2 + \frac{2}{\nu}\int_0^t \|(f-g)(t')\|_{\mathcal{V}'_\sigma}^2\,dt'\right)\exp\left(\frac{C}{\nu^3}\int_0^t \|\nabla u(t')\|_{L^2}^4\,dt'\right).\end{aligned}$$

Remarks

- As in the two-dimensional case, if the domain satisfies the Poincaré inequality, then the estimate becomes

$$\|(u-v)(t)\|_{L^2}^2+\nu\int_0^t\|\nabla(u-v)(t')\|_{L^2}^2\,dt'$$
$$\leq\left(\|u_0-v_0\|_{L^2}^2+\frac{2}{\nu}\int_0^t\|(f-g)(t')\|_{\mathcal{V}'_\sigma}^2\,dt'\right)\exp\left(\frac{C}{\nu^3}\int_0^t\|\nabla u(t')\|_{L^2}^4\,dt'\right).$$

- The proof that such an $L^4([0,T];\mathcal{V}_\sigma)$ solution u exists will be detailed in Section 3.4 in the case of bounded domains, and in Section 3.5 in the case without boundary.

Proof of Theorem 3.3 Thanks to Lemma 2.3, the fact that u belongs to $L^4([0,T];\mathcal{V}_\sigma)$ implies that

$$\begin{aligned}\|Q(u,u)\|_{L^2([0,T];\mathcal{V}')}&\leq C\|u\|_{L^\infty([0,T];L^2)}^{\frac12}\|u\|_{L^3([0,T];H_0^1)}^{\frac32}\\&\leq CT^{\frac18}\|u\|_{L^\infty([0,T];L^2)}^{\frac12}\|u\|_{L^4([0,T];H_0^1)}^{\frac32}.\end{aligned}\tag{3.3.2}$$

Hence the non-linear term $Q(u,u)$ belongs to $L^2([0,T];\mathcal{V}')$. Thus, exactly as in the two-dimensional case, u is the solution of (ES_ν) with initial data u_0 and external force $f+Q(u,u)$. Theorem 3.1 immediately implies that u belongs to $C([0,T];\mathcal{H})$ and satisfies, for any (s,t) such that $0\leq s\leq t$,

$$\frac12\|u(t)\|_{L^2}^2+\nu\int_s^t\|\nabla u(t')\|_{L^2}^2\,dt'$$
$$=\frac12\|u(s)\|_{L^2}^2+\int_s^t\langle f(t'),u(t')\rangle\,dt'+\int_s^t\langle Q(u(t'),u(t')),u(t')\rangle\,dt'.$$

Using Lemma 2.3, we get the energy equality (3.3.1).

The method now used in the proof of the stability is important because we shall follow its lines quite often in this book. As u and v are two Leray solutions, we can write that

$$\begin{aligned}\delta_\nu(t)&\stackrel{\text{def}}{=}\|(u-v)(t)\|_{L^2}^2+2\nu\int_0^t\|\nabla(u-v)(t')\|_{L^2}^2\,dt'\\&=\|u(t)\|_{L^2}^2+2\nu\int_0^t\|\nabla u(t')\|_{L^2}^2\,dt'+\|v(t)\|_{L^2}^2+2\nu\int_0^t\|\nabla v(t')\|_{L^2}^2\,dt'\\&\quad-2(u(t)|v(t))_{L^2}-4\nu\int_0^t(\nabla u(t')|\nabla v(t'))_{L^2}\,dt'\\&\leq\|u_0\|_{L^2}^2+2\int_0^t\langle f(t'),u(t')\rangle\,dt'+\|v_0\|_{L^2}^2+2\int_0^t\langle g(t'),v(t')\rangle\,dt'\\&\quad-2(u(t)|v(t))_{L^2}-4\nu\int_0^t(\nabla u(t')|\nabla v(t'))_{L^2}\,dt'.\end{aligned}\tag{3.3.3}$$

Now the problem consists in evaluating the cross-product terms

$$(u(t)|v(t))_{L^2} + 2\nu \int_0^t (\nabla u(t')|\nabla v(t'))_{L^2}\, dt'$$
$$- \int_0^t \langle g(t'), v(t')\rangle\, dt' - \int_0^t \langle f(t'), u(t')\rangle\, dt'.$$

Let us suppose for a moment that u and v are smooth in space and time; then we can simply scalar-multiply by v the equation satisfied by u, and conversely scalar-multiply by u the equation satisfied by v. We get

$$(\partial_t u|v)_{L^2} + \nu(\nabla u|\nabla v)_{L^2} + (u\cdot\nabla u|v)_{L^2} = (f|v)_{L^2} \quad \text{and}$$

$$(\partial_t v|u)_{L^2} + \nu(\nabla v|\nabla u)_{L^2} + (v\cdot\nabla v|u)_{L^2} = (g|u)_{L^2};$$

hence summing both equalities yields

$$\partial_t(u(t)|v(t))_{L^2} + 2\nu(\nabla u(t)|\nabla v(t))_{L^2} - (f(t)|v(t))_{L^2} - (g(t)|u(t))_{L^2}$$
$$+ (u(t)\cdot\nabla u(t)|v(t))_{L^2} + (v(t)\cdot\nabla v(t)|u(t))_{L^2} = 0.$$

After an easy algebraic computation we find that

$$(u(t)\cdot\nabla u(t)|v(t))_{L^2} + (v(t)\cdot\nabla v(t)|u(t))_{L^2} = ((u-v)(t)\cdot\nabla(u-v)(t)|u(t))_{L^2},$$

hence after integration in time, we obtain

$$(u(t)|v(t))_{L^2} + 2\nu \int_0^t (\nabla u(t')|\nabla v(t'))_{L^2}^2\, dt'$$
$$- \int_0^t (g(t')|v(t'))_{L^2}\, dt' - \int_0^t (f(t')|u(t'))_{L^2}\, dt'$$
$$= -(u_0|v_0)_{L^2} + \int_0^t ((u-v)(t')\cdot\nabla(u-v)(t')|u(t'))_{L^2}\, dt'$$
$$+ \int_0^t (f(t')|(v-u)(t'))\, dt' + \int_0^t (g(t')|(u-v)(t'))\, dt'. \tag{3.3.4}$$

Plugging (3.3.4) into (3.3.3) yields

$$\|(u-v)(t)\|_{L^2}^2 + 2\nu \int_0^t \|\nabla(u-v)(t')\|_{L^2}^2\, dt' \le \|u_0 - v_0\|_{L^2}^2$$
$$+ 2\left|\int_0^t ((f-g)|(u-v))(t')\, dt'\right|$$
$$+ 2\left|\int_0^t ((u-v)\cdot\nabla(u-v)|u)_{L^2}(t')\, dt'\right|.$$

and the Gronwall lemma gives the smallness of δ_ν. However unfortunately the above computations make no sense if no smoothness in space and time is known on u or on v (and in particular the final Gronwall argument does not seem possible to write correctly). So some precautions have to be taken in order to make these computations valid, and to conclude the argument by the Gronwall lemma – in particular we are going to see that the assumption that $u \in L^4([0,T];\mathcal{V}_\sigma)$ is enough to make the above computations valid.

Let us therefore proceed with the rigorous computations. Identity (3.3.3) involves scalar products of u and v, which naturally lead to using the definition of weak solutions choosing for instance u as a test function. Unfortunately, in order to be admissible, test functions need to belong to $C^1(\mathbf{R}^+;\mathcal{V}_\sigma)$. so that some preliminary smoothing in time of u is required. We shall use the following approximation lemma and postpone its proof to the end of the proof of the theorem.

Lemma 3.2 *Let u be a Leray solution which belongs to $L^4([0,T];\mathcal{V}_\sigma)$. A sequence $(\widetilde{u}_k)_{k\in\mathbf{N}}$ of $C^1(\mathbf{R}^+;\mathcal{V}_\sigma)$ exists such that*

- *the sequence $(\widetilde{u}_k)_{k\in\mathbf{N}}$ tends to u in $L^4([0,T];\mathcal{V}_\sigma)\cap L^\infty([0,T];\mathcal{H})$;*
- *for all $k\in\mathbf{N}$, we have*

$$\partial_t\widetilde{u}_k - \nu\Delta\widetilde{u}_k = Q(\widetilde{u}_k,\widetilde{u}_k) + f + R_k + \nabla p_k \tag{3.3.5}$$

with $\lim_{k\to\infty}\|R_k\|_{L^2([0,T];\mathcal{V}'_\sigma)} = 0$.

The function $\widetilde{u}_k$ belongs to $C^1(\mathbf{R}^+;\mathcal{V}_\sigma)$, so it can be used as a test function in Definition 2.5. As v is a Leray solution, we have

$$\begin{aligned}\mathcal{B}_k(t) &\stackrel{\text{def}}{=} (v(t)|\widetilde{u}_k(t))_{L^2}\\ &= (v(0)|\widetilde{u}_k(0))_{L^2} - \nu\int_0^t(\nabla v(t')|\nabla\widetilde{u}_k(t'))_{L^2}\,dt' + \int_0^t\langle g(t'),\widetilde{u}_k(t')\rangle\,dt'\\ &\quad+\int_0^t(v(t')\otimes v(t')|\nabla\widetilde{u}_k(t'))_{L^2}\,dt' + \int_0^t\langle v(t'),\partial_t\widetilde{u}_k(t')\rangle\,dt'.\end{aligned}$$

Thanks to (3.3.5), we get

$$\begin{aligned}\mathcal{B}_k(t) &= (v(0)|\widetilde{u}_k(0))_{L^2} - 2\nu\int_0^t(\nabla v(t')|\nabla\widetilde{u}_k(t'))_{L^2}\,dt' + \int_0^t\langle g(t'),\widetilde{u}_k(t')\rangle\,dt'\\ &\quad+\int_0^t\langle f(t'),v(t')\rangle\,dt' + \int_0^t(v(t')\otimes v(t')|\nabla\widetilde{u}_k(t'))_{L^2}\,dt'\\ &\quad+\int_0^t\langle Q(\widetilde{u}_k(t'),\widetilde{u}_k(t')),v(t')\rangle\,dt' + \int_0^t\langle v(t'),R_k(t')\rangle\,dt'.\end{aligned}$$

Lemma 3.2 implies that $\lim_{k\to\infty} \mathcal{B}_k(t) = (v(t)|u(t))_{L^2}$ and that

$$\lim_{k\to\infty} \Big\{ (v(0)|\widetilde{u}_k(0))_{L^2} - 2\nu \int_0^t (\nabla v(t')|\nabla \widetilde{u}_k(t'))_{L^2}\, dt' + \int_0^t \langle v(t'), R_k(t')\rangle\, dt' + \int_0^t \langle g(t'), \widetilde{u}_k(t')\rangle\, dt' \Big\}$$

is equal to

$$(v(0)|u(0))_{L^2} - 2\nu \int_0^t (\nabla v(t')|\nabla u(t'))_{L^2}\, dt' + \int_0^t \langle g(t'), u(t')\rangle\, dt'.$$

Thus stating

$$\mathcal{N}_k(t) \stackrel{\text{def}}{=} \int_0^t (v(t')\otimes v(t')|\nabla \widetilde{u}_k(t'))_{L^2}\, dt' + \int_0^t \langle Q(\widetilde{u}_k(t'), \widetilde{u}_k(t')), v(t')\rangle\, dt',$$

we obtain

$$\begin{aligned}(v(t)|u(t))_{L^2} &= (v(0)|u(0))_{L^2} - 2\nu \int_0^t (\nabla v(t')|\nabla u(t'))_{L^2}\, dt' \\ &\quad + \int_0^t \langle g(t'), u(t')\rangle\, dt' + \int_0^t \langle f(t'), v(t')\rangle\, dt' + \lim_{k\to\infty} \mathcal{N}_k(t).\end{aligned}$$

Plugging this into (3.3.3) gives

$$\delta_\nu(t) = \|u_0 - v_0\|_{L^2}^2 + 2\int_0^t \langle (f-g)(t'), (u-v)(t')\rangle\, dt' + \lim_{k\to\infty} \mathcal{N}_k(t).$$

It remains to study the term $\mathcal{N}_k(t)$. In order to do this, let us observe that, for any vector field a and b in $\mathcal{V}_\sigma$, we have $(b\otimes b|\nabla a)_{L^2} = \langle Q(b,b), a\rangle$ and thus

$$(b\otimes b|\nabla a)_{L^2} + \langle Q(a,a), b\rangle = \langle Q(b,b), a\rangle + \langle Q(a,a), b\rangle.$$

Using Lemma 2.3, we can write

$$\begin{aligned}\langle Q(b,b), a\rangle + \langle Q(a,a), b\rangle &= \langle Q(b,b), a-b\rangle + \langle Q(a,a), b-a\rangle \\ &= \langle Q(b,a), a-b\rangle + \langle Q(a,a), b-a\rangle.\end{aligned}$$

Thus, it turns out that

$$\begin{aligned}\langle Q(b,b), a\rangle + \langle Q(a,a), b\rangle &= \langle Q(a-b,a), b-a\rangle \\ &= ((a-b)\otimes a|\nabla(b-a))_{L^2}.\end{aligned}$$

Using the Gagliardo–Nirenberg inequality (see Corollary 1.2), we get for any a and c in $\mathcal{V}_\sigma$,

$$\begin{aligned}|(c\otimes a|\nabla c)_{L^2}| &\le C\|a\|_{L^6}\|c\|_{L^3}\|\nabla c\|_{L^2} \\ &\le C\|\nabla a\|_{L^2}\|c\|_{L^2}^{\frac12}\|\nabla c\|_{L^2}^{\frac32}.\end{aligned} \tag{3.3.6}$$

For almost every time t, the vector field $v(t)$ belongs to $\mathcal{V}_\sigma$. It follows that for all $k \in \mathbf{N}$ and $t \geq 0$, taking $a = \widetilde{u}_k(t')$ and $b = v(t')$, $t' \in [0, t]$, we have

$$\mathcal{N}_k(t) \leq C \int_0^t \|\nabla \widetilde{u}_k(t')\|_{L^2} \|(\widetilde{u}_k - v)(t')\|_{L^2}^{\frac{1}{2}} \|\nabla(\widetilde{u}_k - v)(t')\|_{L^2}^{\frac{3}{2}} \, dt'.$$

Using Lemma 3.2, we know that

$$\begin{array}{llll} (\|\nabla \widetilde{u}_k(\cdot)\|_{L^2})_{k\in\mathbf{N}} & \text{tends to} & \|\nabla u(\cdot)\|_{L^2} & \text{in } L^4([0,T]) \\ (\|(\widetilde{u}_k - v)(t')\|_{L^2})_{k\in\mathbf{N}} & \text{tends to} & \|(u-v)(\cdot)\|_{L^2} & \text{in } L^\infty([0,T]) \\ (\|\nabla(\widetilde{u}_k - v)(\cdot)\|_{L^2})_{k\in\mathbf{N}} & \text{tends to} & \|\nabla(u-v)(\cdot)\|_{L^2} & \text{in } L^2([0,T]). \end{array}$$

Therefore, we have

$$\lim_{k\to\infty} \mathcal{N}_k(t) \leq \int_0^t \|\nabla u(t')\|_{L^2} \|(u-v)(t')\|_{L^2}^{\frac{1}{2}} \|\nabla(u-v)(t')\|_{L^2}^{\frac{3}{2}} \, dt'.$$

We conclude that

$$\begin{aligned} \delta_\nu(t) \leq \|u_0 - v_0\|_{L^2}^2 &+ 2\int_0^t \|(f-g)(t')\|_{\mathcal{V}'_\sigma} \|(u-v)(t')\|_{\mathcal{V}} \, dt' \\ &+ C\int_0^t \|\nabla u(t')\|_{L^2} \|\nabla(u-v)(t')\|_{L^2}^{\frac{3}{2}} \|(u-v)(t')\|_{L^2}^{\frac{1}{2}} \, dt'. \end{aligned}$$

Using the convexity inequality (1.3.5) with $\theta = 1/4$ and $\theta = 1/2$, we obtain

$$\begin{aligned} &\|(u-v)(t)\|_{L^2}^2 + \nu \int_0^t \|\nabla(u-v)(t')\|_{L^2}^2 dt' \\ &\leq \|u_0 - v_0\|_{L^2}^2 + \frac{2}{\nu}\int_0^t \|(f-g)(t')\|_{\mathcal{V}'_\sigma}^2 \, dt' \\ &\quad + \int_0^t \left(\frac{C}{\nu^3}\|\nabla u(t')\|_{L^2}^4 + \nu\right) \|(u-v)(t')\|_{L^2}^2 \, dt'. \end{aligned}$$

Gronwall's lemma allows us to conclude the proof of Theorem 3.3 provided we prove Lemma 3.2. □

Proof of Lemma 3.2 Thanks to inequality (3.3.2), $Q(u,u)$ belongs to $L^2([0,T]; \mathcal{V}')$. This implies that if in addition u is a Leray solution, then $\partial_t u$ also belongs to $L^2([0,T]; \mathcal{V}')$. Lebesgue's theorem together with Proposition 2.3, page 40, yields

$$\lim_{k\to\infty} \|P_k u - u\|_{L^4([0,T];\mathcal{V}_\sigma)} = \lim_{k\to\infty} \|P_k \partial_t u - \partial_t u\|_{L^2([0,T];\mathcal{V}'_\sigma)} = 0.$$

Then, as in Lemma 2.2, page 44, we can define by a standard regularization procedure in time, a sequence $(\widetilde{u}_k)_{k\in\mathbf{N}}$ in $C^1(\mathbf{R}^+, \mathcal{V}_\sigma)$ such that $\widetilde{u}_k$ tends to u

in $L^4([0,T];\mathcal{V}_\sigma)\cap L^\infty([0,T],\mathcal{H})$. Moreover $\partial_t\widetilde{u}_k$ tends to $\partial_t u$ in $L^2([0,T];\mathcal{V}')$ and inequality (3.3.2) implies that

$$\lim_{k\to\infty}\|Q(\widetilde{u}_k,\widetilde{u}_k)-Q(u,u)\|_{L^2([0,T];\mathcal{V}'_\sigma)}=0.$$

Lemma 3.2 and thus Theorem 3.3 are now proved. □

3.4 Stable solutions in a bounded domain

The purpose of this section is the proof of the existence of solutions of the system (NS_ν) which are L^4 in time with values in $\mathcal{V}_\sigma$ in the case when the domain Ω is bounded. In order to state (and prove) a sharp theorem, we shall introduce intermediate spaces between the spaces $\mathcal{V}'_\sigma$ and $\mathcal{V}_\sigma$. Then, we shall prove a global existence theorem for small data and then a theorem local in time for large data.

3.4.1 Intermediate spaces

We shall define a family of intermediate spaces between the spaces $\mathcal{V}'_\sigma$ and $\mathcal{V}_\sigma$. This can be done by abstract interpolation theory but we prefer to do it here in an explicit way.

Definition 3.2 *Let s be in $[-1,1]$. We shall denote by $\mathcal{V}^s_\sigma$ the space of vector fields u in $\mathcal{V}'$ such that*

$$\|u\|^2_{\mathcal{V}^s_\sigma}\overset{\text{def}}{=}\sum_{j\in\mathbf{N}}\mu_j^{2s}\langle u,e_j\rangle^2<+\infty.$$

Here $(e_j)_{j\in\mathbf{N}}$ denotes the Hilbert basis on $\mathcal{H}$ given by Theorem 2.1, page 34.

Theorem 2.1 implies that $\mathcal{V}^0_\sigma=\mathcal{H}$ and $\mathcal{V}^1_\sigma=\mathcal{V}_\sigma$. Moreover, it is obvious that, when s is non-negative, $\mathcal{V}^s_\sigma$ endowed with the norm $\|\cdot\|_{\mathcal{V}^s_\sigma}$ is a Hilbert space.

The following proposition will be important in the following two paragraphs.

Proposition 3.1 *The space $\mathcal{V}^{\frac12}_\sigma$ is embedded in L^3 and the space $L^{\frac32}$ is embedded in $\mathcal{V}^{-\frac12}_\sigma$.*

Proof This proposition can be proved using abstract interpolation theory. We prefer to present here a self-contained proof in the spirit of the proof of Theorem 1.2. Let us consider a in $\mathcal{V}^{\frac12}_\sigma$. Without loss of generality, we can assume that $\|a\|_{\mathcal{V}^{\frac12}_\sigma}\leq 1$. Let us define, for a positive real number Λ,

$$a_\Lambda\overset{\text{def}}{=}\sum_{j\,/\,\mu_j<\Lambda}\langle a,e_j\rangle e_j\quad\text{and}\quad b_\Lambda\overset{\text{def}}{=}a-a_\Lambda.$$

Using the fact that

$$\{x \in \Omega \,/\, |a(x)| > \Lambda\} \subset \{x \in \Omega \,/\, |a_\Lambda(x)| > \Lambda/2\} \cup \{x \in \Omega \,/\, |b_\Lambda(x)| > \Lambda/2\},$$

we can write

$$\begin{aligned}\|a\|_{L^3}^3 &\leq 3\int_0^{+\infty} \Lambda^2 \text{meas}\,(\{x \in \Omega \,/\, |a_\Lambda(x)| > \Lambda/2\})\, d\Lambda \\ &\quad + 3\int_0^{+\infty} \Lambda^2 \text{meas}\,(\{x \in \Omega \,/\, |b_\Lambda(x)| > \Lambda/2\})\, d\Lambda \\ &\leq 3 \times 2^6 \int_0^{+\infty} \Lambda^{-4} \|a_\Lambda\|_{L^6}^6 \, d\Lambda + 3 \times 2^2 \int_0^{+\infty} \|b_\Lambda\|_{L^2}^2 \, d\Lambda.\end{aligned}$$

Thanks to Theorem 1.2, we have, by definition of the $\|\cdot\|_{\mathcal{V}_\sigma^s}$ norm,

$$\begin{aligned}\|a_\Lambda\|_{L^6}^2 &\leq C\|a_\Lambda\|_{\mathcal{V}_\sigma}^2 \\ &\leq C \sum_{j \,/\, \mu_j < \Lambda} \mu_j^2 \langle a, e_j\rangle^2 \\ &\leq C\Lambda \sum_{j \,/\, \mu_j < \Lambda} \mu_j \langle a, e_j\rangle^2 \\ &\leq C\Lambda.\end{aligned}$$

Thus we have

$$\begin{aligned}\|a\|_{L^3}^3 &\leq C\int_0^{+\infty} \Lambda^{-2} \|a_\Lambda\|_{\mathcal{V}_\sigma}^2 \, d\Lambda + C\int_0^{+\infty} \|b_\Lambda\|_{L^2}^2 \, d\Lambda \\ &\leq C\sum_{j \in \mathbf{N}} \int_{\mu_j}^{+\infty} \Lambda^{-2} \mu_j^2 \langle a, e_j\rangle^2 \, d\Lambda + C\sum_{j \in \mathbf{N}} \int_0^{\mu_j} \langle a, e_j\rangle^2 \, d\Lambda \\ &\leq C\sum_{j \in \mathbf{N}} \mu_j \langle a, e_j\rangle^2 \\ &\leq C.\end{aligned}$$

This proves the first part of the proposition.

The second part is obtained by a duality argument. By definition, we have, for any a in $\mathcal{V}'$,

$$\begin{aligned}\|a\|_{\mathcal{V}_\sigma^{-\frac{1}{2}}} &= \|(\mu_j^{-\frac{1}{2}} \langle a, e_j\rangle)_{j \in \mathbf{N}}\|_{\ell^2} \\ &= \sup_{\substack{(\alpha_j)_{j \in \mathbf{N}} \\ \|(\alpha_j)_{j \in \mathbf{N}}\|_{\ell^2} \leq 1}} \sum_{j \in \mathbf{N}} \alpha_j \mu_j^{-\frac{1}{2}} \langle a, e_j\rangle. \qquad (3.4.1)\end{aligned}$$

The map L defined by

$$L\begin{cases}\ell^2 & \to \mathcal{V}_\sigma^{\frac{1}{2}}\\ (\alpha_j)_{j\in\mathbf{N}} & \mapsto \displaystyle\sum_{j\in\mathbf{N}} \alpha_j\mu_j^{-\frac{1}{2}}e_j\end{cases}$$

is an onto isometry. Thus, thanks to (3.4.1), we have

$$\|a\|_{\mathcal{V}_\sigma^{-\frac{1}{2}}} = \sup_{\|\varphi\|_{\mathcal{V}_\sigma^{\frac{1}{2}}}\leq 1} \sum_{j\in\mathbf{N}}(L^{-1}\varphi)_j\mu_j^{-\frac{1}{2}}\langle a, e_j\rangle.$$

For any φ in $\mathcal{V}_\sigma$, we have

$$\sum_{j\in\mathbf{N}}(L^{-1}\varphi)_j\mu_j^{-\frac{1}{2}}\langle a, e_j\rangle = \langle a, \varphi\rangle.$$

If we assume that a is in $L^{\frac{3}{2}}$, we have, because φ is in L^3,

$$\langle a, \varphi\rangle = \int_\Omega a(x)\cdot\varphi(x)\,dx.$$

Hölder's inequality and the first part of Proposition 3.1 imply that

$$\begin{aligned}|\langle a, \varphi\rangle| &\leq \|a\|_{L^{\frac{3}{2}}}\|\varphi\|_{L^3}\\ &\leq C\|a\|_{L^{\frac{3}{2}}}\|\varphi\|_{\mathcal{V}_\sigma^{\frac{1}{2}}}.\end{aligned}$$

Thus we have

$$\begin{aligned}\|a\|_{\mathcal{V}_\sigma^{-\frac{1}{2}}} &\leq \sup_{\|\varphi\|_{\mathcal{V}_\sigma^{\frac{1}{2}}}\leq 1}\langle a, \varphi\rangle\\ &\leq C\|a\|_{L^{\frac{3}{2}}}.\end{aligned}$$

This completes the proof of Proposition 3.1. □

3.4.2 The well-posedness result

The aim of this paragraph is the proof of the following existence theorem with data in $\mathcal{V}_\sigma^{\frac{1}{2}}$.

Theorem 3.4 *If the initial data u_0 belongs to $\mathcal{V}_\sigma^{\frac{1}{2}}$ and the bulk force f belongs to $L^2_{\text{loc}}(\mathbf{R}_+;\mathcal{V}_\sigma^{-\frac{1}{2}})$, then a positive time T exists such that a solution u of* (NS_ν) *exists in $L^4([0,T];\mathcal{V}_\sigma)$. This solution is unique and belongs to $C([0,T];\mathcal{V}_\sigma^{\frac{1}{2}})$.*

Moreover, a constant c exists (which can be chosen independently of the domain Ω) such that, if

$$\|u_0\|_{\mathcal{V}_\sigma^{\frac{1}{2}}} + \frac{1}{\nu}\|f\|_{L^2(\mathbf{R}^+;\mathcal{V}_\sigma^{-\frac{1}{2}})} \leq c\nu,$$

then the above solution is global.

Proof For the sake of simplicity, we shall ignore the bulk force in the proof. Let us consider the sequence $(u_k)_{k\in\mathbf{N}}$ used in the proof of Leray's theorem and defined by the ordinary differential equation $(\mathrm{NS}_{\nu,k})$, page 45. The point is to prove that this sequence $(u_k)_{k\in\mathbf{N}}$ is bounded in $L^4([0,T];\mathcal{V}_\sigma)$ for some positive T.

Let us recall the remark on page 56 which tells us that

$$u_k = \sum_{j=0}^{k} U_{j,k}(t)e_j$$

with

$$U_{j,k}(t) \stackrel{\text{def}}{=} (u_0|e_j)_{L^2}e^{-\nu\mu_j^2 t} + \int_0^t e^{-\nu\mu_j^2(t-t')}\langle Q(u_k(t'),u_k(t')),e_j\rangle\,dt'. \tag{3.4.2}$$

Using Proposition 3.1 we claim that for any vector field a and b in $\mathcal{V}_\sigma$ and for all $j\in\mathbf{N}$,

$$\begin{aligned}\|P_j\operatorname{div}(a\otimes b)\|_{\mathcal{V}_\sigma^{-\frac12}} &= \|P_j(a\cdot\nabla b)\|_{\mathcal{V}_\sigma^{-\frac12}}\\ &\le C\|a\cdot\nabla b\|_{L^{\frac32}}\\ &\le C\|a\|_{L^6}\|\nabla b\|_{L^2}.\end{aligned}$$

Using Sobolev embeddings, we deduce that

$$\|P_j\operatorname{div}(a\otimes b)\|_{\mathcal{V}_\sigma^{-\frac12}} \le C\|a\|_{\mathcal{V}_\sigma}\|b\|_{\mathcal{V}_\sigma}. \tag{3.4.3}$$

By definition of the norm on $\mathcal{V}_\sigma^{-\frac12}$, we infer that for all $k\in\mathbf{N}$, a $(c_{j,k}(t))_{j\in\mathbf{N}}$ exists such that

$$|\langle Q(u_k(t),u_k(t)),e_j\rangle| \le Cc_{j,k}(t)\mu_j^{\frac12}\|u_k(t)\|_{\mathcal{V}_\sigma}^2 \tag{3.4.4}$$

with, for any t, $\sum_{j\in\mathbf{N}} c_{j,k}^2(t)=1$. Plugging this inequality into (3.4.2), we get

$$|U_{j,k}(t)| \le |(u_0|e_j)|e^{-\nu\mu_j^2 t} + C\mu_j^{\frac12}\int_0^t e^{-\nu\mu_j^2(t-t')}c_{j,k}(t')\|u_k(t')\|_{\mathcal{V}_\sigma}^2\,dt'. \tag{3.4.5}$$

Thanks to Young's inequality $\|f\star g\|_{L^4}\le\|f\|_{L^{\frac43}}\|g\|_{L^2}$, we have, for any positive T,

$$\|U_{j,k}\|_{L^4([0,T])} \le |(u_0|e_j)|\mu_j^{-\frac12}\left(\frac{1-e^{-4\nu\mu_j^2T}}{4\nu}\right)^{\frac14} + \frac{C}{\nu^{\frac34}}\mu_j^{-1}\left(\int_0^T c_{j,k}^2(t)\|u_k(t)\|_{\mathcal{V}_\sigma}^4\,dt\right)^{\frac12}.$$

Multiplying by μ_j and taking the ℓ^2 norm gives

$$\left(\sum_{j\in\mathbf{N}} \mu_j^2 \|U_{j,k}\|^2_{L^4([0,T])}\right)^{\frac12} \leq \left(\sum_{j\in\mathbf{N}} \mu_j (u_0|e_j)^2 \left(\frac{1-e^{-4\nu\mu_j^2 T}}{\nu}\right)^{\frac12}\right)^{\frac12} + \frac{C}{\nu^{\frac34}}\left(\sum_{j\in\mathbf{N}} \int_0^T c_{j,k}^2(t)\|u_k(t)\|^4_{\mathcal{V}_\sigma}\,dt\right)^{\frac12}.$$

Thanks to (3.4.4), we have

$$\left(\sum_{j\in\mathbf{N}} \mu_j^2 \|U_{j,k}\|^2_{L^4([0,T])}\right)^{\frac12} \leq \left(\sum_{j\in\mathbf{N}} \mu_j (u_0|e_j)^2 \left(\frac{1-e^{-4\nu\mu_j^2 T}}{\nu}\right)^{\frac12}\right)^{\frac12} + \frac{C}{\nu^{\frac34}}\|u_k\|^2_{L^4([0,T];\mathcal{V}_\sigma)}.$$

Now let us observe that, thanks to the Cauchy–Schwarz inequality, for any a in $\ell^2(L^4[0,T])$,

$$\begin{aligned}\int_0^T \|a_j(t)\|^4_{\ell^2(\mathbf{N})}\,dt &= \int_0^T \left(\sum_{j\in\mathbf{N}} a_j^2(t)\right)^2 dt\\ &= \sum_{j\in\mathbf{N},k\in\mathbf{N}} \int_0^T a_j^2(t)a_k^2(t)\,dt\\ &\leq \sum_{j\in\mathbf{N},k\in\mathbf{N}} \|a_j\|^2_{L^4([0,T])}\|a_k\|^2_{L^4([0,T])}\\ &\leq \left\|\left(\|a_j\|_{L^4([0,T])}\right)_{j\in\mathbf{N}}\right\|^4_{\ell^2}\end{aligned}$$

Let us notice that this is a particular case of the Minkowski inequality. Thus we infer that

$$\|u_k\|_{L^4([0,T];\mathcal{V}_\sigma)} \leq \left(\sum_{j\in\mathbf{N}} \mu_j (u_0|e_j)^2 \left(\frac{1-e^{-4\nu\mu_j^2 T}}{\nu}\right)^{\frac12}\right)^{\frac12} + \frac{C}{\nu^{\frac34}}\|u_k\|^2_{L^4([0,T];\mathcal{V}_\sigma)}.$$

Let us define

$$T_k \stackrel{\text{def}}{=} \sup\left\{T>0 /\ \|u_k\|_{L^4([0,T];\mathcal{V}_\sigma)} \leq \frac{\nu^{\frac34}}{2C}\right\}.$$

As u_k belongs to $C^1(\mathbf{R}^+; P_k\mathcal{V}_\sigma)$, the supremum T_k is positive for all k. We have, for all $T \in [0, T_k]$,

$$\|u_k\|_{L^4([0,T];\mathcal{V}_\sigma)} \leq 2\left(\sum_{j\in\mathbf{N}} \mu_j (u_0|e_j)^2 \left(\frac{1-e^{-4\nu\mu_j^2 T}}{\nu}\right)^{\frac{1}{2}}\right)^{\frac{1}{2}}. \tag{3.4.6}$$

In the case of small data, it is enough to observe that, for any positive T,

$$\sum_{j\in\mathbf{N}} \mu_j (u_0|e_j)^2 \left(\frac{1-e^{-4\nu\mu_j^2 T}}{\nu}\right)^{\frac{1}{2}} \leq \frac{1}{\nu^{\frac{1}{2}}}\|u_0\|^2_{\mathcal{V}_\sigma^{\frac{1}{2}}}.$$

Thus, if $\|u_0\|_{\mathcal{V}_\sigma^{\frac{1}{2}}} \leq \dfrac{\nu}{8C}$, we have, for any T smaller than T_k,

$$\|u_k\|_{L^4([0,T];\mathcal{V}_\sigma)} \leq \frac{\nu^{\frac{3}{4}}}{4C}.$$

This implies that $T_k = +\infty$ and that

$$\|u_k\|_{L^4(\mathbf{R}^+;\mathcal{V}_\sigma)} \leq \frac{2}{\nu^{\frac{1}{4}}}\|u_0\|_{\mathcal{V}_\sigma^{\frac{1}{2}}}.$$

In the case of large data, let us define the smallest integer j_0 such that

$$\left(\sum_{j>j_0} \mu_j (u_0|e_j)^2\right)^{\frac{1}{2}} \leq \frac{\nu}{16C}. \tag{3.4.7}$$

Then, using the fact that $1 - e^{-x} \leq x$ for all non-negative x, we can write for all $T < T_k$,

$$\begin{aligned}
U_1(T) &\stackrel{\text{def}}{=} \left(\sum_{j\in\mathbf{N}} \mu_j (u_0|e_j)^2 \left(\frac{1-e^{-4\nu\mu_j^2 T}}{\nu}\right)^{\frac{1}{2}}\right)^{\frac{1}{2}} \\
&\leq \frac{\nu^{\frac{3}{4}}}{16C} + \left(\sum_{j\leq j_0} \mu_j (u_0|e_j)^2 \left(\frac{1-e^{-4\nu\mu_j^2 T}}{\nu}\right)^{\frac{1}{2}}\right)^{\frac{1}{2}} \\
&\leq \frac{\nu^{\frac{3}{4}}}{16C} + \mu_{j_0}\sqrt{2}\,T^{\frac{1}{4}}\|u_0\|_{L^2}.
\end{aligned}$$

Thus, stating

$$T_{u_0} \stackrel{\text{def}}{=} \left(\frac{\nu^{\frac{3}{4}}}{16\sqrt{2}C\mu_{j_0}\|u_0\|_{L^2}}\right)^4,$$

we have, for any positive T less than $\min\{T_k; T_{u_0}\}$,

$$\|u_k\|_{L^4([0,T);\mathcal{V}_\sigma)} \leq \frac{\nu^{\frac{3}{4}}}{2C}.$$

Thus, for all k, $T_k \geq T_{u_0}$. This implies that $(u_k)_{k\in\mathbf{N}}$ is a bounded sequence of $L^4([0,T];\mathcal{V}_\sigma)$. We infer that a Leray solution u of (NS_ν) exists such that u belongs to $L^4([0,T];\mathcal{V}_\sigma)$. Thanks to Theorem 3.3, this solution is unique on $[0,T]$, continuous from $[0,T]$ into $\mathcal{H}$ and satisfies the energy equality on $[0,T]$.

The only thing we have to prove now is the continuity of u from $[0,T]$ into $\mathcal{V}_\sigma^{\frac{1}{2}}$. As u belongs to $L^4([0,T];\mathcal{V}_\sigma)$, we infer from (3.4.3) that $Q(u,u)$ is in $L^2([0,T];\mathcal{V}_\sigma^{-\frac{1}{2}})$. Using Theorem 3.1, we get

$$(u(t)|e_j) = (u_0|e_j)e^{-\nu\mu_j^2 t} + \int_0^t e^{-\nu\mu_j^2(t-t')}\langle Q(u(t'),u(t')),e_j\rangle\,dt'.$$

Using (3.4.3) again, we infer by definition of the norm on $\mathcal{V}_\sigma^{-\frac{1}{2}}$ that there exists a sequence $(c_j(t))_{j\in\mathbf{N}}$, such that

$$|(u(t)|e_j)| \leq |(u_0|e_j)|e^{-\nu\mu_j^2 t} + C\mu_j^{\frac{1}{2}}\int_0^t e^{-\nu\mu_j^2(t-t')}c_j(t')\|u(t')\|_{\mathcal{V}_\sigma}^2\,dt'$$

with $\sum_{j\in\mathbf{N}} c_j^2(t) = 1$. Using the Cauchy–Schwarz inequality, we have

$$\|(u(t)|e_j)\|_{L^\infty([0,T])} \leq |(u_0|e_j)| + \frac{C}{\nu^{\frac{1}{2}}}\mu_j^{-\frac{1}{2}}\left(\int_0^T c_j^2(t)\|u(t)\|_{\mathcal{V}_\sigma}^4\,dt\right)^{\frac{1}{2}}.$$

Multiplying by $\mu_j^{\frac{1}{2}}$ and taking the ℓ^2 norm gives

$$\begin{aligned} U_2(T) &\stackrel{\text{def}}{=} \left(\sum_{j\in\mathbf{N}} \mu_j\|(u(\cdot)|e_j)\|_{L^\infty([0,T])}^2\right)^{\frac{1}{2}} \\ &\leq \sqrt{2}\|u_0\|_{\mathcal{V}_\sigma^{\frac{1}{2}}} + \sqrt{2}\frac{C}{\nu^{\frac{1}{2}}}\left(\sum_{j\in\mathbf{N}}\int_0^T c_j^2(t)\|u(t)\|_{\mathcal{V}_\sigma}^4\,dt\right)^{\frac{1}{2}} \\ &\leq \sqrt{2}\|u_0\|_{\mathcal{V}_\sigma^{\frac{1}{2}}} + \frac{C}{\nu^{\frac{1}{2}}}\|u\|_{L^4([0,T];\mathcal{V}_\sigma)}^2. \end{aligned}$$

This gives that u is in $L^\infty([0,T];\mathcal{V}_\sigma^{\frac{1}{2}})$. In fact, it will imply continuity using the following argument. Let η be any positive number. An integer j_0 exists such that

$$\left(\sum_{j>j_0} \mu_j\|(u(\cdot)|e_j)\|_{L^\infty([0,T])}^2\right)^{\frac{1}{2}} < \frac{\eta}{2}.$$

Now, it turns out that for all $(t_1,t_2)\in[0,T]^2$, one has

$$\begin{aligned}\|u(t_1)-u(t_2)\|_{\mathcal{V}_\sigma^{\frac12}} &\leq \Big(\sum_{j>j_0}\mu_j\|(u(\cdot)|e_j)\|^2_{L^\infty([0,T])}\Big)^{\frac12}+\Big(\sum_{j\leq j_0}\mu_j(u(t_1)-u(t_2)|e_j)^2\Big)^{\frac12}\\ &\leq \frac{\eta}{2}+\mu_{j_0}^{\frac12}\|u(t_1)-u(t_2)\|_{L^2}.\end{aligned}$$

Theorem 3.1 tells us that u is continuous from $[0,T]$ into $\mathcal{H}$. Thus the whole Theorem 3.4 is proved. □

3.4.3 Some remarks about stable solutions

In this paragraph, we shall assume that the bulk force f is identically 0. We shall establish some results about the maximal existence time of the solution constructed in the preceding paragraph.

Proposition 3.2 *Let us assume that the initial data u_0 belongs to $\mathcal{V}_\sigma$. Then the maximal time of existence $T^\star$ of the solution u in the space*

$$C([0,T^\star[;\mathcal{V}_\sigma^{\frac12})\cap L^4_{\text{loc}}([0,T^\star[;\mathcal{V}_\sigma)$$

satisfies

$$T^\star\geq\frac{c\nu^3}{\|\nabla u_0\|^4_{L^2}}.$$

Proof Thanks to (3.4.6), the maximal time of existence $T^\star$ is bounded from below by T such that

$$4\sum_j\mu_j(u_0|e_j)^2\left(\frac{1-e^{-\nu\mu_j^2T}}{\nu}\right)^{\frac12}\leq c\nu^{\frac32}.$$

As $1-e^{-\nu\mu_j^2T}\leq\nu\mu_j^2T$, we infer that

$$\begin{aligned}4\sum_j\mu_j(u_0|e_j)^2\left(\frac{1-e^{-\nu\mu_j^2T}}{\nu}\right)^{\frac12}&\leq 4T^{\frac12}\sum_j\mu_j^2(u_0|e_j)^2\\&\leq 4T^{\frac12}\|u_0\|^2_{\mathcal{V}_\sigma}.\end{aligned}$$

This proves the proposition. □

From this proposition, we infer the following corollary.

Corollary 3.1 *Let $T^\star$ be the maximal time of existence for a solution u of the system* (NS$_\nu$) *in the space $C([0,T^\star);\mathcal{V}_\sigma^{\frac12})\cap L^4_{\text{loc}}([0,T^\star);\mathcal{V}_\sigma)$. If $T^\star$ is finite, then*

$$\int_0^{T^\star}\|\nabla u(t)\|^4_{L^2}\,dt=+\infty\quad\textit{and}\quad T^\star\leq\frac{c}{\nu^5}\|u_0\|^4_{L^2}.$$

Proof For almost every t, $u(t)$ belongs to $\mathcal{V}_\sigma$. Then, thanks to the above proposition, the maximal time of existence of the solution starting at time t, which is of course $T^\star - t$, satisfies

$$T^\star - t \geq \frac{c\nu^3}{\|\nabla u(t)\|_{L^2}^4} \cdot$$

This can be written as

$$\|\nabla u(t)\|_{L^2}^4 \geq \frac{c\nu^3}{T^\star - t} \cdot$$

This gives the first part of the corollary. Taking the square root of the above inequality gives, thanks to the energy estimate,

$$c\nu^{\frac{5}{2}} \int_0^{T^\star} \frac{dt}{(T^\star - t)^{\frac{1}{2}}} \leq \frac{1}{2}\|u_0\|_{L^2}^2.$$

The corollary is proved. □

3.5 Stable solutions in a domain without boundary

The case of two dimensions was dealt with in Section 3.2 in complete generality. The purpose of this section reduces to the study of well-posedness of the incompressible Navier–Stokes equations in the whole space $\mathbf{R}^3$ or in a periodic box $\mathbf{T}^3$.

The basic remark about the level of regularity which is necessary to get uniqueness and stability is related to the scaling of the Navier–Stokes equation. If u is a solution of the Navier–Stokes equations in the whole space $\mathbf{R}^d$, then the vector field u_λ defined by

$$u_\lambda(t,x) \stackrel{\text{def}}{=} \lambda u(\lambda^2 t, \lambda x)$$

is also a solution with initial data $\lambda u_0(\lambda\cdot)$. It turns out that non-linear estimates will be based in this section on scaling invariant norms. It is very easy to check that in two dimensions, the energy norm

$$\|u(t)\|_{L^2}^2 + 2\nu \int_0^t \|\nabla u(t')\|_{L^2}^2 \, dt'$$

is scaling invariant. This is not the case in three dimensions. In three dimensions, the analogous scaling invariant norm is

$$\|u(t)\|_{\dot{H}^{\frac{1}{2}}}^2 + 2\nu \int_0^t \|\nabla u(t')\|_{\dot{H}^{\frac{1}{2}}}^2 \, dt'. \tag{3.5.1}$$

Unfortunately, no conservation or global a priori control is known about that quantity.

Theorem 3.5 *Let u_0 be a divergence-free vector field in $H^{\frac{1}{2}}$ and let f be an external force in $L^2_{\rm loc}(\mathbf{R}^+ : \dot H^{-\frac{1}{2}})$. There is a positive time T such that there is a solution of* (NS_ν) *satisfying*

$$u \in C([0,T]; H^{\frac{1}{2}}) \quad and \quad \nabla u \in L^2([0,T]; H^{\frac{1}{2}}).$$

Moreover a constant c exists such that

$$\|u_0\|_{\dot H^{\frac{1}{2}}} + \frac{1}{\nu}\|f\|_{L^2(\mathbf{R}^+;\dot H^{-\frac{1}{2}})} \leq c\nu \Longrightarrow T = \infty,$$

in which case one also has $\lim_{t\to+\infty} \|u(t)\|_{\dot H^{\frac{1}{2}}} = 0$.

Remark The convexity inequality on Sobolev norms implies that

$$C([0,T]; \dot H^{\frac{1}{2}}) \cap L^2([0,T]; \dot H^{\frac{3}{2}}) \subset L^4([0,T]; \dot H^1).$$

Thus the solutions constructed by the above theorem are stable.

Proof of Theorem 3.5

Step 1: The case of small data

We shall start by examining the case when the initial data u_0 are small in $\dot H^{\frac{1}{2}}$. We shall of course be using the sequence of approximate solutions $(u_k)_{k\in\mathbf{N}}$ introduced in Subsection 2.2.1 as solutions of the system $(\mathrm{NS}_{\nu,k})$ defined on page 45. Let us write the energy estimate in the space $\dot H^{\frac{1}{2}}$. Taking the $\dot H^{\frac{1}{2}}$ scalar product of $(\mathrm{NS}_{\nu,k})$ with u_k, it turns out, due to the divergence-free condition, that

$$\frac{d}{dt}\|u_k(t)\|^2_{\dot H^{\frac{1}{2}}} + 2\nu\|\nabla u_k(t)\|^2_{\dot H^{\frac{1}{2}}} = 2\left(Q(u_k(t), u_k(t))|u_k(t)\right)_{\dot H^{\frac{1}{2}}}.$$

By definition of the scalar product on $\dot H^{\frac{1}{2}}$, we get

$$\left|\left(Q(u_k(t), u_k(t))|u_k(t)\right)_{\dot H^{\frac{1}{2}}}\right| \leq \|Q(u_k(t), u_k(t))\|_{\dot H^{-\frac{1}{2}}} \|\nabla u_k(t)\|_{\dot H^{\frac{1}{2}}}.$$

The Sobolev embeddings proved in Theorem 1.2, page 23, and Corollary 1.1, page 25, imply that

$$\begin{aligned}\|Q(a,b)\|_{\dot H^{-\frac{1}{2}}} &\leq C\|a\cdot\nabla b\|_{L^{\frac{3}{2}}}\\ &\leq C\|a\|_{L^6}\|\nabla b\|_{L^2}\\ &\leq C\|\nabla a\|_{L^2}\|\nabla b\|_{L^2}.\end{aligned} \tag{3.5.2}$$

By the interpolation inequality between $\dot H^{\frac{1}{2}}$ and $\dot H^{\frac{3}{2}}$, we infer

$$\frac{d}{dt}\|u_k(t)\|^2_{\dot H^{\frac{1}{2}}} + 2\nu\|\nabla u_k(t)\|^2_{\dot H^{\frac{1}{2}}} \leq C\|u_k(t)\|_{\dot H^{\frac{1}{2}}}\|\nabla u_k(t)\|^2_{\dot H^{\frac{1}{2}}}. \tag{3.5.3}$$

A quick examination of that inequality shows that it is of little use when the norm $\|u_k(t)\|_{\dot H^{\frac{1}{2}}}$ is large. On the other hand, it is very good when that norm is small enough. This is a typical phenomenon of global existence theorems for small data. From now on we suppose that $\|u_k(0)\|_{\dot H^{\frac12}} \leq c\nu$; let us define T_k in $[0,+\infty]$ by

$$T_k \stackrel{\text{def}}{=} \sup\{t \,/\, \forall t' \leq t\,,\ \|u_k(t')\|_{\dot H^{\frac12}} \leq c\nu\} \quad \text{with } c \stackrel{\text{def}}{=} \frac{1}{C}\,,$$

the constant C being that in (3.5.3) By estimate (3.5.3), we have

$$\left(\frac{d}{dt}\|u_k(t)\|^2_{\dot H^{\frac12}}\right)_{|t=0} < 0.$$

Thus T_k is positive and there is some positive t_k such that

$$\forall t \in]0,t_k]\,,\quad \|u_k(t)\|_{\dot H^{\frac12}} < c\nu.$$

Moreover, still by inequality (3.5.3), the function $\|u_k(t)\|_{\dot H^{\frac12}}$ decreases on the interval $[0,T_k[$. So for any $t \in [t_k, T_k[$ we have $\|u_k(t)\|_{\dot H^{\frac12}} < c\nu$. Hence $T_k = +\infty$. We deduce directly from inequality (3.5.3) that

$$\forall t \geq 0\,,\quad \frac{d}{dt}\|u_k(t)\|^2_{\dot H^{\frac12}} + \nu\|\nabla u_k(t)\|^2_{\dot H^{\frac12}} \leq 0.$$

By integration it follows that for any real number t, we have

$$\|u_k(t)\|^2_{\dot H^{\frac12}} + \nu\int_0^t \|\nabla u_k(t')\|^2_{\dot H^{\frac12}}\,dt' \leq \|u_k(0)\|^2_{\dot H^{\frac12}}\,.$$

Extracting a subsequence which converges weakly towards a Leray solution u, the above estimate implies that this Leray solution u satisfies

$$u \in L^\infty(\mathbf{R}_+; H^{\frac12}) \quad \text{and} \quad \nabla u \in L^2(\mathbf{R}_+; H^{\frac12}).$$

Now let us prove that the solution u goes to zero for large times, in $\dot H^{\frac12}$: this result is simply due to the fact that as u is a Leray solution it is in the space $L^\infty(\mathbf{R}^+;L^2)\cap L^2(\mathbf{R}^+,\dot H^1)$, hence by interpolation in $L^4(\mathbf{R}^+;\dot H^{\frac12})$. It follows that for any $\eta > 0$ one can find a time T_η such that $\|u(T_\eta,\cdot)\|_{\dot H^{\frac12}} \leq \eta$, which yields the result recalling that the $\dot H^{\frac12}$ norm decreases in time.

Step 2: The case of large data

Let us now consider the case of large data. We start by decomposing the initial data into a high-frequency part and a low-frequency part. Let k_0 be a positive real number which will be chosen later on and let us consider u_L the solution of the evolution Stokes problem

$$(ES_\nu)\begin{cases}\partial_t u_L - \nu\Delta u_L = -\nabla p\\ \operatorname{div} u_L = 0\\ u_{L|t=0} = \mathbf{P}_{k_0}u_0.\end{cases}$$

Stating $w_k \stackrel{\text{def}}{=} u_k - u_L$, it is obvious after very elementary computations that w_k is the solution of the evolution Stokes problem with initial data $\mathbf{P}_k(u_0 - \mathbf{P}_{k_0}u_0)$ and external force

$$g_k \stackrel{\text{def}}{=} \mathbf{P}_k\left(Q(w_k, w_k) + Q(w_k, u_L) + Q(u_L, w_k) + Q(u_L, u_L)\right).$$

Using again an $\dot{H}^{\frac{1}{2}}$ energy estimate, we get that

$$\begin{aligned}\frac{d}{dt}\|w_k(t)\|^2_{\dot{H}^{\frac{1}{2}}} + 2\nu\|\nabla w_k(t)\|^2_{\dot{H}^{\frac{1}{2}}} &= 2(g_k(t)|w_k(t))_{\dot{H}^{\frac{1}{2}}}\\ &\leq \|g_k(t)\|_{\dot{H}^{-\frac{1}{2}}}\|\nabla w_k(t)\|_{\dot{H}^{\frac{1}{2}}}.\end{aligned}$$

Using estimate (3.5.2), we get that

$$\|g_k(t)\|_{\dot{H}^{-\frac{1}{2}}} \leq C\left(\|\nabla w_k(t)\|^2_{L^2} + \|\nabla u_L(t)\|^2_{L^2}\right).$$

Then by an interpolation inequality between $\dot{H}^{\frac{1}{2}}$ and $\dot{H}^{\frac{3}{2}}$, we deduce that

$$2(g_k(t)|w_k(t))_{\dot{H}^{\frac{1}{2}}} \leq \left(C_1\|w_k(t)\|_{\dot{H}^{\frac{1}{2}}} + \frac{\nu}{2}\right)\|\nabla w_k(t)\|^2_{\dot{H}^{\frac{1}{2}}} + \frac{C_2}{\nu}\|\nabla u_L(t)\|^4_{L^2}.$$

Thus we get

$$\frac{d}{dt}\|w_k(t)\|^2_{\dot{H}^{\frac{1}{2}}} + \frac{3}{2}\nu\|\nabla w_k(t)\|^2_{\dot{H}^{\frac{1}{2}}} \leq C_1\|w_k(t)\|_{\dot{H}^{\frac{1}{2}}}\|\nabla w_k(t)\|^2_{\dot{H}^{\frac{1}{2}}} + \frac{C_2}{\nu}\|\nabla u_L(t)\|^4_{L^2}.$$

Now let us prove that w_k, which can be made arbitrarily small initially, will remain so for a long enough time. More precisely, let us define T_k by

$$T_k \stackrel{\text{def}}{=} \sup\left\{t \,/\, t' \leq t\,,\ \|w_k(t')\|_{\dot{H}^{\frac{1}{2}}} \leq \frac{\nu}{2C_1}\right\}.$$

Let us choose k_0 the smallest integer such that

$$\|u_0 - \mathbf{P}_{k_0}u_0\|_{\dot{H}^{\frac{1}{2}}} \leq \frac{\nu}{4C_1}. \tag{3.5.4}$$

Let us prove that a positive time T exists such that for any integer k, we have $T_k \geq T$. For any time t smaller than or equal to T_k, we get

$$\frac{d}{dt}\|w_k(t)\|^2_{\dot{H}^{\frac{1}{2}}} + \nu\|\nabla w_k(t)\|^2_{\dot{H}^{\frac{1}{2}}} \leq \frac{C_2}{\nu}\|\nabla u_L(t)\|^4_{L^2}.$$

By time integration, we infer, for any $t \leq T_k$,

$$\begin{aligned}&\|w_k(t)\|^2_{\dot{H}^{\frac{1}{2}}} + \nu\int_0^t \|\nabla w_k(t')\|^2_{\dot{H}^{\frac{1}{2}}}\,dt'\\ &\qquad \leq \left(\frac{\nu}{4C_1}\right)^2 + \frac{C}{\nu}\int_0^t \|\nabla u_L(t')\|^4_{L^2}\,dt'.\end{aligned} \tag{3.5.5}$$

As u_L belongs to $\mathbf{P}_{k_0}\mathcal{H}$, we have

$$\|\nabla u_L(t)\|^4_{L^2} \leq k_0^2\|u_L(t)\|^4_{L^2} \leq k_0^2\|u_0\|^4_{L^2}$$

Thus, thanks to (3.5.4), we get that

$$\|w_k(t)\|^2_{\dot H^{\frac12}} + \nu\int_0^t \|\nabla w_k(t')\|^2_{\dot H^{\frac12}}\,dt' \le \left(\frac{\nu}{4C_1}\right)^2 + \frac{C}{\nu}tk_0^2\|u_0\|^4_{L^2}.$$

Now let us state

$$T \stackrel{\text{def}}{=} \frac{\nu^3}{16C_1^2C}\frac{1}{k_0^2\|u_0\|^4_{L^2}}. \tag{3.5.6}$$

Then for any k we get $T_k \ge T$.

Similarly to the case of small data, one can extract from $(w_k)_{k\in\mathbf{N}}$ a subsequence which converges weakly towards $u - u_L$, where u is a Leray solution. By the above estimate on w_k, u satisfies

$$u \in L^\infty([0,T]; H^{\frac12}) \quad \text{and} \quad \nabla u \in L^2([0,T]; H^{\frac12}).$$

To end the proof of the theorem, it remains therefore to prove that u belongs to $C([0,T]; \dot H^{\frac12})$. In order to do so, we recall that the above bounds on u imply in particular that $u \in L^4([0,T]; H^1)$. Hence u is stable in the sense of Theorem 3.3, page 58, and we can write, as noted in formula (3.1.3),

$$u(t,x) = (2\pi)^{-d}\int_{\mathbf{R}^d} e^{ix\cdot\xi}\widehat{u}(t,\xi)\,d\xi \quad \text{with}$$

$$\widehat{u}(t,\xi) \stackrel{\text{def}}{=} e^{-\nu|\xi|^2t}\widehat{u}_0(\xi) + \int_0^t e^{-\nu|\xi|^2(t-t')}\mathcal{F}Q(u(t'),u(t'))(\xi)\,dt'.$$

Note that we have supposed here that the domain is $\mathbf{R}^3$, but the computations are identical in the case of $\mathbf{T}^3$, simply replacing everywhere the integrals in $\xi \in \mathbf{R}^3$ by sums on $k \in \mathbf{Z}^3$. We leave the details to the reader. Let us state the following proposition, which we will prove at the end of this section.

Proposition 3.3 *If v is the solution of the Stokes evolution system* (ES$_\nu$), *with an initial data v_0 in $H^{\frac12}$ and an external force in $L^2_{\text{loc}}(\mathbf{R}^+; H^{-\frac12})$, then*

$$\int_{\mathbf{R}^3} |\xi|\|\widehat{v}(\cdot,\xi)\|^2_{L^\infty([0,T])}\,d\xi \le \|v_0\|_{\dot H^{\frac12}} + \frac{1}{2\nu^{\frac12}}\|f\|_{L^2([0,T];\dot H^{-\frac12})}.$$

Let us recall that

$$\|Q(u,u)\|_{\dot H^{-\frac12}} \le C\|\nabla u\|^2_{L^2}.$$

Thus we may apply Proposition 3.3. This implies directly the fact that u is continuous in time on $[0,T]$ with values in $\dot H^{\frac12}$. Indeed let η be any positive number. One can find, according to Proposition 3.3, a positive integer N_0 such that

$$\int_{|\xi|\ge N_0} |\xi|\|\widehat{u}(\cdot,\xi)\|^2_{L^\infty([0,T])}\,d\xi < \frac{\eta}{2}.$$

Now consider t_1 and t_2 in the time interval $[0,T]$. We have

$$\|u(t_1)-u(t_2)\|_{\dot H^{\frac12}} \le \left(\int_{|\xi|\le N_0} |\xi||\widehat{u}(t_1,\xi)-\widehat{u}(t_2,\xi)|^2\,d\xi\right)^{\frac12} + \left(\int_{|\xi|\ge N_0} |\xi|\|\widehat{u}(\cdot,\xi)\|^2_{L^\infty([0,T])}\right)^{\frac12}$$
$$\le N_0^{\frac12}\|u(t_1,\cdot)-u(t_2,\cdot)\|_{L^2}+\frac{\eta}{2}\cdot$$

But we know from Theorem 3.3, page 58, that u is continuous in time with values in L^2, so the result follows, up to the proof of Proposition 3.3.

Proof of Proposition 3.3 We have

$$|\xi|^{\frac12}\,|\widehat{u}(t,\xi)| \le |\xi|^{\frac12}|\widehat{u}_0(\xi)| + \left|\int_0^t e^{-\nu|\xi|^2(t-t')}|\xi|^{\frac12}\widehat{f}(t',\xi)\,dt'\right|.$$

Young's inequality enables us to infer that

$$\||\xi|^{\frac12}|\widehat{u}(t,\xi)|\|_{L^\infty([0,T])} \le |\xi|^{\frac12}|\widehat{u}_0(\xi)| + \frac{1}{2\nu^{\frac12}}\||\xi|^{-\frac12}\widehat{f}(t,\xi)\|_{L^2([0,T])}\cdot$$

Taking the L^2 norm gives

$$\left(\int_{\mathbf{R}^d}|\xi|\|\widehat{v}(\cdot,\xi)\|^2_{L^\infty([0,T])}\,d\xi\right)^{\frac12} \le \left(\int_{\mathbf{R}^d}|\xi||\widehat{u}_0(\xi)|^2\,d\xi\right)^{\frac12} + \frac{1}{2\nu^{\frac12}}\left(\int_0^T\int_{\mathbf{R}^d}|\xi|^{-1}|\widehat{f}(t,\xi)|^2\,dtd\xi\right)^{\frac12}.$$

The proposition follows. □

3.6 Blow-up condition and propagation of regularity

The aim of the first subsection is a version of Proposition 3.2 and Corollary 3.1 for the case of domains without boundary. The purpose of the second subsection is the proof of the propagation of regularity result in the case of dimension two which will be useful in Chapter 6 in the periodic case.

3.6.1 Blow-up condition

Proposition 3.4 *Let u_0 be in $H^1(\mathbf{R}^3)$. Then the maximal time of existence $T^\star$ of the solution u in $C([0,T^\star[;H^{\frac12})\cap L^2_{\rm loc}([0,T^\star[;\dot H^{\frac32})$ satisfies*

$$T^\star \ge \frac{c\nu^3}{\|\nabla u_0\|^4_{L^2}}\cdot$$

Proof If u_0 is in H^1, we have

$$\begin{aligned}\|u_0 - P_{k_0}u_0\|_{\dot H^{\frac12}}^2 &\leq \int_{|\xi|\geq k_0} |\xi|\,|\widehat{u}_0(\xi)|^2\,d\xi\\ &\leq k_0^{-1}\int |\xi|^2|\widehat{u}_0(\xi)|^2\,d\xi\\ &\leq k_0^{-1}\|\nabla u_0\|_{L^2}^2.\end{aligned}$$

Thus choosing $k_0 = (4C_1\|\nabla u_0\|_{L^2}^2)/\nu^2$ ensures (3.5.4). Now, estimate (3.5.5) gives, thanks to the conservation of the H^1 norm by the heat flow,

$$\|w_k(t)\|_{\dot H^{\frac12}}^2 + \nu\int_0^t \|\nabla w_k(t')\|_{\dot H^{\frac12}}^2\,dt' \leq \left(\frac{\nu}{4C_1}\right)^2 + \frac{C}{\nu}t\|\nabla u_0\|_{L^2}^4.$$

Proposition 3.4 is now proved. □

We can now state an analog of Corollary 3.1, the proof of which is left to the reader as an exercise.

Corollary 3.2 *Let $T^\star$ be the maximal time of existence for a solution of the system* (NS_ν) *in the space* $C([0,T^\star);H^{\frac12})\cap L^4_{\mathrm{loc}}([0,T^\star);H^1)$. *If $T^\star$ is finite then*

$$\int_0^{T^\star}\|\nabla v(t)\|_{L^2}^4\,dt = \int_0^{T^\star}\|\nabla u(t)\|_{\dot H^{\frac12}}^2\,dt = +\infty \quad \textit{and} \quad T^\star \leq \frac{c}{\nu^5}\|u_0\|_{L^2}^4.$$

This property will have the following useful application (see in particular the scheme of the proof of the forthcoming Theorem 6.2, page 119).

Theorem 3.6 *Two real numbers c and C exist which satisfy the following property. Let u be the solution of* (NS_ν) *in* $\mathbf{T}^3$ *associated with initial data u_0 in $H^{\frac12}$ and an external force f in $L^2(\mathbf{R}^+;H^{-\frac12})$. Let us assume that u is global and that*

$$\|u\|_{\frac12}^2 \stackrel{\text{def}}{=} \sup_{t\geq0}\left(\|u(t)\|_{H^{\frac12}}^2 + 2\nu\int_0^t\|\nabla u(t')\|_{H^{\frac12}}^2\,dt'\right) < +\infty.$$

Then, for any v_0 in $H^{\frac12}$ and any g in $L^2(\mathbf{R}^+;H^{-\frac12})$ such that

$$\|v_0-u_0\|_{H^{\frac12}}^2 + \frac{4}{\nu}\|f-g\|_{L^2(\mathbf{R}^+;H^{-\frac12})}^2 \leq c\nu\exp\left(-\frac{C}{\nu^4}\|u\|_{\frac12}^4\right),$$

the solution for $(\mathrm{NS})_\nu$ *associated with v_0 and g is global and belongs to the space $E_{\frac12}$.*

Proof Let us consider the maximal solution v given by Theorem 3.5. It belongs to the space $C([0,T^\star[;H^{\frac12})\cap L^2_{\mathrm{loc}}([0,T^\star[;H^{\frac32})$. Let us state $w \stackrel{\text{def}}{=} v-u$. It is solution of

$$\begin{cases}\partial_t w - \nu\Delta w = Q(w,w)+Q(u,w)+Q(w,u)+h\\ \operatorname{div} w = 0\\ w_{|t=0} = v_0-u_0\end{cases}$$

with $h \stackrel{\text{def}}{=} g - f$. Inequality (3.5.2), together with the interpolation inequality between $H^{\frac{1}{2}}$ and $H^{\frac{3}{2}}$, tells us that

$$Q(w,w) + Q(u,w) + Q(w,u) \in L^2_{\text{loc}}([0,T^\star[;H^{-\frac{1}{2}}).$$

Again inequality (3.5.2), together with the interpolation inequality between $H^{\frac{1}{2}}$, and $H^{\frac{3}{2}}$ gives

$$\begin{aligned}
&\|w(t)\|^2_{H^{\frac{1}{2}}} + 2\nu \int_0^t \|\nabla w(t')\|^2_{H^{\frac{1}{2}}}\,dt' \\
&\quad \leq \|w_0\|^2_{H^{\frac{1}{2}}} + 2\int_0^t \|h(t')\|_{H^{-\frac{1}{2}}} \|\nabla w(t')\|_{H^{\frac{1}{2}}}\,dt' \\
&\qquad + C\int_0^t \|w(t')\|_{H^{\frac{1}{2}}} \|\nabla w(t')\|^2_{H^{\frac{1}{2}}}\,dt' \\
&\qquad + C\int_0^t \|\nabla u(t')\|_{L^2} \|\nabla w(t')\|^{\frac{1}{2}}_{H^{\frac{1}{2}}} \|\nabla w(t')\|^{\frac{3}{2}}_{H^{\frac{1}{2}}}\,dt'.
\end{aligned}$$

Using the convexity inequality (1.3.5), page 26, gives

$$\begin{aligned}
\|w(t)\|^2_{H^{\frac{1}{2}}} + \nu \int_0^t \|\nabla w(t')\|^2_{H^{\frac{1}{2}}}\,dt' &\leq \|w_0\|^2_{H^{\frac{1}{2}}} + \frac{4}{\nu}\|h\|^2_{L^2(\mathbf{R}^+;H^{-\frac{1}{2}})} \\
+ C\int_0^t \|w(t')\|_{H^{\frac{1}{2}}} \|\nabla w(t')\|^2_{H^{\frac{1}{2}}}\,dt' &+ C\int_0^t \|\nabla u(t')\|^4_{L^2} \|w(t')\|^2_{H^{\frac{1}{2}}}\,dt'.
\end{aligned}$$

Let us assume that

$$\|w_0\|^2_{H^{\frac{1}{2}}} + \frac{4}{\nu}\|h\|^2_{L^2(\mathbf{R}^+;H^{-\frac{1}{2}})} \leq \left(\frac{\nu}{4C}\right)^2.$$

Let us define $\overline{T} \stackrel{\text{def}}{=} \sup\left\{t < T^\star \;/\; \forall t' \leq t,\ \|w(t')\|_{H^{\frac{1}{2}}} \leq \dfrac{\nu}{2C}\right\}$. As w is continuous with values in $H^{\frac{1}{2}}$, the time $\overline{T}$ is positive. For any $t < \overline{T}$, we have

$$\begin{aligned}
\|w(t)\|^2_{H^{\frac{1}{2}}} + \frac{\nu}{2}\int_0^t \|\nabla w(t')\|^2_{H^{\frac{1}{2}}}\,dt' &\leq \|w_0\|^2_{H^{\frac{1}{2}}} + \frac{4}{\nu}\|h\|^2_{L^2(\mathbf{R}^+;H^{-\frac{1}{2}})} \\
&\quad + C\int_0^t \|\nabla u(t')\|^4_{L^2} \|w(t')\|^2_{H^{\frac{1}{2}}}\,dt'.
\end{aligned}$$

Gronwall's lemma gives, for any $t < \overline{T}$,

$$\begin{aligned}
\|w(t)\|^2_{H^{\frac{1}{2}}} + \frac{\nu}{2}\int_0^t \|\nabla w(t')\|^2_{H^{\frac{1}{2}}}\,dt' &\leq \left(\|w_0\|^2_{H^{\frac{1}{2}}} + \frac{4}{\nu}\|h\|^2_{L^2(\mathbf{R}^+;H^{-\frac{1}{2}})}\right) \\
&\quad \times \exp\left(\frac{C}{\nu^3}\int_0^\infty \|\nabla u(t)\|^4_{L^2}\,dt'\right).
\end{aligned}$$

Using the interpolation inequality between $H^{\frac{1}{2}}$ and $H^{\frac{3}{2}}$, we get, for any t less than T.

$$\|w(t)\|^2_{H^{\frac{1}{2}}} + \frac{\nu}{2}\int_0^t \|\nabla w(t')\|^2_{H^{\frac{1}{2}}}\,dt' \le \left(\|w_0\|^2_{H^{\frac{1}{2}}} + \frac{4}{\nu}\|h\|^2_{L^2(\mathbf{R}^+;H^{-\frac{1}{2}})}\right)$$
$$\times \exp\left(\frac{C}{\nu^4}\|\nabla u\|^4_{\frac{1}{2}}\right). \qquad (3.6.1)$$

Thus, if

$$\|w_0\|^2_{H^{\frac{1}{2}}} + \frac{4}{\nu}\|h\|^2_{L^2(\mathbf{R}^+;H^{-\frac{1}{2}})} \le \left(\frac{\nu}{4C}\right)^2 \exp\left(-2\frac{C}{\nu^4}\|\nabla u\|^4_{\frac{1}{2}}\right),$$

then, for any $t \le \overline{T}$, we have $\|w(t)\|_{H^{\frac{1}{2}}} \le \nu/2C$ and thus $\overline{T} = T^\star$. Then inequality (3.6.1) together with the blow-up condition given by Proposition 3.4 implies that $T^\star = +\infty$. Theorem 3.6 is now proved. □

3.6.2 Propagation of regularity

We shall investigate this problem only in the two-dimensional periodic case. The following theorem holds.

Theorem 3.7 *Let u_0 be in $H^{\frac{1}{2}}(\mathbf{T}^2)$ and $f \in L^2_{\rm loc}(\mathbf{R}^+;H^{-\frac{1}{2}}(\mathbf{T}^2))$. Then the stable Leray solution u belongs to $C(\mathbf{R}^+;H^{\frac{1}{2}}(\mathbf{T}^2)) \cap L^2_{\rm loc}(\mathbf{R}^+;H^{\frac{3}{2}}(\mathbf{T}^2))$ and satisfies*

$$\|u(t)\|^2_{H^{\frac{1}{2}}} + \nu\int_0^t \|\nabla u(t')\|^2_{H^{\frac{1}{2}}}dt' \le \exp\left(\frac{C}{\nu}\int_0^t \|\nabla u(t')\|^2_{L^2}dt'\right)$$
$$\times\left(\|u_0\|^2_{H^{\frac{1}{2}}} + \frac{C}{\nu}\int_0^t \|f(t')\|^2_{H^{-\frac{1}{2}}} \exp\left(-\frac{C}{\nu}\int_{t'}^t \|\nabla u(t'')\|^2_{L^2}dt''\right)dt'\right).$$

As usual, we prove a priori bounds on the sequence $(u_k)_{k\in\mathbf{N}}$ of solutions of the approximated problem. Taking the $H^{\frac{1}{2}}$ scalar product of the system $(\mathrm{NS}_{\nu,k})$ gives

$$\frac{1}{2}\frac{d}{dt}\|u_k(t)\|^2_{H^{\frac{1}{2}}} + \nu\|u_k(t)\|^2_{H^{\frac{3}{2}}} = -\left(u_k\cdot\nabla u_k \mid u_k\right)_{H^{\frac{1}{2}}} + \left(f_k \mid u_k\right)_{H^{\frac{1}{2}}}. \qquad (3.6.2)$$

By definition of the $H^{\frac{1}{2}}$ scalar product, we have

$$(a|b)_{H^{\frac{1}{2}}} = \sum_{n\in\mathbf{Z}^2} \widehat{a}(n)|n|\widehat{b}(n).$$

The Cauchy–Schwarz inequality and the Fourier–Plancherel theorem imply that

$$(a|b)_{H^{\frac{1}{2}}} \le \|a\|_{L^2}\|\nabla b\|_{L^2}.$$

Thus we get

$$\left|\left(u_k\cdot\nabla u_k \mid u_k\right)_{H^{\frac{1}{2}}}\right| \le C\|u_k\cdot\nabla u_k\|_{L^2}\|\nabla u_k\|_{L^2}.$$

The Sobolev embedding $H^{\frac{1}{2}}(\mathbf{T}^2) \hookrightarrow L^4(\mathbf{T}^2)$ together with the Hölder estimate gives

$$\begin{aligned}|(u_k \cdot \nabla u_k \mid u_k)_{H^{\frac{1}{2}}}| &\leq C\|u_k\|_{L^4}\|\nabla u_k\|_{L^4}\|\nabla u_k\|_{L^2}\\ &\leq C\|u_k\|_{H^{\frac{1}{2}}}\|\nabla u_k\|_{H^{\frac{1}{2}}}\|\nabla u_k\|_{L^2}.\end{aligned}$$

Using the fact that $(f_k \mid u_k)_{H^{\frac{1}{2}}} \leq \|\nabla u_k\|_{H^{\frac{1}{2}}}\|f_k\|_{H^{-\frac{1}{2}}}$ and the convexity inequality (1.3.5), we infer

$$\frac{d}{dt}\|u_k(t)\|^2_{H^{\frac{1}{2}}} + \nu\|u_k(t)\|^2_{H^{\frac{3}{2}}} \leq \frac{C}{\nu}\left(\|u_k\|^2_{H^{\frac{1}{2}}}\|\nabla u_k\|^2_{L^2} + \|f_k\|^2_{H^{-\frac{1}{2}}}\right)$$

The Gronwall lemma implies that

$$\begin{aligned}\|u_k(t)\|^2_{H^{\frac{1}{2}}} + \nu\int_0^t \|\nabla u_k(t')\|^2_{H^{\frac{1}{2}}}\,dt' &\leq \exp\left(\frac{C}{\nu}\int_0^t \|\nabla u_k(t')\|^2_{L^2}\,dt'\right)\\ \times\left(\|u_0\|^2_{H^{\frac{1}{2}}} + \frac{C}{\nu}\int_0^t \exp\left(-\frac{C}{\nu}\int_{t'}^t \|\nabla u_k(t'')\|^2_{L^2}\,dt''\right)\|f_k(t')\|^2_{H^{-\frac{1}{2}}}\,dt'\right).\end{aligned}$$

Thus the Leray solution belongs to $L^\infty_{\mathrm{loc}}(\mathbf{R}^+; H^{\frac{1}{2}}) \cap L^2_{\mathrm{loc}}(\mathbf{R}^+; H^{\frac{3}{2}})$ and satisfies the estimate of the theorem. Now let us prove that it is continuous with values in $H^{\frac{1}{2}}$. Thanks to Sobolev embeddings (see Theorem 1.2, page 23, and Corollary 1.1, page 25), one has

$$\begin{aligned}\|a \cdot \nabla b\|_{H^{-\frac{1}{2}}} &\leq C\|a \cdot \nabla b\|_{L^{\frac{4}{3}}}\\ &\leq C\|a\|_{L^2}\|\nabla b\|_{L^4}\\ &\leq C\|a\|_{L^2}\|\nabla b\|_{H^{\frac{1}{2}}}.\end{aligned}$$

As u belongs in particular to $L^\infty_{\mathrm{loc}}(\mathbf{R}^+; L^2) \cap L^2_{\mathrm{loc}}(\mathbf{R}^+; H^{\frac{3}{2}})$, it solves the Stokes problem with initial data in $H^{\frac{1}{2}}$ and an external force in the space $L^2_{\mathrm{loc}}(\mathbf{R}^+; H^{-\frac{1}{2}})$. Thus following Proposition 3.3 allows us to conclude the proof of the theorem. □

4

References and remarks on the Navier–Stokes equations

The purpose of this chapter is to give some historical landmarks to the reader. The concept of weak solutions certainly has its origin in mechanics; the article by C. Oseen [100] is referred to in the seminal paper [87] by J. Leray. In that famous article, J. Leray proved the global existence of solutions of (NS_ν) in the sense of Definition 2.5, page 42, in the case when $\Omega = \mathbf{R}^3$. The case when Ω is a bounded domain was studied by E. Hopf in [74]. The study of the regularity properties of those weak solutions has been the purpose of a number of works. Among them, we recommend to the reader the fundamental paper of L. Caffarelli, R. Kohn and L. Nirenberg [21]. In two space dimensions, J.-L. Lions and G. Prodi proved in [91] the uniqueness of weak solutions (this corresponds to Theorem 3.2, page 56, of this book). Theorem 3.3, page 58, of this book shows that regularity and uniqueness are two closely related issues. In the case of the whole space $\mathbf{R}^3$, theorems of that type have been proved by J. Leray in [87]. For generalizations of that theorem we refer to [113], [121], [58] and [62]. In the article [87], J. Leray also proved the global regularity of weak solutions (and their global uniqueness) for small initial data, namely initial data satisfying

$$\|u_0\|_{L^2}^2\|u_0\|_{L^\infty} \leq c^3\nu^3 \quad \text{or} \quad \|u_0\|_{L^2}^2\|\nabla u_0\|_{L^2} \leq c^2\nu^2.$$

Theorem 3.4, page 66, and Theorem 3.5, page 73, which are generalizations of the first smallness condition above, were proved by H. Fujita and T. Kato in [57]. In the case if the whole space $\mathbf{R}^3$, the smallness condition has been generalized in terms of the Lebesgue space L^3 by T. Kato in [79], of Besov spaces by M. Cannone, Y. Meyer and F. Planchon in [25], and of the BMO-type space BMO^{-1} by H. Koch and D. Tataru in [84]. The set of results presented in this part contains the material required for the further study of rotating fluids. The reader who wants to learn more about the theory of the incompressible Navier–Stokes system can read the following monographs:

- M. Cannone: *Ondelettes, paraproduits et Navier–Stokes* [24]
- J.-Y. Chemin: *Localization in Fourier space and Navier–Stokes system* [30]
- P. Constantin and C. Foias: *Navier–Stokes Equations* [38]
- P.-G. Lemarié-Rieusset: *Recent Developments in the Navier–Stokes problem* [86]
- P.-L. Lions: *Mathematical topics in fluid mechanics* [92]

- Y. Meyer: *Wavelets, Paraproducts and Navier–Stokes* [98]
- L. Tartar: *Topics in non linear analysis* [117]
- R. Temam: *Navier–Stokes Equations, Theory and Numerical Analysis* [118].

Let us conclude this chapter by noticing that the proof of Sobolev embeddings presented here comes from [35]. Moreover, Section 1.4 follows mainly [117] by L. Tartar.

PART III

Rotating fluids

In this part we intend to study the rotating-fluid equations (NSC_ε) presented in the introduction. Let us recall the system:

$$(\mathrm{NSC}_\varepsilon) \quad \begin{cases} \partial_t u + u \cdot \nabla u - \nu \Delta u + \dfrac{e^3 \wedge u}{\varepsilon} + \nabla p = f \\ \operatorname{div} u = 0. \end{cases}$$

We recall that the parameter ε is the Rossby number, which will be considered very small: one of the aims of this part is to study the asymptotics of the solutions as ε goes to zero. As explained in the introduction, the situation depends strongly on the boundary conditions.

In Chapter 5 we consider the case when the system (NSC_ε) is written in the whole space $\mathbf{R}^3$. In that situation we are faced with dispersion – this concept is studied in an abstract setting in Section 5.1, and applied to the system (NSC_ε) in Sections 5.2 and 5.3.

Chapter 6 deals with the case of periodic boundary conditions. In that situation there is no longer any dispersion, and the possible interaction of oscillatory Rossby waves for all times has to be taken into account.

Finally in Chapter 7 we deal with the more difficult case where the fluid evolves between two fixed, horizontal plates. The assumption made on the boundary is that the fluid is stopped (this corresponds to Dirichlet boundary conditions), which, as explained in the introduction, immediately creates boundary layers. Depending on the horizontal boundary conditions (in the whole space or periodic), these boundary layers are coupled with dispersion, or oscillatory phenomena.

5

Dispersive cases

5.1 A brief overview of dispersive phenomena

It is well known that dispersive phenomena play a significant role in the study of partial differential equations. Historically, the use of dispersive effects appeared in the study of the wave equation in the whole space $\mathbf{R}^d$ with the proof of the so-called Strichartz estimates. The idea is that even though the wave equation is time reversible and preserves the energy, it induces a time decay in L^p norms, of course for exponents p greater than 2. In particular, the energy of the waves over a bounded subdomain vanishes as time goes to infinity. These decay properties also yield smoothing effects, which have been the beginning of a long series of works in which the aforementioned smoothing is used in the analysis of non-linear wave equations to improve the classical well-posedness results. Similar developments have been applied to the non-linear Schrödinger equations.

Let us give a flavor of the proof of dispersion estimates in the case of simple systems. As a first example, the free transport equation describing the evolution of a system of free particles in $\mathbf{R}^d$ is expressed in terms of a non-negative microscopic density $f(t, x, v)$ as

$$\partial_t f + v \cdot \nabla_x f = 0, \quad f_{|t=0}(x, v) = f_0(x, v). \tag{5.1.1}$$

where $x \in \mathbf{R}^d$ and $v \in \mathbf{R}^d$, respectively, denote the particles' position and velocity. The associated macroscopic density is given by

$$\rho(t, x) = \int_{\mathbf{R}^d_v} f(t, x, v) dv, \tag{5.1.2}$$

so that the total mass conservation property reduces to

$$\frac{d}{dt} \int_{\mathbf{R}^d_x} \rho(t, x) dx = 0, \quad \text{i.e.} \quad \|f(t, \cdot, \cdot)\|_{L^1(\mathbf{R}^d_x \times L^1(\mathbf{R}^d_v))} = \|f_0(\cdot, \cdot)\|_{L^1(\mathbf{R}^d_x \times \mathbf{R}^d_v)}.$$

On the other hand, the exact expression of f in terms of f_0 is given by integration along the characteristics $f(t, x, v) = f_0(x - vt, v)$, so that

$$\begin{aligned} \|\rho(t, \cdot)\|_{L^\infty(\mathbf{R}^d)} &= \|f(t, \cdot, \cdot)\|_{L^\infty(\mathbf{R}^d_x; L^1(\mathbf{R}^d_v))} \\ &\leq |t|^{-d} \sup_{x \in \mathbf{R}^d} \int_{\mathbf{R}^d_w} f_0\left(x - w, \frac{w}{t}\right) dw \end{aligned}$$

$$\leq |t|^{-d} \sup_{x\in\mathbf{R}^d} \int_{\mathbf{R}^d_w} \|f_0(x-w,\cdot)\|_{L^\infty(\mathbf{R}^d)} dw$$

$$\leq |t|^{-d} \|f_0\|_{L^1(\mathbf{R}^d_x;L^\infty(\mathbf{R}^d_v))}, \tag{5.1.3}$$

which means in view of (5.1.2) that the macroscopic density ρ decays in the L^∞ norm, even though the total mass is preserved.

Another simple illustration is provided by the linear Schrödinger equation in the whole space $\mathbf{R}^d$

$$i\partial_t \Psi = \frac{1}{2}\Delta\Psi \quad \text{in} \quad \mathbf{R}^d, \qquad \Psi_{|t=0} = \Psi_0, \tag{5.1.4}$$

where $\Psi(t,\cdot)$ denotes the wave (complex valued) function at time t. Its Fourier transform $\widehat{\Psi}$ can be expressed in terms of the Fourier transform of Ψ_0

$$\widehat{\Psi}(t,\xi) = \widehat{\Psi}_0(\xi) e^{it|\xi|^2/2},$$

which yields the L^2 conservation $\|\Psi(t,\cdot)\|_{L^2(\mathbf{R}^d)} = \|\Psi_0\|_{L^2(\mathbf{R}^d)}$. Moreover, easy computations show

$$\Psi(t,x) = \int_{\mathbf{R}^d} \Psi_0(x-y) K(t,y)\, dy$$

where

$$K(t,y) \stackrel{\text{def}}{=} (2\pi t)^{-d/2} e^{i\pi d/4} e^{-i|y|^2/t},$$

so that we deduce from convolution estimates that the L^∞ norm of $\Psi(t,\cdot)$ decays when $|t|$ tends to $+\infty$

$$\|\Psi(t,\cdot)\|_{L^\infty(\mathbf{R}^d)} \leq |2\pi t|^{-d/2} \|\Psi_0\|_{L^1(\mathbf{R}^d)}. \tag{5.1.5}$$

This estimate is one of the key tools for proving well-posedness properties for non-linear Schrödinger equations.

The analysis of the wave equation is another classical framework for the application of dispersion estimates. The linear wave equation is as follows:

$$\partial_t^2 u - \Delta u = 0 \quad \text{in} \quad \mathbf{R}\times\mathbf{R}^d. \tag{5.1.6}$$

It reduces to the study of

$$\partial_t u_\pm \pm i|D|u_\pm = 0 \quad \text{in} \quad \mathbf{R}\times\mathbf{R}^d, \quad \text{with} \quad |D|v := \mathcal{F}^{-1}(|\xi|\widehat{v}(\xi)), \tag{5.1.7}$$

where $\mathcal{F}$ denotes the Fourier transform. Thus, the solution is of the form

$$u(t) = \mathcal{F}^{-1}\left(e^{it|\xi|}\gamma^+(\xi) + e^{-it|\xi|}\gamma^-(\xi)\right).$$

Let us suppose that the support of the Fourier transform of the initial data γ^+ and γ^- is included in a fixed ring $\mathcal{C}$ of $\mathbf{R}^d$. The Strichartz estimate is based on the so-called "dispersive estimate"

$$\|u(t)\|_{L^\infty(\mathbf{R}^d)} \leq \frac{C}{|t|^{\frac{d-1}{2}}} \left(\|\gamma^+\|_{L^1(\mathbf{R}^d)} + \|\gamma^-\|_{L^1(\mathbf{R}^d)}\right). \tag{5.1.8}$$

Then, functional analysis arguments based upon interpolation results (the so-called TT^* result) imply that

$$\|u\|_{L^p(\mathbf{R};L^q(\mathbf{R}^d))} \leq C(\|\gamma^+\|_{L^1(\mathbf{R}^d)} + \|\gamma^-\|_{L^1(\mathbf{R}^d)}) \tag{5.1.9}$$

for suitable p and q, which is clearly a decay property. Using scaling arguments and Sobolev embeddings, it can be proved that

$$\|\nabla u\|_{L^2([0,T];L^\infty(\mathbf{R}^d))} \leq C_T \left(\|\gamma^+\|_{H^s(\mathbf{R}^d)} + \|\gamma^-\|_{H^s(\mathbf{R}^d)}\right)$$

for some $s < d/2$ depending on the dimension of the space. This kind of smoothing estimate is used to solve non-linear wave equations locally in time.

The aim of this section is to emphasize basic properties which allow us to derive dispersion estimates in a much more general framework, as long as waves propagate in a physical medium. Indeed, the main tool for proving time decay for solutions of wave equations is multiple integration by parts like in the stationary phase theorem.

In the next subsection, we shall explain the general way to derive Strichartz estimates and shall illustrate how this method works on the wave equation in Subsection 5.1.2.

In Section 5.2, we apply these ideas to the rotating incompressible Navier–Stokes equations. Dispersion takes place in the direction transverse to the rotation vector, which influences the time decay of the associated waves in the L^∞ norm. Finally, the application to the non-linear case of the rotating Navier–Stokes equations is given in Section 5.3.

5.1.1 *Strichartz-type estimates*

We now intend to describe mathematically dispersion phenomena in terms of the time decay in the L^∞ norm in the case of frequency localized functions. It turns out that dispersion is obtained by integration by parts just like in the proof of the stationary phase theorem. Let $d \geq 1$ and $m \geq 1$ be two integers, B a bounded open subset of $\mathbf{R}^d$, and Ω an open subset of $\mathbf{R}^m$. Let $\Psi \in \mathcal{D}^\infty(B;\mathbf{C})$ and $a \in C^\infty(\mathbf{R}^d \times \Omega;\mathbf{R})$. We define $K : \mathbf{R} \times \Omega \to \mathbf{C}$ by

$$K(\tau, z) = \int_{\mathbf{R}^d} \Psi(\xi) \exp(i\tau a(\xi, z))\, d\xi. \tag{5.1.10}$$

We also introduce the stationary set

$$X = \{(\xi, z) \in B \times \Omega \;/\; \nabla_\xi a(\xi, z) = 0\}.$$

We assume that the phase a satisfies the bounds

$$\nabla a \in L^\infty(B \times \Omega) \quad \text{and} \quad \nabla^2 a \in L^\infty(B \times \Omega). \tag{5.1.11}$$

We intend to prove the following two theorems.

Theorem 5.1 *Let us consider the non-stationary case $X = \emptyset$. Assuming that*

$$\inf_{B\times\Omega} |\nabla_\xi a| \geq \beta > 0, \tag{5.1.12}$$

then K decays in τ at any order: for any $k \in \mathbf{N}$, there exists $C_k > 0$ such that for any τ in $\mathbf{R}^+$,

$$\|K(\tau,\cdot)\|_{L^\infty(\Omega)} \leq C_k(1+\tau)^{-k}. \tag{5.1.13}$$

In the stationary case, we shall prove the following result.

Theorem 5.2 *Assume that a function ξ_0 exists in $C^\infty(\Omega; B)$ such that*

$$X = \{(\xi_0(z), z)\ ,\ z \in \Omega\} \quad \text{and} \quad \inf_{|x|=1,\ z\in\Omega} |D^2 a(\xi_0(z), z)\cdot x| \geq \delta > 0. \tag{5.1.14}$$

Then, there exists $C > 0$ such that for all $\tau \in \mathbf{R}^*$,

$$\|K(\tau,\cdot)\|_{L^\infty(\Omega)} \leq C|\tau|^{-d/2}. \tag{5.1.15}$$

Proof of Theorems 5.1 and 5.2 The main idea in proving the above two theorems is to perform multiple integrations by parts. As a matter of fact, we introduce the following operator

$$\mathcal{L} = \frac{1 - i\alpha\partial_\xi}{1+\tau|\alpha|^2}, \quad \text{where} \quad \alpha(\xi, z) \stackrel{\text{def}}{=} \partial_\xi a(\xi, z). \tag{5.1.16}$$

It can be easily checked that $\mathcal{L}\exp(i\tau a) = \exp(i\tau a)$. On the other hand, the transposed operator ${}^t\mathcal{L}$ is defined by

$$\int_{\mathbf{R}^d} {}^t\mathcal{L}f\cdot g\,dx = \int_{\mathbf{R}^d} f\cdot\mathcal{L}g\,dx \quad \text{for any} \quad (f,g)\in\mathcal{D}(\mathbf{R}^d;\mathbf{C})^2.$$

Straightforward computations now yield

$${}^t\mathcal{L} = \frac{1}{1+\tau\alpha^2} + i\frac{(\partial_\xi\alpha - \tau\alpha^2\partial_\xi\alpha)}{(1+\tau\alpha^2)^2} + \frac{i\alpha}{1+\tau\alpha^2}\partial_\xi\,. \tag{5.1.17}$$

It follows from the assumptions of Theorem 5.1 that for all $\Psi \in \mathcal{D}(B)$,

$$\begin{aligned}\left\|{}^t\mathcal{L}\Psi(\tau,\cdot)\right\|_{L^\infty(B\times\Omega)} \leq \frac{C}{1+\beta^2\tau}&\left(\|\Psi\|_{L^\infty(B)}\|\partial_\xi\alpha\|_{L^\infty(\Omega)}\right.\\ &\left.+\|\nabla\Psi\|_{L^\infty(B)}\|\alpha\|_{L^\infty(\Omega)}\right).\end{aligned}$$

As a result, it can be deduced by induction that for all $k \in \mathbf{N}^*$, there exists a positive C_k such that for all $\tau \geq 0$, we have

$$\left\|({}^t\mathcal{L})^k(\tau,\cdot,\cdot)\Psi\right\|_{L^\infty(B\times\Omega)} \leq \frac{C_k}{(1+\tau)^k}\|\Psi\|_{W^{k,\infty}(B)}. \tag{5.1.18}$$

Writing

$$K(\tau, z) = \int_{\mathbf{R}^d} \Psi(\xi)\exp(i\tau a(\xi,z))\,d\xi = \int_{\mathbf{R}^d} {}^t\mathcal{L}^k\Psi\exp(i\tau a(\xi,z))\,d\xi,$$

we deduce from (5.1.18) that

$$\|K(\tau,\cdot)\|_{L^\infty(\Omega)} \le \mathrm{Vol}(B)\frac{C_k}{(1+\tau)^k}\|\Psi\|_{W^{k,\infty}(\Omega)},$$

which completes the proof of Theorem 5.1. In order to prove Theorem 5.2, it only remains to deal with the case when (ξ, z) is localized in a neighborhood $B' \times \Omega$ of the stationary set X, otherwise Theorem 5.1 would apply and yield the claimed time decay. For (ξ, z) in $B' \times \Omega$, we may use the non-degeneracy assumption on the phase along the stationary set to prove that

$$|\alpha(\xi,z)| \ge |\xi - \xi_0(z)|\frac{\delta}{2}\cdot$$

As a result, we deduce similarly by induction that for $k \in \mathbf{N}$ and (ξ, z) in $B' \times \Omega$, we have

$$\left|({}^t\mathcal{L})^k\Psi(\tau,\xi,z)\right| \le \frac{C_k}{(1+\tau|\xi-\xi_0(z)|^2)^k}\|\psi\|_{W^{k,\infty}(B)}. \tag{5.1.19}$$

For a given integer k greater than $d/2$, we conclude by making the change of variables $\zeta = \sqrt{|\tau|}(\xi - \xi_0(z))$ and using the fact that $\zeta \to 1/(1+|\zeta|^2)^k$ is integrable over $\mathbf{R}^d$. □

5.1.2 *Illustration of the wave equation*

Let us consider again the wave equation (5.1.6). Solutions can be expressed in terms of

$$u^\pm(t,x) = \int_{\mathbf{R}^d} \gamma^\pm(\xi)\exp(i\xi\cdot x \pm it|\xi|)\, d\xi. \tag{5.1.20}$$

The above integral can be rewritten as a function of $z = x/t$ in order to match the assumptions of Theorem 5.2. It corresponds to the case when

$$a(\xi,z) = \pm|\xi| + z\cdot\xi$$

and to amplitude functions localized in $\mathbf{R}^d_\xi \backslash \{0\}$. Observing that

$$\nabla_\xi a(\xi,z) = z \pm \frac{\xi}{|\xi|},$$

we deduce that the phase is likely to be stationary when ξ is directed along z. The idea is then to write the integration over $\mathbf{R}^d$ as an integration over $\mathbf{R}\times\mathbf{R}^{d-1}$ for suitable one and $d-1$ dimensional spaces. Let us assume that the direction of z is the first vector basis e_1. Then the phase can be written

$$a(\xi,z) = \xi_1 z_1 \pm (|\xi'|^2 + |\xi_1|^2)^{1/2} \quad \text{with} \quad \xi = (\xi_1,\xi') \in \mathbf{R}\times\mathbf{R}^{d-1}$$

and its $d-1$ dimensional gradient as

$$\nabla_{\xi'} a(z,\xi) = \pm\frac{\xi'}{|\xi|},$$

so that the stationary set over $\mathbf{R}^{d-1}_{\xi'}$ is given by $X=\{(\xi'=0, z_1 e_1)\}$. The second derivative in ξ' of the phase is expressed as

$$\nabla^2_{\xi'} a(\xi_1, \xi', z_1, 0) = \left(I_{\xi'} - \frac{\xi'}{|\xi|} \otimes \frac{\xi'}{|\xi|} \right) \frac{1}{|\xi|},$$

and hence reduces to $I_{\xi'}/|\xi_1|$ along the stationary set (we recall that $\xi_1 = 0$ is forbidden since the amplitude function is supported outside $\{0\}$). As a result, Theorem 5.2 applies and yields the claimed time decay (5.1.8) in $t^{-(d-1)/2}$. The classical estimate (5.1.9) follows from the so–called TT^* argument (see [80]).

There are many applications of the above estimates to fluid mechanics problems such as the low Mach number limit of compressible flows (otherwise known as the incompressible limit). In the inviscid case, the Euler equations can be considered in the isentropic case to simplify to

$$(\text{ALE}_{\text{comp}}) \qquad \begin{cases} \partial_t \rho + \operatorname{div}(\rho u) = 0 \\ \partial_t(\rho u) + \dfrac{\nabla \rho^\gamma}{\gamma \varepsilon^2} = 0, \end{cases}$$

where $\gamma > 1$ denotes the exponent of the pressure law and ε the Mach number of the flow. As the Mach number vanishes, the density tends to a constant, say 1, and we find $\operatorname{div} u = 0$. Since general initial data do not necessarily have constant density and incompressible velocity, waves propagate at the very high speed $1/\varepsilon$. Introducing the density fluctuation

$$\psi = \frac{\rho - 1}{\varepsilon},$$

we get the acoustic wave equations

$$(\text{WE}_\varepsilon) \qquad \begin{cases} \partial_t u + \dfrac{\nabla \psi}{\varepsilon} = 0 \\ \partial_t \psi + \dfrac{\operatorname{div} u}{\varepsilon} = 0 \end{cases}$$

and hence $\partial_t^2 \psi - \Delta\psi/\varepsilon^2 = 0$. From the time decay (5.1.8) in $L^\infty(\mathbf{R}^d)$ deduced from Theorem 5.2, classical duality arguments allow us to derive bounds such as (5.1.9) for suitable p and q such that $q > 2$. As a result, rescaling the time according to the speed of sound $1/\varepsilon$ gives convergence to zero at a rate ε. In the case of the Euler equations for finite time or in the viscous case globally in time, convergence to the incompressible limit system can be proved for general initial data, namely the local energy of the potential part of the flow, which is carried out by the acoustic waves, vanishes as the Mach number tends to zero.

Let us emphasize that a number of physically relevant systems can be studied similarly, as long as waves propagate in an infinite medium. Indeed, a linearization procedure followed by a travelling wave analysis allows us to derive dispersion relations expressing the pulsation ω as a function of the wavenumber ξ. Then, the superposition of such waves can be analyzed like the above simple example.

The next section will illustrate once again dispersion effects in a non-linear system such as the rotating Navier–Stokes equations, which is the main purpose of this book.

5.2 The particular case of the Rossby operator in $\mathbf{R}^3$

In this section, we give another illustration of the preceding dispersion estimates on a linearized version of the rotating Navier–Stokes equations. In the sequel, we shall denote $e^3 = (0,0,1)$ the unit vector directed along the x_3-coordinate. $\varepsilon > 0$ the Rossby number and $\nu \geq 0$ the viscosity of the fluid. The inviscid case $\nu = 0$ can be treated by using Theorem 5.2 and yields a $\tau^{-1/2}$ time decay in the L^∞ norm. In the Navier–Stokes-like case $\nu > 0$, the viscosity provides additional regularity which can be used for nonlinear applications, so that the proof is detailed in the sequel. The model equations read as

$$(\mathrm{VC}_\varepsilon) \quad \begin{cases} \partial_t v - \nu \Delta v + \dfrac{e^3 \wedge v}{\varepsilon} + \nabla p = f \\ \operatorname{div} v = 0 \\ v_{|t=0} = v_0, \end{cases}$$

which yields in Fourier variables $\xi \in \mathbf{R}^3$

$$(\mathrm{FVC}_\varepsilon) \quad \begin{cases} \partial_t \widehat{v} + \nu |\xi|^2 \widehat{v} + \dfrac{\xi_3 \xi \wedge \widehat{v}}{\varepsilon |\xi|^2} = \widehat{f} \\ \widehat{v}_{|t=0} = \widehat{v}_0. \end{cases}$$

The matrix $Mv \stackrel{\text{def}}{=} \dfrac{\xi_3 \xi \wedge v}{|\xi|^2}$ has three eigenvalues, 0 and $\pm i \dfrac{\xi_3}{|\xi|} \cdot$ The associated eigenvectors are

$$e^0(\xi) = {}^t(0,0,1)$$

and

$$e^\pm(\xi) = \frac{1}{\sqrt{2}|\xi||\xi_h|} {}^t\left(\xi_1\xi_3 \mp i\xi_2|\xi|,\, \xi_2\xi_3 \pm i\xi_1|\xi|, -|\xi_h|^2\right).$$

The precise value of these vectors is not needed for our study; all we need to know is that the last two are divergence-free, in the sense that $\xi \cdot e^\pm(\xi) = 0$. Furthermore they are orthogonal, and we leave the proof of the following easy property to the reader.

Lemma 5.1 *Let $v \in \mathcal{H}(\mathbf{R}^3)$ be given, and define*

$$v^\pm \stackrel{def}{=} \mathcal{F}^{-1}\left((\widehat{v}^\pm(\xi) \cdot e^\pm(\xi)) e^\pm(\xi)\right). \tag{5.2.1}$$

Then

$$\|v\|_{L^2}^2 = \|v^+\|_{L^2}^2 + \|v^-\|_{L^2}^2.$$

We are now led to study

$$\mathcal{G}_\nu^{\varepsilon,\pm}(\tau)\colon\, g \mapsto \int_{\mathbf{R}^3_\xi} \widehat{g}(\xi)e^{\pm i\tau\frac{\xi_3}{|\xi|}-\nu\tau\varepsilon|\xi|^2+ix\cdot\xi}\,d\xi$$

$$= \int_{\mathbf{R}^3_\xi\times\mathbf{R}^3_y} g(y)e^{\pm i\tau\frac{\xi_3}{|\xi|}-\nu\tau\varepsilon|\xi|^2+i(x-y)\cdot\xi}\,d\xi dy,$$

first considering the case when $\widehat{g}$ is supported in $\mathcal{C}_{r,R}$ for some $r<R$, defined by

$$\mathcal{C}_{r,R} = \{\xi\in\mathbf{R}^3 \,/\, |\xi_3|\geq r \text{ and } |\xi|\leq R\}. \tag{5.2.2}$$

Let us introduce

$$K^\pm(t,\tau,z) \stackrel{\text{def}}{=} \int_{\mathbf{R}^3_\xi} \psi(\xi)e^{\pm i\tau a(\xi,z)+iz\cdot\xi-\nu t|\xi|^2}\,d\xi, \tag{5.2.3}$$

where

$$a(\xi,z)\stackrel{\text{def}}{=}\frac{\xi_3}{|\xi|},$$

and ψ is a function of $\mathcal{D}(\mathbf{R}^3\backslash\{0\})$, such that $\psi\equiv 1$ in a neighborhood of $\mathcal{C}_{r,R}$, which in addition is radial with respect to the horizontal variable $\xi_h=(\xi_1,\xi_2)$.

As we shall need anisotropic-type estimates in the next section (when we apply those estimates to the full non-linear rotating fluid equations), it is convenient to introduce also

$$I^\pm(t,\tau,z_h,\xi_3)\stackrel{\text{def}}{=}\int_{\mathbf{R}^2_{\xi_h}} \psi(\xi)e^{\pm i\tau a(\xi,z_h)+iz_h\cdot\xi_h-\nu t|\xi|^2}\,d\xi_h. \tag{5.2.4}$$

Note that

$$K^\pm(t,\tau,z) = \int_{\mathbf{R}} e^{iz_3\xi_3} I^\pm(t,\tau,z_h,\xi_3)\,d\xi_3. \tag{5.2.5}$$

Lemma 5.2 *For any (r,R) such that $0<r<R$, a constant $C_{r,R}$ exists such that $\forall z\in\mathbf{R}^3$, $\forall\xi_3\in\mathbf{R}$,*

$$|K^\pm(t,\tau,z)|+|I^\pm(t,\tau,z_h,\xi_3)| \leq C_{r,R}\min\{1,\tau^{-\frac{1}{2}}\}e^{-\frac{\nu}{2}r^2t}. \tag{5.2.6}$$

Proof Due to (5.2.5) and to the fact that ξ_3 is restricted to $r\leq|\xi_3|\leq R$, it is enough to prove the result on $I^\pm$ and the estimate on $K^\pm$ will follow.

The proof of the estimate on $I^\pm$ follows the lines of Theorem 5.2 in the previous section, in a very simple way; the only difference is that we have to take care of the dependence upon the viscosity. Moreover for the sake of simplicity we will only consider I^+, the term I^- being dealt with exactly in the same way.

First using the rotation invariance in (ξ_1,ξ_2), we restrict ourselves to the case when $z_2=0$. Next, denoting $\alpha(\xi)\stackrel{\text{def}}{=}-\partial_{\xi_2}a(\xi)=\xi_2\xi_3/|\xi|^3$, we introduce the following differential operator:

$$\mathcal{L}\stackrel{\text{def}}{=}\frac{1}{1+\tau\alpha^2(\xi)}\left(1+i\alpha(\xi)\partial_{\xi_2}\right),$$

which acts on the ξ_2 variable, and satisfies $\mathcal{L}(e^{i\tau a}) = e^{i\tau a}$. Integrating by parts, we obtain

$$I^+(t,\tau,z_h,\xi_3) = \int_{\mathbf{R}^2} e^{i\tau a(\xi,z_h)+iz_1\xi_1} \left({}^t\mathcal{L}(\psi(\xi)e^{-\nu t|\xi|^2})\right) d\xi_h.$$

Easy computations yield

$$\begin{aligned}{}^t\mathcal{L}\left(\psi(\xi)e^{-\nu t|\xi|^2}\right) &= \left(\frac{1}{1+\tau\alpha^2} - i(\partial_{\xi_2}\alpha)\frac{1-\tau\alpha^2}{(1+\tau\alpha^2)^2}\right)\psi(\xi)e^{-\nu t|\xi|^2}\\ &\quad - \frac{i\alpha}{1+\tau\alpha^2}\partial_{\xi_2}\left(e^{-\nu t|\xi|^2}\psi(\xi)\right).\end{aligned}$$

As ξ belongs to the set $\mathcal{C}_{r,R}$ defined by (5.2.2), we have clearly

$$\frac{|\xi_2|r}{R^3} \leq |\alpha(\xi)| \leq \frac{R^2}{r^3},$$

hence

$$\frac{1}{1+\tau\alpha^2} + \frac{|1-\tau\alpha^2|}{1+\tau\alpha^2} + \frac{|\alpha|}{1+\tau\alpha^2} \leq \frac{C_{r,R}}{1+\tau\xi_2^2}.$$

An easy computation also shows that $|\partial_{\xi_2}\alpha(\xi)| \leq C_{r,R}$. Finally, since $\psi \in \mathcal{D}(\mathbf{R}^3)$, we have

$$\begin{aligned}\left|\partial_{\xi_2}\left(e^{-\nu t|\xi|^2}\psi(\xi)\right)\right| &\leq |\partial_{\xi_2}\psi(\xi)|e^{-\nu tr^2} + \nu t|\xi_2||\psi(\xi)|e^{-\nu tr^2}\\ &\leq C_{r,R}e^{-\frac{\nu}{2}tr^2}.\end{aligned}$$

Putting all those estimates together we infer that

$$\left|{}^t\mathcal{L}\left(\psi(\xi)e^{-\nu t|\xi|^2}\right)\right| \leq \frac{C_{r,R}}{1+\tau\xi_2^2}e^{-\frac{\nu}{2}tr^2}$$

so since $r \leq |\xi_1| \leq R$ and $|e^{i\tau a(\xi,z_h)+iz_1\xi_1}| = 1$, we obtain, for all $z_h \in \mathbf{R}^2$ and all $\xi_3 \in \mathbf{R}$,

$$|I^+(t,\tau,z_h,\xi_3)| \leq C_{r,R}e^{-\frac{\nu}{2}tr^2}\int_{\mathbf{R}}\frac{d\xi_2}{1+\tau\xi_2^2},$$

which proves Lemma 5.2. □

Lemma 5.2 yields the following theorem.

Theorem 5.3 *For any positive constants r and R such that $r < R$, let $\mathcal{C}_{r,R}$ be the frequency domain defined in (5.2.2). Then a constant $C_{r,R}$ exists such that if $v_0 \in L^2(\mathbf{R}^3)$ and $f \in L^1(\mathbf{R}^+; L^2(\mathbf{R}^3))$ are two vector fields such that*

$$\operatorname{Supp}\widehat{v}_0 \cup \bigcup_{t\geq 0}\operatorname{Supp}\widehat{f}(t,\cdot) \subset \mathcal{C}_{r,R},$$

and if v is the solution of the linear equation (VC_ε) *with forcing term f and initial data v_0, then for all p in* $[1,+\infty]$,

$$\|v\|_{L^p(\mathbf{R}^+;L^\infty(\mathbf{R}^3))} \leq C_{r,R}\,\varepsilon^{\frac{1}{4p}} \left(\|v_0\|_{L^2(\mathbf{R}^3)} + \|f\|_{L^1(\mathbf{R}^+;L^2(\mathbf{R}^3))}\right) \tag{5.2.7}$$

and

$$\|v\|_{L^p(\mathbf{R}^+;L^{\infty,2}_{x_h,x_3})} \leq C_{r,R}\,\varepsilon^{\frac{1}{4p}} \left(\|v_0\|_{L^2(\mathbf{R}^3)} + \|f\|_{L^1(\mathbf{R}^+;L^2(\mathbf{R}^3))}\right), \tag{5.2.8}$$

where we have noted, for $(\alpha,\beta)\in[1,+\infty]^2$,

$$\|f\|_{L^{\alpha,\beta}_{x_h,x_3}} \stackrel{def}{=} \Big\|\|f(x_h,\cdot)\|_{L^\beta(\mathbf{R}_{x_3})}\Big\|_{L^\alpha(\mathbf{R}^2_{x_h})}.$$

Proof Let us start by proving estimate (5.2.7). Its anisotropic counterpart (estimate (5.2.8)) will be obtained in a similar though slightly more complicated way.

Before proving the result (5.2.7), we note that Duhamel's formula enables us to restrict our attention to the case $f=0$. Indeed if f is non-zero then we write $v=v^+ + v^-$ with

$$v^\pm(t) \stackrel{\text{def}}{=} \mathcal{G}^{\varepsilon,\pm}_\nu\left(\frac{t}{\varepsilon}\right) v_0^\pm + \int_0^t \mathcal{G}^{\varepsilon,\pm}_\nu\left(\frac{t-t'}{\varepsilon}\right) f^\pm(t')\,dt',$$

where we have used notation (5.2.1) above. Then by Lemma 5.1 it is enough to prove the result for v^+ remembering that

$$\|v^\pm\|_{L^2(\mathbf{R}^3)} \leq \|v\|_{L^2(\mathbf{R}^3)} \quad\text{and}\quad \|f^\pm\|_{L^1(\mathbf{R}^+,L^2(\mathbf{R}^2))} \leq \|f\|_{L^1(\mathbf{R}^+,L^2(\mathbf{R}^2))}.$$

Note that the eigenvalue 0 does not appear in this formula since the corresponding eigenvector is not divergence-free.

If the result (5.2.7) holds when $f^+=0$, then of course

$$\|\mathcal{G}^{\varepsilon,+}_\nu\left(\frac{t}{\varepsilon}\right) v_0^+\|_{L^p(\mathbf{R}^+;L^\infty(\mathbf{R}^3))} \leq C_{r,R}\,\varepsilon^{\frac{1}{4p}}\|v_0\|_{L^2(\mathbf{R}^3)}.$$

Then we write, if $p=1$,

$$\begin{aligned}
\mathcal{G}^{\varepsilon,+}_1(f) &\stackrel{\text{def}}{=} \left\|\int_0^t \mathcal{G}^{\varepsilon,+}_\nu\left(\frac{t-t'}{\varepsilon}\right) f^+(t')\,dt'\right\|_{L^1(\mathbf{R}^+;L^\infty)}\\
&\leq \int\!\!\int_0^t \left\|\mathcal{G}^{\varepsilon,+}_\nu\left(\frac{t-t'}{\varepsilon}\right) f^+(t')\right\|_{L^\infty} dt'dt\\
&\leq \int\!\!\int_{t'}^\infty \left\|\mathcal{G}^{\varepsilon,+}_\nu\left(\frac{t-t'}{\varepsilon}\right) f^+(t')\right\|_{L^\infty} dtdt'\\
&\leq C_{r,R}\,\varepsilon^{\frac{1}{4}}\int \|f(t')\|_{L^2}dt'
\end{aligned}$$

which yields the result if $p = 1$. Similarly if $p = +\infty$,

$$\mathcal{G}_1^{\varepsilon,+}(f) \leq \sup_{t\geq 0} \int_0^t \left\| \mathcal{G}_\nu^{\varepsilon,+}\left(\frac{t-t'}{\varepsilon}\right) f^+(t') \right\|_{L^\infty} dt'$$
$$\leq C_{r,R} \sup_{t\geq 0} \int_0^t \|f(t')\|_{L^2}\, dt'$$

and estimate (5.2.7), for all $p \in [1,\infty]$, follows by interpolation. So from now on we shall suppose that $f = 0$.

As a consequence of Lemma 5.2, considering g such that $\widehat{g}$ is supported in $\mathcal{C}_{r,R}$ and $\psi \in \mathcal{D}(\mathbf{R}^3 \setminus \{0\})$ radial with respect to ξ_h, such that $\psi \equiv 1$ in a neighborhood of $\mathcal{C}_{r,R}$, one has

$$\mathcal{G}_\nu^{\varepsilon,+}(\tau)g(x) = \int_{\mathbf{R}^3_\xi \times \mathbf{R}^3_y} \psi(\xi)g(y)e^{i\tau\frac{\xi_3}{|\xi|} - \nu\tau\varepsilon|\xi|^2 + i(x-y)\cdot\xi}\, d\xi dy$$
$$= \int_{\mathbf{R}^3_y} K^+(\varepsilon\tau, \tau, x-y)\, g(y)\, dy,$$

where K^+ is defined by formula (5.2.3). Moreover, the following estimate holds:

$$\left\| \mathcal{G}_\nu^{\varepsilon,+}\left(\frac{t}{\varepsilon}\right) g \right\|_{L^\infty} \leq C_{r,R} \frac{\varepsilon^{\frac{1}{2}}}{t^{\frac{1}{2}}} e^{-ct} \|g\|_{L^1}.$$

Now we shall use a duality argument, otherwise known as the TT^* argument. Once we have observed that

$$\|b\|_{L^1(\mathbf{R}^+;L^\infty(\mathbf{R}^3))} = \sup_{\varphi\in\mathcal{B}} \int_{\mathbf{R}^+\times\mathbf{R}^3} b(t,x)\varphi(t,x)\, dxdt,$$

with

$$\mathcal{B} = \left\{ \varphi \in \mathcal{D}(\mathbf{R}^+ \times \mathbf{R}^3),\ \|\varphi\|_{L^\infty(\mathbf{R}^+;L^1(\mathbf{R}^3))} \leq 1 \right\},$$

we can write

$$\left\| \mathcal{G}_\nu^{\varepsilon,+}\left(\frac{t}{\varepsilon}\right) g \right\|_{L^1(\mathbf{R}^+;L^\infty)}$$
$$= \sup_{\varphi\in\mathcal{B}} \int_{\mathbf{R}^+\times\mathbf{R}^6} K^+\left(t, \frac{t}{\varepsilon}, x-y\right) g(y)\varphi(t,x)\, dtdxdy$$

which can be written

$$\left\| \mathcal{G}_\nu^{\varepsilon,+}\left(\frac{t}{\varepsilon}\right) g \right\|_{L^1(\mathbf{R}^+;L^\infty)}$$
$$= \sup_{\varphi\in\mathcal{B}} \int_{\mathbf{R}^+\times\mathbf{R}^3} g(y) \left(\int_{\mathbf{R}^3} K^+\left(t, \frac{t}{\varepsilon}, x-y\right) \varphi(t,x) dx \right) dtdy.$$

A Cauchy–Schwarz inequality then yields, denoting $\check{b}(x) = b(-x)$,

$$\left\| \mathcal{G}_\nu^{\varepsilon,+}\left(\frac{t}{\varepsilon}\right) g \right\|_{L^1(\mathbf{R}^+;L^\infty)} \leq \|g\|_{L^2}\Phi$$

with

$$\Phi \stackrel{\text{def}}{=} \sup_{\varphi\in\mathcal{B}} \left\| \int_{\mathbf{R}^+} \check{K}^+\left(t, \frac{t}{\varepsilon}, \cdot\right) * \varphi(t,\cdot)\, dt \right\|_{L^2}.$$

By the Fourier–Plancherel theorem, we have

$$\begin{aligned}\Phi^2 &= \frac{1}{(2\pi)^3} \left\| \int_{\mathbf{R}^+} \mathcal{F}\check{K}^+\left(t, \frac{t}{\varepsilon}, \cdot\right) \widehat{\varphi}(t,\cdot)\, dt \right\|_{L^2}^2 \\ &\leq C \int_{\mathbf{R}^+\times\mathbf{R}^+\times\mathbf{R}^3} \widehat{K}^+\left(t, \frac{t}{\varepsilon}, -\xi\right) \overline{\widehat{\varphi}}(t,\xi) \overline{\widehat{K}}^+\left(s, \frac{s}{\varepsilon}, -\xi\right) \widehat{\varphi}(s,\xi)\, d\xi dt ds.\end{aligned}$$

By definition (5.2.3) of K, we have

$$\widehat{K}^+(t,\tau,-\xi) = \psi(-\xi) e^{i\tau a(-\xi,z)} e^{-\tau t|\xi|^2}.$$

Thus the following identity holds

$$\widehat{K}^+\left(t, \frac{t}{\varepsilon}, -\xi\right) \overline{\widehat{K}}^+\left(s, \frac{s}{\varepsilon}, -\xi\right) = \widehat{K}^+\left(t+s, \frac{t-s}{\varepsilon}, -\xi\right) \psi(-\xi).$$

It follows that

$$\Phi^2 \leq C \int_{(\mathbf{R}^+)^2\times\mathbf{R}^3} \psi(\xi) \mathcal{F}\left(\check{K}^+\left(t+s, \frac{t-s}{\varepsilon}, \cdot\right) * \varphi(t,\cdot)\right) \overline{\widehat{\varphi}}(s,\xi)\, d\xi dt ds.$$

We now use the Fourier–Plancherel theorem again to get

$$\begin{aligned}\Phi^2 &\leq C \int_{(\mathbf{R}^+)^2\times\mathbf{R}^3} \psi(\xi) \left(\check{K}^+\left(t+s, \frac{t-s}{\varepsilon}, \cdot\right) * \varphi(t,\cdot)\right)(x) \varphi(s,-x)\, dx dt ds. \\ &\leq C \int_{(\mathbf{R}^+)^2} \left\| \check{K}^+\left(t+s, \frac{t-s}{\varepsilon}, \cdot\right) * \varphi(t,\cdot) \right\|_{L^\infty} \|\varphi(s,\cdot)\|_{L^1}\, dt ds, \\ &\leq C \int_{(\mathbf{R}^+)^2} \left\| \check{K}^+\left(t+s, \frac{t-s}{\varepsilon}, \cdot\right) \right\|_{L^\infty(\mathbf{R}^3)} \|\varphi(t,\cdot)\|_{L^1} \|\varphi(s,\cdot)\|_{L^1(\mathbf{R}^3)}\, dt ds.\end{aligned}$$

The dispersion estimate (5.2.6) on K^+ yields

$$\Phi^2 \leq C_{r,R} \int_{(\mathbf{R}^+)^2} \frac{\varepsilon^{\frac{1}{2}}}{(t-s)^{\frac{1}{2}}} e^{-\nu r^2(t+s)} \|\varphi(t,\cdot)\|_{L^1(\mathbf{R}^3)} \|\varphi(s,\cdot)\|_{L^1(\mathbf{R}^3)}\, dt ds.$$

Now we conclude simply by writing

$$\Phi^2 \leq C_{r,R}\varepsilon^{\frac{1}{2}} \|\varphi\|_{L^\infty(\mathbf{R}^+;L^1(\mathbf{R}^3))}^2 \int_{(\mathbf{R}^+)^2} \frac{1}{(t-s)^{\frac{1}{2}}} ds dt.$$

As the integral

$$\int_{(\mathbf{R}^+)^2} \frac{1}{(t-s)^{\frac{1}{2}}} e^{-\nu r^2 (t+s)} \, dtds$$

is finite, this yields the expected estimate in the case $p = 1$. To get the whole interval $p \in [1, +\infty]$, we simply notice that due to the skew-symmetry of the rotation operator, the $L^\infty(\mathbf{R}^+; L^2(\mathbf{R}^3))$ norm of v is bounded, uniformly in ε. As its frequencies are bounded by R, the same goes for the $L^\infty(\mathbf{R}^+ \times \mathbf{R}^3)$ norm since according to Lemma 1.2, page 26,

$$\|v\|_{L^\infty(\mathbf{R}^+ \times \mathbf{R}^3)} \leq C_R \|v\|_{L^\infty(\mathbf{R}^+; L^2(\mathbf{R}^3))} \leq \|v_0\|_{L^2(\mathbf{R}^3)}.$$

Interpolation of that uniform bound with the $L^1(\mathbf{R}^+; L^\infty(\mathbf{R}^3))$ estimate found above yields estimate (5.2.7).

Now let us turn to the anisotropic estimate (5.2.8), which is obtained in a very similar manner to the isotropic estimate above. If we denote, for any function g defined on $\mathbf{R}^3$, by $\widetilde{g}$ its vertical Fourier transform

$$\widetilde{g}(x_h, \xi_3) \stackrel{\text{def}}{=} \int_{\mathbf{R}} e^{-ix_3\xi_3} g(x_h, x_3) \, dx_3,$$

then we have of course

$$\left\| \mathcal{G}_\nu^{\varepsilon,\pm}(\tau) g \right\|_{L^{\infty,2}_{x_h,x_3}} = C \left\| \widetilde{\mathcal{G}}_\nu^{\varepsilon,\pm}(\tau) \widetilde{g} \right\|_{L^{\infty,2}_{x_h,\xi_3}}.$$

where we have defined

$$\widetilde{\mathcal{G}}_\nu^{\varepsilon,\pm}(\tau) \widetilde{g}(x_h, \xi_3) \stackrel{\text{def}}{=} \int_{\mathbf{R}^2_{y_h}} I^\pm(\varepsilon\tau, \tau, x_h - y_h, \xi_3) \widetilde{g}(y_h, \xi_3) dy_h$$

where $I^\pm$ is defined by (5.2.4). By Lemma 5.2 we have

$$\left\| \widetilde{\mathcal{G}}_\nu^{\varepsilon,\pm}\left(\frac{t}{\varepsilon}\right) \widetilde{g} \right\|_{L^{\infty,2}_{x_h,\xi_3}} \leq C_{r,R} \frac{\varepsilon^{\frac{1}{2}}}{t^{\frac{1}{2}}} e^{-ct} \|\widetilde{g}\|_{L^{1,2}_{x_h,\xi_3}}.$$

We shall now be using exactly the same TT^* argument as in the isotropic case: we reproduce it here for the reader's convenience. We have

$$\|\widetilde{b}\|_{L^1(\mathbf{R}^+; L^{\infty,2}_{x_h,\xi_3})} = \sup_{\widetilde{\varphi} \in \widetilde{\mathcal{B}}} \int_{\mathbf{R}^+ \times \mathbf{R}^3} \widetilde{b}(t, x_h, \xi_3) \widetilde{\varphi}(t, x_h, \xi_3) \, dx_h d\xi_3 dt,$$

with $\widetilde{\mathcal{B}} = \{\widetilde{\varphi} \in \mathcal{D}(\mathbf{R}^+ \times \mathbf{R}^3),\ \|\widetilde{\varphi}\|_{L^\infty(\mathbf{R}^+;L^{1,2}_{x_h,\xi_3})} \leq 1\}$. So we can write

$$\begin{aligned}
\widetilde{G} &\stackrel{\text{def}}{=} \left\| \widetilde{\mathcal{G}}^{\varepsilon,+}_\nu \left(\frac{t}{\varepsilon}\right) \widetilde{g} \right\|_{L^1(\mathbf{R}^+;L^{\infty,2}_{x_h,\xi_3})} \\
&= \sup_{\widetilde{\varphi}\in\widetilde{\mathcal{B}}} \int_{\mathbf{R}^+\times\mathbf{R}^5} I^+\left(t, \frac{t}{\varepsilon}, x_h - y_h, \xi_3\right) \widetilde{g}(y_h,\xi_3)\widetilde{\varphi}(t,x_h,\xi_3)\, dt dx_h dy_h d\xi_3 \\
&= \sup_{\widetilde{\varphi}\in\widetilde{\mathcal{B}}} \int_{\mathbf{R}^+\times\mathbf{R}^3} \widetilde{g}(y_h,\xi_3) \\
&\qquad \times \left(\int_{\mathbf{R}^2} I^+\left(t, \frac{t}{\varepsilon}, x_h - y_h, \xi_3\right) \widetilde{\varphi}(t,x_h,\xi_3)\, dx_h \right) dt dy_h d\xi_3.
\end{aligned}$$

We infer that

$$\widetilde{G} \leq \|\widetilde{g}\|_{L^2(\mathbf{R}^3)} \sup_{\widetilde{\varphi}\in\widetilde{\mathcal{B}}} \left\| \int_{\mathbf{R}^+} \check{I}^+\left(t, \frac{t}{\varepsilon}, \cdot, \xi_3\right) * \varphi(t,\cdot,\xi_3)\, dt \right\|_{L^2(\mathbf{R}^3)}.$$

Similarly to the isotropic case, we have

$$\begin{aligned}
&\left\| \int_{\mathbf{R}^+} \check{I}^+\left(t, \frac{t}{\varepsilon}, \cdot, \xi_3\right) * \varphi(t,\cdot,\xi_3)\, dt \right\|^2_{L^2} \\
&\quad \leq C_{r,R} \int_{(\mathbf{R}^+)^2} \frac{\varepsilon^{\frac{1}{2}}}{(t-s)^{\frac{1}{2}}} e^{-\nu r^2(t+s)} \|\widetilde{\varphi}(t,\cdot)\|_{L^{1,2}_{x_h,\xi_3}} \|\widetilde{\varphi}(s,\cdot)\|_{L^{1,2}_{x_h,\xi_3}}\, dt ds
\end{aligned}$$

and the result follows as previously: to get the L^p estimate in time for all p, we now notice that the $L^\infty(\mathbf{R}^+, L^{\infty,2}_{x_h,x_3})$ norm of the solution is bounded, again by frequency localization and the fact that the rotation operator is an isometry on $L^2(\mathbf{R}^3)$, and the corollary is proved. □

5.3 Application to rotating fluids in $\mathbf{R}^3$

In this section, we focus on the small Rossby number limit of solutions to the incompressible Navier–Stokes–Coriolis system, namely

$$(\mathrm{NSC}_\varepsilon) \begin{cases} \partial_t u^\varepsilon + u^\varepsilon \cdot \nabla u^\varepsilon - \nu\Delta u^\varepsilon + \dfrac{e^3 \wedge u^\varepsilon}{\varepsilon} + \nabla p^\varepsilon = 0 \\ \operatorname{div} u^\varepsilon = 0 \\ u^\varepsilon_{|t=0} = u_0. \end{cases}$$

We refer to the introduction of this book for physical discussions of the model. In the sequel, we will focus on the viscous case $\nu > 0$. We have neglected the presence of bulk forces to simplify the presentation. Let us introduce some notation: let $\mathbf{P}$ be the Leray projector onto divergence-free vector fields; the fact that the operator $\mathbf{P}(e^3 \wedge u^\varepsilon)$ is skew-symmetric implies that if the initial velocity u_0 belongs to $L^2(\mathbf{R}^3)$, then we obtain a sequence $(u^\varepsilon)_{\varepsilon>0}$ of Leray's weak solutions, uniformly bounded in the space $L^\infty(\mathbf{R}^+; L^2(\mathbf{R}^3)) \cap L^2(\mathbf{R}^+; \dot{H}^1(\mathbf{R}^3))$. We leave as an exercise to the reader, the adaptation of the proof of Theorem 2.3, page 42, to obtain the existence of such solutions.

We now wish to analyze that system in the limit when ε goes to zero. In the introduction (Part I), we claimed that the weak limit of u^ε does not depend on the vertical variable x_3. The only element of $L^2(\mathbf{R}^3)$ which does not depend on x_3 is zero, so in order to get some more relevant results on the asymptotics of u^ε, we are going to study the existence (and convergence) of Leray-type solutions for initial data of the type

$$u_0^j = \overline{u}_0^j(x_h) + w_0^j(x_h, x_3), \quad j \in \{1,2,3\}, \tag{5.3.1}$$

with $\operatorname{div}_h u_0^h = \operatorname{div} w_0 = 0$.

Let us denote by $\widetilde{\mathcal{H}}(\mathbf{R}^2)$ the space of vector fields $\overline{u}$ with three components in $L^2(\mathbf{R}^2)$, such that $\partial_1 \overline{u}^1 + \partial_2 \overline{u}^2 = 0$. We have, of course, for $\overline{u}$ in $\widetilde{\mathcal{H}}(\mathbf{R}^2)$,

$$\overline{u} = (\overline{u}^h, 0) + \overline{u}^3(0,0,1)$$

with $\overline{u}^h$ in $\mathcal{H}(\mathbf{R}^2)$ and $\overline{u}^3$ in $L^2(\mathbf{R}^2)$.

We will denote by $(\overline{\mathrm{NS}})$ the two-dimensional Navier–Stokes equations, when the velocity field has three components and not two:

$$(\overline{\mathrm{NS}}) \begin{cases} \partial_t \overline{u} + \overline{u}^h \cdot \nabla^h \overline{u} - \nu \Delta_h \overline{u} + (\nabla^h p, 0) = 0 \\ \operatorname{div}_h \overline{u}^h = 0 \\ \overline{u}_{|t=0} = \overline{u}_0 \in \widetilde{\mathcal{H}}. \end{cases}$$

Our aim is to prove the convergence of the solutions of the rotating fluid equations $(\mathrm{NSC}_\varepsilon)$ associated with data of the type (5.3.1) towards the solution of $(\overline{\mathrm{NS}})$. So we first need to define what a solution of $(\overline{\mathrm{NS}})$ is, and to prove an existence theorem for that system; that will be done in Section 5.3.1 below. Then we will investigate the existence and convergence of solutions to $(\mathrm{NSC}_\varepsilon)$ in a "Leray" framework (in Section 5.3.2) and we will discuss their stability and global well-posedness in time in Section 5.3.3.

5.3.1 *Study of the limit system*

In this brief section we shall discuss the existence of solutions to system $(\overline{\mathrm{NS}})$.

Definition 5.1 *We shall say that a vector field $\overline{u}$ in the space*

$$L^\infty_{\mathrm{loc}}(\mathbf{R}^+; \widetilde{\mathcal{H}}(\mathbf{R}^2)) \cap L^2_{\mathrm{loc}}(\mathbf{R}^+; H^1(\mathbf{R}^2))$$

is a weak solution of $(\overline{\mathrm{NS}})$ with initial data u_0 in $\widetilde{\mathcal{H}}(\mathbf{R}^2)$ if and only if $\overline{u}^h$ is a solution of the two-dimensional Navier–Stokes equations in the sense of Definition 2.5, page 42, and if for any function Ψ in $C^1(\mathbf{R}^+; H^1(\mathbf{R}^2))$,

$$\langle \overline{u}^3(t), \Psi(t)\rangle = \langle \overline{u}^3(0), \Psi(0)\rangle + \int_0^t \int_{\mathbf{R}^2} \left(\nu \nabla^h \overline{u}^3 \cdot \nabla^h \Psi - \overline{u}^3 \cdot \partial_t \Psi\right)(t', x_h)\, dx_h dt'.$$

We state without proof the following theorem, which is a trivial adaptation of Theorem 3.2, page 56.

Theorem 5.4 *Let $\overline{u}_0$ be a vector field in $\widetilde{\mathcal{H}}(\mathbf{R}^2)$. Then there is a unique solution $\overline{u}$ to $(\overline{\mathrm{NS}})$ in the sense of Definition 5.1. Moreover, this solution belongs to $C(\mathbf{R}^+; \widetilde{\mathcal{H}}(\mathbf{R}^2))$ and satisfies the energy equality*

$$\frac{1}{2}\|\overline{u}(t)\|_{L^2}^2 + \nu \int_0^t \|\nabla^h \overline{u}(t')\|_{L^2}^2 \, dt' = \frac{1}{2}\|\overline{u}_0\|_{L^2}^2.$$

5.3.2 *Existence and convergence of solutions to the rotating-fluid equations*

The goal of this section is to study the well-posedness of the rotating-fluid equations in the functional framework presented above. We will then investigate their asymptotics as the parameter ε goes to zero.

Due to the skew-symmetry of the rotation operator, studying $(\mathrm{NSC}_\varepsilon)$ in such a framework means in fact studying three-dimensional perturbations of the two-dimensional Navier–Stokes equations. Let us first establish the equation on such a perturbed flow. We consider $\overline{u}_0$ in $\widetilde{H}(\mathbf{R}^2)$ and the associate solution $\overline{u}$ of $(\overline{\mathrm{NS}})$ given by Theorem 5.4. Let us now define $w \stackrel{\text{def}}{=} u - \overline{u}$, where u is assumed to solve the three-dimensional Navier–Stokes equation. Then we can write formally

$$(\mathrm{PNS}_\nu) \begin{cases} \partial_t w + w \cdot \nabla w + \overline{u} \cdot \nabla w + w \cdot \nabla \overline{u} - \nu \Delta w = -\nabla p \\ \operatorname{div} w = 0. \end{cases}$$

The definition of a solution is simply a duplication of Definition 2.5, page 42.

Definition 5.2 *Let w_0 be a vector field in $\mathcal{H}(\mathbf{R}^3)$. We shall say that w is a weak solution of (PNS_ν) with initial data w_0 if and only if w belongs to the space $L^\infty_{\mathrm{loc}}(\mathbf{R}^+; \mathcal{H}) \cap L^2_{\mathrm{loc}}(\mathbf{R}^+; \mathcal{V}_\sigma)$ and for any function Ψ in $C^1(\mathbf{R}^+; \mathcal{V}_\sigma)$,*

$$\int_{\mathbf{R}^3} w(t,x) \cdot \Psi(t,x)\, dx - \int_0^t \int_{\mathbf{R}^3} (w \cdot \partial_t \Psi + \nu \nabla w : \nabla \Psi)(t', x)\, dx dt'$$
$$- \int_0^t \int_{\mathbf{R}^3} (w \otimes (\overline{u} + w) + \overline{u} \otimes w) : \nabla \Psi(t', x)\, dx dt' = \int_{\mathbf{R}^3} w_0(x) \cdot \Psi(0, x)\, dx,$$

where $\overline{u}$ is the unique solution of $(\overline{\mathrm{NS}})$ associated with $\overline{u}_0$, given by Theorem 5.4.

We have the following result.

Theorem 5.5 *Let $\overline{u}_0$ and w_0 be two vector fields, respectively in $\widetilde{\mathcal{H}}(\mathbf{R}^2)$ and in $\mathcal{H}(\mathbf{R}^3)$. Let $\overline{u}$ be the unique solution of $(\overline{\mathrm{NS}})$ associated with $\overline{u}_0$, given by*

Theorem 5.4. Then, there exists a global weak solution w to the system (PNS$_\nu$) *in the sense of Definition 5.2. Moreover, this solution satisfies the energy inequality*

$$\int_{\mathbf{R}^3} |w(t,x)|^2\,dx + \nu \int_0^t \int_{\mathbf{R}^3} |\nabla w(t',x)|^2\,dxdt' \leq \|w_0\|_{L^2}^2 \exp\left(\frac{C}{\nu^2}\|\overline{u}_0\|_{L^2}^2\right).$$

Proof We shall not give all the details of the proof here, as it is very similar to the proof of the Leray Theorem 2.3, page 42. The only difference is the presence of the additional vector field $\overline{u}$, which could cause some trouble, were it not for the fact that it only depends on two variables and not on three. So the key to the proof of that theorem is to establish some estimates on the product of two vector fields, one of which only depends on two variables.

Let us therefore consider the sequence $(w_k)_{k\in\mathbf{N}}$ defined by

$$\dot{w}_k(t) = \nu\mathbf{P}_k\Delta w_k(t) + F_k(w_k(t)) + G_k(w_k(t),\overline{u}(t)), \tag{5.3.2}$$

where we recall that $F_k(a) = \mathbf{P}_k Q(a,a)$, and where we have defined

$$G_k(a,b) = \mathbf{P}_k G(a,b) \stackrel{\text{def}}{=} \mathbf{P}_k\left(Q(a,b) + Q(b,a)\right). \tag{5.3.3}$$

The only step of the proof of Theorem 2.3 we shall retrace here is the proof of the analog of the energy bound (2.2.4), page 46. An integration by parts yields

$$\begin{aligned}\frac{1}{2}\frac{d}{dt}\|w_k(t)\|_{L^2}^2 + \nu\|\nabla w_k(t)\|_{L^2}^2 = &-\int_{\mathbf{R}^3} (\overline{u}(t,x_h)\cdot\nabla w_k(t,x))\cdot w_k(t,x)\,dx\\ &-\int_{\mathbf{R}^3} (w_k^h(t,x)\cdot\nabla^h\overline{u}(t,x_h))\cdot w_k(t,x)\,dx\end{aligned}$$

and again

$$\frac{1}{2}\frac{d}{dt}\|w_k(t)\|_{L^2}^2 + \nu\|\nabla w_k(t)\|_{L^2}^2 = -\int_{\mathbf{R}^3} (w_k^h(t,x)\cdot\nabla^h\overline{u}(t,x_h))\cdot w_k(t,x)\,dx.$$

Let us define

$$I_k(t) \stackrel{\text{def}}{=} \int_{\mathbf{R}^3} (w_k^h(t,x)\cdot\nabla^h\overline{u}(t,x_h))\cdot w_k(t,x)\,dx.$$

We can write

$$|I_k(t)| \leq \left(\int_{\mathbf{R}} \|w_k(t,\cdot,x_3)\|_{L^4(\mathbf{R}^2)}^2\,dx_3\right)\|\nabla^h\overline{u}(t,\cdot)\|_{L^2(\mathbf{R}^2)}$$

and the Gagliardo–Nirenberg inequality (see Corollary 1.2), page 25 yields

$$|I_k(t)| \leq C\|\nabla^h\overline{u}(t)\|_{L^2(\mathbf{R}^2)}\int_{\mathbf{R}} \|w_k(t,\cdot,x_3)\|_{L^2(\mathbf{R}^2)}\|\nabla^h w_k(t,\cdot,x_3)\|_{L^2(\mathbf{R}^2)}\,dx_3.$$

It is then simply a matter of using the Cauchy–Schwarz inequality to find that

$$|I_k(t)| \leq \frac{\nu}{2}\|\nabla w_k(t)\|_{L^2(\mathbf{R}^3)}^2 + \frac{C}{\nu}\|w_k(t)\|_{L^2(\mathbf{R}^3)}^2\|\nabla^h\overline{u}(t)\|_{L^2(\mathbf{R}^2)}^2.$$

Going back to the estimate on w_k we find that

$$\frac{d}{dt}\|w_k(t)\|_{L^2}^2 + \nu\|\nabla w_k(t)\|_{L^2}^2 \le \frac{C}{\nu}\|w_k(t)\|_{L^2(\mathbf{R}^3)}^2\|\nabla^h\overline{u}(t)\|_{L^2(\mathbf{R}^2)}^2$$

which by Gronwall's lemma yields

$$\|w_k(t)\|_{L^2}^2 + \nu\int_0^t \|\nabla w_k(t')\|_{L^2}^2\,dt' \le \|w_k(0)\|_{L^2}^2 \exp\left(\frac{C}{\nu}\int_0^t \|\nabla^h\overline{u}(t')\|_{L^2}^2\,dt'\right).$$

The energy estimate on $\overline{u}$ enables us to infer finally that

$$\|w_k(t)\|_{L^2}^2 + \nu\int_0^t \|\nabla w_k(t')\|_{L^2}^2\,dt' \le \|w_k(0)\|_{L^2}^2 \exp\left(\frac{C}{\nu^2}\|\overline{u}_0\|_{L^2}^2\right).$$

The end of the proof of Theorem 5.5 is then identical to the case treated in Section 2.2, page 42, so we will not give any more detail here. □

We are now ready to study the asymptotics of Leray solutions of (NSC_ε). We start by defining the Coriolis version, denoted by ($\mathrm{PNSC}_\varepsilon$) of the perturbed system (PNS_ν):

$$\begin{cases} \partial_t w^\varepsilon + w^\varepsilon\cdot\nabla w^\varepsilon + \overline{u}\cdot\nabla w^\varepsilon + w^\varepsilon\cdot\nabla\overline{u} - \nu\Delta w^\varepsilon + \dfrac{e^3\wedge w^\varepsilon}{\varepsilon} = -\dfrac{\nabla p^\varepsilon}{\varepsilon} \\ \operatorname{div} w^\varepsilon = 0. \end{cases}$$

Let us note that $\overline{u}$ belongs to the kernel of the Coriolis operator because $e^3\wedge\overline{u}$ is a gradient.

Definition 5.3 *Let w_0 be a vector field in $\mathcal{H}(\mathbf{R}^3)$, and let $\varepsilon > 0$ be given. We shall say that w^ε is a weak solution of ($\mathrm{PNSC}_\varepsilon$) with initial data w_0 if and only if w^ε belongs to the space $L^\infty_{\mathrm{loc}}(\mathbf{R}^+;\mathcal{H})\cap L^2_{\mathrm{loc}}(\mathbf{R}^+;\mathcal{V}_\sigma)$ and for any function Ψ in $C^1(\mathbf{R}^+;\mathcal{V}_\sigma)$,*

$$\begin{aligned} &\int_{\mathbf{R}^3} w^\varepsilon(t,x)\cdot\Psi(t,x)\,dx - \int_0^t\int_{\mathbf{R}^3}(w^\varepsilon\cdot\partial_t\Psi + \nu\nabla w^\varepsilon:\nabla\Psi)(t',x)\,dxdt' \\ &\quad + \int_0^t\int_{\mathbf{R}^3}\left(\frac{e^3\wedge w^\varepsilon}{\varepsilon}\cdot\Psi - (w^\varepsilon\otimes(\overline{u}+w^\varepsilon) + \overline{u}\otimes w^\varepsilon):\nabla\Psi\right)(t',x)\,dxdt' \\ &\quad = \int_{\mathbf{R}^3} w_0(x)\cdot\Psi(0,x)\,dx, \end{aligned}$$

where $\overline{u}$ is the unique solution of ($\overline{\mathrm{NS}}$) associated with $\overline{u}_0$, given by Theorem 5.4.

Theorem 5.6 *Let $\overline{u}_0$ and w_0 be two vector fields, respectively in $\widetilde{\mathcal{H}}(\mathbf{R}^2)$ and in $\mathcal{H}(\mathbf{R}^3)$. Let $\overline{u}$ be the unique solution of ($\overline{\mathrm{NS}}$) associated with $\overline{u}_0$, given by*

Theorem 5.4. Then, there exists a global weak solution w^ε *to the system* ($\mathrm{PNSC}_\varepsilon$). *satisfying*

$$\int_{\mathbf{R}^3} |w^\varepsilon(t,x)|^2\,dx + \nu\int_0^t\int_{\mathbf{R}^3} |\nabla w^\varepsilon(t',x)|^2\,dxdt' \le \|w_0\|_{L^2}^2 \exp\left(\frac{C}{\nu^2}\|\overline{u}_0\|_{L^2}^2\right).$$

Moreover, for any $q \in]2,6[$, *for any time* T, *we have*

$$\lim_{\varepsilon\to 0}\int_0^T \|w^\varepsilon(t)\|_{L^q(\mathbf{R}^3)}^2\,dt = 0.$$

This means that one can solve the system (NSC_ε) with initial data $\overline{u}_0 + w_0$, and any such solution converges towards the unique solution $\overline{u}$ of ($\overline{\mathrm{NS}}$) associated with $\overline{u}_0$.

Proof of Theorem 5.6 We shall omit the proof of the existence of w^ε satisfying the energy estimate, as it is identical to the proof of Theorem 5.5 due to the skew-symmetry of the Coriolis operator. So what we must concentrate on now is the proof of the convergence of w^ε to zero. It is here that the Strichartz estimates proved in Section 5.2 will be used. In order to use those estimates, we have to get rid of high frequencies and low vertical frequencies. Let us define the following operator

$$\mathcal{P}_R f \stackrel{\text{def}}{=} \chi\left(\frac{|D|}{R}\right) f, \quad \text{where } \chi\in\mathcal{D}(]-2,2[),\ \chi(x) = 1 \text{ for } |x|\le 1.$$

Let us observe that, thanks to Sobolev embeddings (see Theorem 1.2, page 23) and the energy estimate, we have, for any $q \in [2,6[$,

$$\begin{aligned}
\|w^\varepsilon - \mathcal{P}_R w^\varepsilon\|_{L^2(\mathbf{R}^+;L^q(\mathbf{R}^3))} &\le C\|w^\varepsilon - \mathcal{P}_R w^\varepsilon\|_{L^2(\mathbf{R}^+;\dot H^{3(\frac12-\frac1q)})}\\
&\le CR^{-\alpha_q}\|w^\varepsilon\|_{L^2(\mathbf{R}^+;\dot H^1)}\\
&\le CR^{-\alpha_q}\|w_0\|_{L^2}\exp\left(\frac{C}{\nu^2}\|\overline{u}_0\|_{L^2(\mathbf{R}^2)}^2\right) \qquad (5.3.4)
\end{aligned}$$

with $\alpha_q \stackrel{\text{def}}{=} (3/q) - (1/2)$. Now let us define

$$\chi\left(\frac{D_3}{r}\right) a \stackrel{\text{def}}{=} \mathcal{F}^{-1}\left(\chi\left(\frac{\xi_3}{r}\right)\widehat{a}(\xi)\right).$$

As the support of $\chi(D_3/r)\mathcal{P}_R w^\varepsilon$ is included in $\mathcal{B}_{r,R} \stackrel{\text{def}}{=} \{\xi\in B(0,R)\,/\,|\xi_3|\le 2r\}$, we have, thanks to Lemma 1.1, page 24,

$$\begin{aligned}
\left\|\chi\left(\frac{D_3}{r}\right)\mathcal{P}_R w^\varepsilon(t)\right\|_{L^\infty} &\le \left(\int_{\mathcal{B}_{r,R}}\frac{d\xi}{|\xi|^2}\right)^{\frac12}\left\|\chi\left(\frac{D_3}{r}\right)\mathcal{P}_R w^\varepsilon(t)\right\|_{\dot H^1}\\
&\le Cr^{\frac12}\left(\log(1+R^2)\right)^{\frac12}\|w^\varepsilon(t)\|_{\dot H^1}.
\end{aligned}$$

Then, by the energy estimate, we have

$$\left\|\chi\left(\frac{D_3}{r}\right)\mathcal{P}_R w^\varepsilon\right\|_{L^2(\mathbf{R}^+;L^\infty)} \leq C_R\, r^{\frac{1}{2}}\|w_0\|_{L^2}\exp\left(\frac{C}{\nu^2}\|\overline{u}_0\|^2_{L^2(\mathbf{R}^2)}\right). \qquad (5.3.5)$$

Let us define

$$\mathcal{P}_{r,R}f \stackrel{\text{def}}{=} \left(\text{Id} - \chi\left(\frac{D_3}{r}\right)\right)\mathcal{P}_R f. \qquad (5.3.6)$$

The following lemma, which we admit for the moment, describes the dispersive effects due to fast rotation.

Lemma 5.3 *For any positive real numbers r, R and T, and for any q in $]2,+\infty[$,*

$$\forall \varepsilon > 0\,, \quad \|\mathcal{P}_{r,R}\, w^\varepsilon\|_{L^2([0,T];L^q(\mathbf{R}^3))} \leq C\varepsilon^{\frac{1}{8}\left(1-\frac{2}{q}\right)},$$

the constant C above depending on r, q, R, T, $\|\overline{u}_0\|_{L^2}$ and $\|w_0\|_{L^2}$ but not on ε.

Together with inequalities (5.3.4) and (5.3.5), this lemma implies that, for any positive r, R and T, for $q\in]2,6[$,

$$\forall \varepsilon > 0\,,\ \|w^\varepsilon\|_{L^2([0,T];L^q)} \leq CR^{-\alpha_q} + C_R\, r^{\frac{1}{2}} + C_3\varepsilon^{\frac{1}{8}\left(1-\frac{2}{q}\right)},$$

the constant C_3 above depending on r, R, T, $\|\overline{u}_0\|_{L^2}$ and $\|w_0\|_{L^2}$ but not on ε. We deduce that, for any positive r, R and T, for $q\in]2,6[$,

$$\limsup_{\varepsilon\to 0}\|w^\varepsilon\|_{L^2([0,T];L^q)} \leq CR^{-\alpha_q} + C_R\, r^{\frac{1}{2}}.$$

Passing to the limit when r tends to 0 and then when R tends to ∞ gives Theorem 5.6, provided of course we prove Lemma 5.3. □

Proof of Lemma 5.3 Thanks to Duhamel's formula we have,

$$\mathcal{P}_{r,R}w^\varepsilon(t) = \sum_{j=1}^{3}\mathcal{P}^j_{r,R}w^\varepsilon(t) \quad \text{with}$$

$$\mathcal{P}^1_{r,R}w^\varepsilon(t) \stackrel{\text{def}}{=} \mathcal{G}^\varepsilon_\nu\left(\frac{t}{\varepsilon}\right)\mathcal{P}_{r,R}w_0,$$

$$\mathcal{P}^2_{r,R}w^\varepsilon(t) \stackrel{\text{def}}{=} \int_0^t \mathcal{G}^\varepsilon_\nu\left(\frac{t-t'}{\varepsilon}\right)\mathcal{P}_{r,R}Q(w^\varepsilon(t'),w^\varepsilon(t'))\,dt' \quad \text{and}$$

$$\mathcal{P}^3_{r,R}w^\varepsilon(t) \stackrel{\text{def}}{=} \int_0^t \mathcal{G}^\varepsilon_\nu\left(\frac{t-t'}{\varepsilon}\right)\mathcal{P}_{r,R}\left(Q(w^\varepsilon(t'),\overline{u}(t')) + Q(\overline{u}(t'),w^\varepsilon(t'))\right)\,dt'.$$

Theorem 5.3 implies

$$\|\mathcal{P}^1_{r,R}w^\varepsilon\|_{L^2(\mathbf{R}^+;L^\infty)} \leq C_{r,R}\,\varepsilon^{\frac{1}{8}}\|w_0\|_{L^2}.$$

By interpolation with the energy bound, we infer that

$$\|\mathcal{P}^1_{r,R}w^\varepsilon\|_{L^2([0,T];L^q)} \leq C_{r,R,T}\varepsilon^{\frac{1}{8}(1-\frac{2}{q})}\|w_0\|_{L^2} \exp\left(\frac{C}{\nu^2}\|\overline{u}_0\|^2_{L^2(\mathbf{R}^2)}\right). \quad (5.3.7)$$

Using again Theorem 5.3, we have

$$\|\mathcal{P}^2_{r,R}w^\varepsilon\|_{L^2([0,T];L^\infty)} \leq C_{r,R}\,\varepsilon^{\frac{1}{8}}\|\mathcal{P}_R Q(w^\varepsilon, w^\varepsilon)\|_{L^1([0,T];L^2)}.$$

Lemma 1.2, page 26, together with the energy estimate implies that

$$\begin{aligned}\|\mathcal{P}_R Q(w^\varepsilon, w^\varepsilon)\|_{L^1([0,T];L^2)} &\leq CR\|\mathcal{P}_R(w^\varepsilon\otimes w^\varepsilon)\|_{L^1([0,T];L^2)}\\ &\leq CR^{1+\frac{3}{2}}\|w^\varepsilon\otimes w^\varepsilon\|_{L^1([0,T];L^1)}\\ &\leq C_R T\|w_0\|^2_{L^2(\mathbf{R}^3)}\exp\left(\frac{C}{\nu^2}\|\overline{u}_0\|^2_{L^2(\mathbf{R}^2)}\right).\end{aligned}$$

Thus we have

$$\|\mathcal{P}^2_{r,R}w^\varepsilon\|_{L^2([0,T];L^\infty)} \leq C_{r,R}T\,\varepsilon^{\frac{1}{8}}\|w_0\|^2_{L^2(\mathbf{R}^3)}\exp\left(\frac{C}{\nu^2}\|\overline{u}_0\|^2_{L^2(\mathbf{R}^2)}\right).$$

By interpolation with the energy bound, we infer

$$\|\mathcal{P}^2_{r,R}w^\varepsilon\|_{L^2([0,T];L^q)} \leq C_{r,R,T}\,\varepsilon^{\frac{1}{8}(1-\frac{2}{q})}\|w_0\|_{L^2}^{2(1-\frac{1}{q})}\exp\left(\frac{C}{\nu^2}\|\overline{u}_0\|^2_{L^2(\mathbf{R}^2)}\right). \quad (5.3.8)$$

Still using Theorem 5.3, we have

$$\|\mathcal{P}^3_{r,R}w^\varepsilon\|_{L^2([0,T];L^\infty)} \leq C_{r,R}\,\varepsilon^{\frac{1}{8}}\|\mathcal{P}_R(Q(w^\varepsilon,\overline{u})+Q(\overline{u},w^\varepsilon))\|_{L^1([0,T];L^2)}. \quad (5.3.9)$$

Lemma 1.2, page 26, implies that

$$\|\mathcal{P}_R Q(\overline{u}, w^\varepsilon)\|_{L^1([0,T];L^2)} \leq CR\|\mathcal{P}_R(\overline{u}\otimes w^\varepsilon)\|_{L^1([0,T];L^2)}.$$

We shall prove the following lemma, which is an anisotropic version of Lemma 1.2, page 26.

Lemma 5.4 *For any function $f \in L^{1,2}_{x_h,x_3}$, we have*

$$\|\mathcal{P}_R f\|_{L^2(\mathbf{R}^3)} \leq CR\|f\|_{L^{1,2}_{x_h,x_3}}.$$

Proof Let $x_3 \in \mathbf{R}$ be given. Then Lemma 1.2 in two space dimensions implies that

$$\|\mathcal{P}_R f(\cdot,x_3)\|_{L^2(\mathbf{R}^2)} \leq CR\|f(\cdot,x_3)\|_{L^1(\mathbf{R}^2)}.$$

Taking the L^2 norm in x_3 therefore yields

$$\|\mathcal{P}_R f\|_{L^2(\mathbf{R}^3)} \leq CR\left(\int_{\mathbf{R}}\|f(\cdot,x_3)\|^2_{L^1(\mathbf{R}^2)}\,dx_3\right)^{\frac{1}{2}},$$

and the result follows from the fact that

$$\int_{\mathbf{R}} \|f(\cdot, x_3)\|^2_{L^1(\mathbf{R}^2)}\, dx_3 \leq \|f\|^2_{L^{1,2}_{x_h,x_3}}. \qquad \square$$

This lemma, together with Hölder's estimate and the now classical energy bound, implies that

$$\begin{aligned}\|\mathcal{P}_R Q(\overline{u}, w^\varepsilon)\|_{L^1([0,T];L^2(\mathbf{R}^3))} &\leq CR^2 \|\overline{u} \otimes w^\varepsilon\|_{L^1([0,T];L^{1,2}_{x_h,x_3})} \\ &\leq CR^2 \|\overline{u}\|_{L^2([0,T];L^2(\mathbf{R}^2))} \|w^\varepsilon\|_{L^2([0,T];L^2(\mathbf{R}^3))} \\ &\leq C_{R,T} \|\overline{u}_0\|_{L^2(\mathbf{R}^2)} \|w_0\|_{L^2(\mathbf{R}^3)} \\ &\qquad\qquad \times \exp\left(\frac{C}{\nu^2} \|\overline{u}_0\|^2_{L^2(\mathbf{R}^2)}\right).\end{aligned}$$

Clearly, the term $\|\mathcal{P}_R Q(w^\varepsilon, \overline{u})\|_{L^1([0,T];L^2(\mathbf{R}^3))}$ can be estimated in the same way. Thus, inequality (5.3.9) becomes

$$\|\mathcal{P}^3_{r,R} w^\varepsilon\|_{L^2([0,T];L^\infty)} \leq C_{r,R,T}\, \varepsilon^{\frac{1}{8}} \|\overline{u}_0\|_{L^2} \|w_0\|_{L^2} \exp\left(\frac{C}{\nu^2}\|\overline{u}_0\|^2_{L^2}\right).$$

By interpolation with the energy bound, we get

$$\|\mathcal{P}^3_{r,R} w^\varepsilon\|_{L^2([0,T];L^q)} \leq C_{r,R,T}\, \varepsilon^{\frac{1}{8}\left(1-\frac{2}{q}\right)} \|\overline{u}_0\|^{1-\frac{2}{q}}_{L^2} \|w_0\|_{L^2} \exp\left(\frac{C}{\nu^2}\|\overline{u}_0\|^2_{L^2}\right).$$

Combining inequalities (5.3.7)–(5.3.9) concludes the proof of the lemma and thus of Theorem 5.6. □

5.3.3 Global well-posedness

In the same way as in Part II, once solutions have been obtained it is natural to address the question of their stability. The stability arguments of Section 3.5, page 72, still hold in the setting of rotating fluids, due as usual to the skew-symmetry of the rotation operator. What we are interested in, therefore, is proving the existence, in the framework set up in the introduction of this section, of a solution in $L^4([0,T]; \mathcal{V}_\sigma)$ with uniform bounds. Actually we will do better than that since we will be able to prove a result global in time, with no smallness assumption on the initial data. That will be of course due to the presence of strong enough rotation in the equation.

The theorem we shall prove is the following.

Theorem 5.7 *Let $\overline{u}_0$ and w_0 be two divergence-free vector fields, respectively, in the spaces $\widetilde{\mathcal{H}}$ and $H^{\frac{1}{2}}(\mathbf{R}^3)$. Then a positive ε_0 exists such that for all $\varepsilon \leq \varepsilon_0$, there is a unique global solution u^ε to the system* (NSC$_\varepsilon$). *More precisely, denoting by $\overline{u}$ the (unique) solution of* $(\overline{\text{NS}})$ *associated with $\overline{u}_0$, by v^ε_F the solution* *of* $(\mathrm{VC}_\varepsilon)$ *with initial data* w_0 *(with* $f = 0$*), and defining* $w^\varepsilon \stackrel{def}{=} u^\varepsilon - \overline{u}$*, we have*

$$\begin{array}{lll} w^\varepsilon \in C_b^0(\mathbf{R}^+; H^{\frac{1}{2}}(\mathbf{R}^3)) & \textit{and} & \nabla w^\varepsilon \in L^2(\mathbf{R}^+; H^{\frac{1}{2}}(\mathbf{R}^3)), \\ w^\varepsilon - v_F^\varepsilon \to 0 & \textit{in} & L^\infty(\mathbf{R}^+; \dot{H}^{\frac{1}{2}}(\mathbf{R}^3)) \\ \textit{and } \nabla(w^\varepsilon - v_F^\varepsilon) \to 0 & \textit{in} & L^2(\mathbf{R}^+; \dot{H}^{\frac{1}{2}}(\mathbf{R}^3)) \end{array}$$

as ε *goes to zero.*

Proof Let us start by proving the uniqueness of such solutions. In order to do so, it is enough to prove the uniqueness of w^ε solving$(\mathrm{PNSC}_\varepsilon)$. So let us consider two vector fields w_1^ε and w_2^ε solving $(\mathrm{PNSC}_\varepsilon)$, with the same initial data w_0 in $H^{\frac{1}{2}}$ and the same $\overline{u}$ in $L^\infty(\mathbf{R}^+; \widetilde{\mathcal{H}}) \cap L^2(\mathbf{R}^+; \dot{H}^1)$ solution of $(\overline{\mathrm{NS}})$. As in Section 3.5, page 72, we are going to use results on the time-dependent Stokes system. If $\widetilde{w}^\varepsilon$ denotes the difference $\widetilde{w}^\varepsilon \stackrel{\mathrm{def}}{=} w_1^\varepsilon - w_2^\varepsilon$, then $\widetilde{w}^\varepsilon$ solves the time-dependent Stokes system with force

$$Q(w_1^\varepsilon, w_1^\varepsilon) - Q(w_2^\varepsilon, w_2^\varepsilon) + G(\overline{u}, \widetilde{w}^\varepsilon),$$

using notation (5.3.3). As seen in Chapter 3, page 53, for any vector field a which belongs to $L^4([0,T]; \mathcal{V}_\sigma)$, $Q(a,a)$ is in the space $L^2([0,T]; \mathcal{V}'_\sigma)$, so the term $Q(w_1^\varepsilon, w_1^\varepsilon) - Q(w_2^\varepsilon, w_2^\varepsilon)$ is dealt with exactly as for the three-dimensional stability result (Theorem 3.3, page 58– it is in fact even easier here since both w_1^ε and w_2^ε are supposed to be in $L^4([0,T]; \mathcal{V}_\sigma)$). So the only term we must control is $G(\overline{u}, \widetilde{w}^\varepsilon)$. Let us prove that it is an element of $L^2([0,T]; \mathcal{V}'_\sigma)$. In order to do so, it is enough to prove that $\overline{u} \otimes \widetilde{w}^\varepsilon$ is in $L^2([0,T]; L^2)$. But we have, by the Gagliardo–Nirenberg inequality (1.3.4), page 25,

$$\begin{aligned} \|\overline{u} \otimes \widetilde{w}^\varepsilon\|_{L^2(\mathbf{R}^3)}^2 &= \int_{\mathbf{R}} \|(\overline{u} \otimes \widetilde{w}^\varepsilon)(\cdot, x_3)\|_{L^2(\mathbf{R}^2)}^2 \, dx_3 \\ &\leq \|u\|_{L^4(\mathbf{R}^2)}^2 \int_{\mathbf{R}} \|\widetilde{w}^\varepsilon(\cdot, x_3)\|_{L^4(\mathbf{R}^2)}^2 \, dx_3 \\ &\leq C \|\overline{u}\|_{L^2(\mathbf{R}^2)} \|\nabla \overline{u}\|_{L^2(\mathbf{R}^2)} \|\widetilde{w}^\varepsilon\|_{L^2(\mathbf{R}^3)} \|\nabla \widetilde{w}^\varepsilon\|_{L^2(\mathbf{R}^3)}, \end{aligned}$$

so the result follows; for $G(\overline{u}, \widetilde{w}^\varepsilon)$ to be in $L^2([0,T]; \mathcal{V}'_\sigma)$ it is in fact enough to suppose that $\widetilde{w}^\varepsilon$ belongs to $L^\infty([0,T]; \mathcal{H}) \cap L^2([0,T]; \mathcal{V}_\sigma)$, contrary to the purely three-dimensional case where a $L^4([0,T]; \mathcal{V}_\sigma)$ bound is required.

Let us now prove the global existence of w^ε as stated in the theorem. We can immediately note that it is hopeless to try to prove the global existence of w^ε simply by considering $\dot{H}^{\frac{1}{2}}$ estimates on (5.3.2): by skew-symmetry, the rotation would immediately disappear and unless the initial data w_0 are small, it is impossible to find an estimate global in time using the methods introduced in Section 3.5. So the idea is to subtract from (5.3.2) the solution v_F^ε of $(\mathrm{VC}_\varepsilon)$, which we know goes to zero (at least for low frequencies) by the Strichartz estimates. We will then be led to solving a system of the type (5.3.2) on the difference $w_k^\varepsilon - v_F^\varepsilon$, which will have small data and small source terms. The methods of Section 3.5 should then enable us to find the expected result, up to the additional difficulty consisting in coping with the interaction of two-dimensional and three-dimensional vector fields.

Let us be more precise. We can introduce v^ε, the solution of $(\mathrm{VC}_\varepsilon)$ associated with the initial data $\mathcal{P}_{r,R}w_0$ where $\mathcal{P}_{r,R}$ is the cut-off operator defined in (5.3.6). Notice that $\mathcal{P}_{r,R}w_0$ converges towards w_0 in $H^{\frac{1}{2}}$ as r goes to zero and R goes to infinity. So the proof of the fact that $w^\varepsilon - v^\varepsilon$ can be made arbitrarily small, in the function spaces given in Theorem 5.7, implies that $w^\varepsilon - v^\varepsilon_{_F}$ can also be made arbitrarily small in those same spaces; therefore, from now on we shall restrict our attention to v^ε, and we are ready to consider the following approximate system:

$$\dot{\delta}_k^\varepsilon(t) = -\frac{e^3 \wedge \mathbf{P}_k \delta_k^\varepsilon(t)}{\varepsilon} + \nu \mathbf{P}_k \Delta \delta_k^\varepsilon(t) + F_k(\delta_k^\varepsilon(t)) + F_k(v^\varepsilon(t))$$
$$+ G_k(\delta_k^\varepsilon(t), \overline{u}(t) + v^\varepsilon(t)) + G_k(\overline{u}(t), v^\varepsilon(t)),$$

where F_k and G_k were defined in (5.3.3). The initial data for δ_k^ε are

$$\delta^\varepsilon_{k|t=0} = \mathbf{P}_k(\mathrm{Id} \; - \; \mathcal{P}_{r,R})w_0.$$

We omit in the notation of δ_k^ε and v^ε the dependence on r and R, although that fact will of course be crucial in the estimates. We notice that δ_k^ε remains bounded in $L^2(\mathbf{R}^3)$ and is defined as a smooth solution for any time t. The point is now to prove a global bound in $\dot{H}^{\frac{1}{2}}(\mathbf{R}^3)$, as well as the fact that δ_k^ε can be made arbitrarily small. Indeed $w_k^\varepsilon \stackrel{\text{def}}{=} v^\varepsilon + \delta_k^\varepsilon$ solves

$$\partial_t w_k^\varepsilon + \mathbf{P}_k(w_k^\varepsilon \cdot \nabla w_k^\varepsilon + \overline{u} \cdot \nabla w_k^\varepsilon + w_k^\varepsilon \cdot \nabla \overline{u}) - \nu \mathbf{P}_k \Delta w_k^\varepsilon + \frac{\mathbf{P}_k(e^3 \wedge w_k^\varepsilon)}{\varepsilon} = 0,$$

with initial data $\mathbf{P}_k \mathcal{P}_{r,R} w_0$.

The main step in the proof of the theorem reduces to the following proposition.

Proposition 5.1 *For any positive η, one can find three positive real numbers r_0, R_0 and ε_0 such that for any $k \in \mathbf{N}$ and for any $\varepsilon < \varepsilon_0$, we have*

$$\sup_{t \geq 0} \|\delta_k^\varepsilon(t)\|^2_{\dot{H}^{\frac{1}{2}}(\mathbf{R}^3)} + \frac{\nu}{2} \int_{\mathbf{R}^+} \|\nabla \delta_k^\varepsilon(t')\|^2_{\dot{H}^{\frac{1}{2}}(\mathbf{R}^3)} \, dt' \leq \eta.$$

We leave the reader to complete the proof of Theorem 5.7 using that proposition, since it consists simply in copying the end of the proof of Theorem 3.5 page 73.

□

Proof of Proposition 5.1 Let us consider from now on three positive real numbers $\eta < 1$, r and R. We define the time T_k (depending of course on those three numbers), as the biggest time t for which

$$\sup_{t' \leq t} \|\delta_k^\varepsilon(t')\|^2_{\dot{H}^{\frac{1}{2}}} + \frac{\nu}{2} \int_0^t \|\nabla \delta_k^\varepsilon(t')\|^2_{\dot{H}^{\frac{1}{2}}} \, dt' \leq \eta.$$

Thanks to the Lebesgue theorem, r and R can be chosen in order to have

$$\|\delta_k^\varepsilon(0)\|_{\dot H^{\frac12}}^2 \le \|(\mathrm{Id} - \mathcal{P}_{r,R})w_0\|_{\dot H^{\frac12}}^2 \le \eta^2, \tag{5.3.10}$$

which implies in particular that T_k is positive for each k, since $\eta^2 < \eta$.

Let us now write an energy estimate in $\dot H^{\frac12}(\mathbf{R}^3)$ on δ_k^ε. We have

$$\begin{aligned}\frac12\|\delta_k^\varepsilon(t)\|_{\dot H^{\frac12}}^2 &+ \nu\int_0^t \|\nabla\delta_k^\varepsilon(t')\|_{\dot H^{\frac12}}^2\,dt' \\ &\le \frac12\|\delta_k^\varepsilon(0)\|_{\dot H^{\frac12}}^2 + \int_0^t (F_k(\delta_k^\varepsilon(t'))|\delta_k^\varepsilon(t'))_{\dot H^{\frac12}}\,dt' + \int_0^t (\mathcal{F}_k^\varepsilon(t')|\delta_k^\varepsilon(t'))_{\dot H^{\frac12}}\,dt'\end{aligned}$$

where

$$\mathcal{F}_k^\varepsilon \stackrel{\mathrm{def}}{=} G_k(\delta_k^\varepsilon, \overline{u} + v^\varepsilon) + G_k(\overline{u}, v^\varepsilon) + F_k(v^\varepsilon).$$

By (3.5.2) we have

$$\|F_k(\delta_k^\varepsilon)\|_{\dot H^{-\frac12}} \le C\|\nabla\delta_k^\varepsilon\|_{L^2}^2 \le C\|\delta_k^\varepsilon\|_{\dot H^{\frac12}}\|\nabla\delta_k^\varepsilon\|_{\dot H^{\frac12}},$$

so we infer that

$$\begin{aligned}\frac12\|\delta_k^\varepsilon(t)\|_{\dot H^{\frac12}}^2 &+ \nu\int_0^t \|\nabla\delta_k^\varepsilon(t')\|_{\dot H^{\frac12}}^2\,dt' \\ &\le \frac12\|\delta_k^\varepsilon(0)\|_{\dot H^{\frac12}}^2 + C\|\delta_k^\varepsilon(t)\|_{\dot H^{\frac12}}\|\nabla\delta_k^\varepsilon(t)\|_{\dot H^{\frac12}}^2 + \int_0^t (\mathcal{F}_k^\varepsilon(t')|\delta_k^\varepsilon(t'))_{\dot H^{\frac12}}\,dt'.\end{aligned}$$

Now let us suppose that $\eta \le (1/2C)$, where C is the constant appearing in the estimate above. Then by (5.3.10) we have, for all times $t \le T_k$,

$$\|\delta_k^\varepsilon(t)\|_{\dot H^{\frac12}}^2 + \nu\int_0^t \|\nabla\delta_k^\varepsilon(t')\|_{\dot H^{\frac12}}^2\,dt' \le \eta^2 + \int_0^t (\mathcal{F}_k^\varepsilon(t')|\delta_k^\varepsilon(t'))_{H^{\frac12}}\,dt'. \tag{5.3.11}$$

We have the following estimate for $\mathcal{F}_k^\varepsilon$; we postpone the proof to the end of this section.

Proposition 5.2 *With the previous choices of η, r_0 and R_0, there is a $\varepsilon_0 > 0$ and a family of functions $(f^\varepsilon)_{\varepsilon\le\varepsilon_0}$, uniformly bounded in $L^1(\mathbf{R}^+)$ by a constant depending on $\|\overline{u}_0\|_{L^2}$ and on $\|w_0\|_{\dot H^{\frac12}}$, such that, for all $t \le T_k$, we can write, for any $\varepsilon \le \varepsilon_0$,*

$$\left|\int_0^t (\mathcal{F}_k^\varepsilon|\delta_k^\varepsilon)_{\dot H^{\frac12}}\,dt'\right| \le \frac{\nu}{2}\int_0^t \|\nabla\delta_k^\varepsilon(t')\|_{\dot H^{\frac12}}^2\,dt' + \int_0^t f^\varepsilon(t')\|\delta_k^\varepsilon(t')\|_{\dot H^{\frac12}}^2\,dt' + \eta^2.$$

Let us end the proof of Proposition 5.1. Applying Gronwall's lemma to the estimate (5.3.11) and using Proposition 5.2, yields

$$\|\delta_k^\varepsilon(t)\|_{\dot H^{\frac12}(\mathbf{R}^3)}^2 + \frac{\nu}{2}\int_0^t \|\nabla\delta_k^\varepsilon(t')\|_{\dot H^{\frac12}(\mathbf{R}^3)}^2\,dt' \le 2\eta^2\exp\|f^\varepsilon\|_{L^1(\mathbf{R}^+)}.$$

It is therefore now a matter of choosing η small enough so that

$$2\eta^2 \exp\left(\sup_{\varepsilon\leq\varepsilon_0} \|f^\varepsilon\|_{L^1(\mathbf{R}^+)}\right) \leq \frac{\eta}{2},$$

to infer that $T_k = +\infty$; Proposition 5.1 is proved. □

Proof of Proposition 5.2 Recalling that

$$\mathcal{F}_k^\varepsilon \stackrel{\text{def}}{=} G_k(\delta_k^\varepsilon, v^\varepsilon) + G_k(\delta_k^\varepsilon, \overline{u}) + G_k(v^\varepsilon, \overline{u}) + F_k(v^\varepsilon).$$

we have four terms to estimate. First of all we can write, again using (3.5.2),

$$\left|(G_k(\delta_k^\varepsilon, v^\varepsilon)|\delta_k^\varepsilon)_{\dot H^{\frac12}}\right| \leq C\|\nabla v^\varepsilon\|_{L^2}\|\nabla\delta_k^\varepsilon\|_{L^2}\|\nabla\delta_k^\varepsilon\|_{\dot H^{\frac12}}$$

so

$$\int_0^t \left|(G_k(\delta_k^\varepsilon, v^\varepsilon)|\delta_k^\varepsilon)_{\dot H^{\frac12}}\right| dt' \leq \frac{\nu}{8}\int_0^t \|\nabla\delta_k^\varepsilon(t')\|^2_{\dot H^{\frac12}}\, dt' + \frac{C}{\nu^3}\int_0^t \|\delta_k^\varepsilon(t')\|^2_{\dot H^{\frac12}}\|\nabla v^\varepsilon(t')\|^4_{L^2}\, dt'. \quad (5.3.12)$$

Then to conclude we just need to notice that

$$\int_{\mathbf{R}^+} \|\nabla v^\varepsilon\|^4_{L^2}\, dt' \leq \frac{C}{\nu}\|w_0\|^4_{\dot H^{\frac12}}. \quad (5.3.13)$$

The second term is more delicate: it is here that we have to deal with interactions between two-dimensional and three-dimensional vector fields. This is done by the following lemma.

Lemma 5.5 *A constant C exists such that for all vector fields a and b*

$$|(a|b)_{\dot H^{\frac12}}| \leq C\|a\|_{L^2(\mathbf{R}_{x_3};L^{\frac43}(\mathbf{R}^2))}\|\nabla b\|_{L^2(\mathbf{R}_{x_3};\dot H^{\frac12}(\mathbf{R}^2))}.$$

Proof By definition of the scalar product on $\dot H^{\frac12}$, we have

$$\begin{aligned}(a|b)_{\dot H^{\frac12}} &= (2\pi)^{-d}\int_{\mathbf{R}^3} |\xi|\widehat{a}(\xi)\check{\widehat{b}}(\xi)\, d\xi\\ &= (2\pi)^{-d}\int_{\mathbf{R}^3} |\xi_h|^{-\frac12}\widehat{a}(\xi)|\xi_h|^{\frac12}|\xi|\check{\widehat{b}}(\xi)\, d\xi.\end{aligned}$$

We have denoted $\check{\widehat{b}} = \mathcal{F}(b(-\cdot))$. By the Cauchy–Schwarz inequality, we get

$$|(a|b)_{\dot H^{\frac12}}| \leq (2\pi)^{-d}\left(\int_{\mathbf{R}^3} |\xi_h|^{-1}|\widehat{a}(\xi)|^2\, d\xi\right)^{\frac12}\left(\int_{\mathbf{R}^3} |\xi_h|\,|\mathcal{F}(\nabla b)(\xi)|^2\, d\xi\right)^{\frac12}.$$

Using the Fourier–Plancherel theorem in $\mathbf{R}_{x_3}$ we infer

$$(a|b)_{\dot H^{\frac12}} \leq (2\pi)^{-d}\left(\int_{\mathbf{R}} \|a(\cdot,x_3)\|^2_{\dot H^{-\frac12}(\mathbf{R}^2)}dx_3\right)^{\frac12}\left(\int_{\mathbf{R}} \|\nabla b(\cdot,x_3)\|^2_{\dot H^{\frac12}(\mathbf{R}^2)}dx_3\right)^{\frac12}.$$

Thanks to the dual Sobolev estimate stated in Corollary 1.1, page 25, we have

$$\forall x_3 \in \mathbf{R}, \quad \|a(\cdot, x_3)\|^2_{\dot{H}^{-\frac{1}{2}}(\mathbf{R}^2)} \leq C\|a(\cdot, x_3)\|^2_{L^{\frac{4}{3}}(\mathbf{R}^2)}.$$

Lemma 5.5 is proved.

As for any function a in $\dot{H}^{\frac{1}{2}}(\mathbf{R}^3)$,

$$\|a\|_{L^2(\mathbf{R}_{x_3};\dot{H}^{\frac{1}{2}}(\mathbf{R}^2))} \leq C\|a\|_{\dot{H}^{\frac{1}{2}}(\mathbf{R}^3)},$$

Lemma 5.5 enables to find

$$\int_0^t \left|(G_k(\delta_k^\varepsilon, \overline{u})|\delta_k^\varepsilon)_{\dot{H}^{\frac{1}{2}}}\right| dt' \leq \int_0^t \Big(\|\delta_k^\varepsilon(t') \cdot \nabla\overline{u}(t')\|_{L^2(\mathbf{R}_{x_3};L^{\frac{4}{3}}(\mathbf{R}^2))}$$
$$+ \|\overline{u}(t') \cdot \nabla\delta_k^\varepsilon(t')\|_{L^2(\mathbf{R}_{x_3};L^{\frac{4}{3}}(\mathbf{R}^2))}\Big) \|\nabla\delta_k^\varepsilon(t')\|_{\dot{H}^{\frac{1}{2}}} dt'. \tag{5.3.14}$$

On the one hand, we have by Hölder's inequality

$$\|\overline{u} \cdot \nabla\delta_k^\varepsilon\|_{L^2(\mathbf{R}_{x_3};L^{\frac{4}{3}}(\mathbf{R}^2))} \leq \|\overline{u}\|_{L^4(\mathbf{R}^2)}\|\nabla\delta_k^\varepsilon\|_{L^2(\mathbf{R}^3)},$$

hence using the Gagliardo–Nirenberg inequality (1.3.4), page 25,

$$\int_0^t \|\overline{u}(t') \cdot \nabla\delta_k^\varepsilon(t')\|_{L^2(\mathbf{R}_{x_3};L^{\frac{4}{3}}(\mathbf{R}^2))}\|\nabla\delta_k^\varepsilon(t')\|_{\dot{H}^{\frac{1}{2}}} dt'$$
$$\leq \frac{\nu}{16}\int_0^t \|\nabla\delta_k^\varepsilon(t')\|^2_{\dot{H}^{\frac{1}{2}}} dt' + \frac{C}{\nu^3}\int_0^t \|\delta_k^\varepsilon(t')\|^2_{\dot{H}^{\frac{1}{2}}}\|\overline{u}(t')\|^4_{L^4(\mathbf{R}^2)} dt'. \tag{5.3.15}$$

On the other hand, we have

$$\|\delta_k^\varepsilon \cdot \nabla\overline{u}\|_{L^2(\mathbf{R}_{x_3};L^{\frac{4}{3}}(\mathbf{R}^2))} \leq \|\nabla\overline{u}\|_{L^2(\mathbf{R}^2)}\|\delta_k^\varepsilon\|_{\dot{H}^{\frac{1}{2}}(\mathbf{R}^3)}.$$

Thus we infer

$$\int_0^t \|\delta_k^\varepsilon(t') \cdot \nabla\overline{u}(t')\|_{L^2(\mathbf{R}_{x_3};L^{\frac{4}{3}}(\mathbf{R}^2))}\|\nabla\delta_k^\varepsilon(t')\|_{\dot{H}^{\frac{1}{2}}} dt'$$
$$\leq \frac{\nu}{16}\int_0^t \|\nabla\delta_k^\varepsilon(t')\|^2_{\dot{H}^{\frac{1}{2}}} dt' + \frac{C}{\nu}\int_0^t \|\delta_k^\varepsilon(t')\|^2_{\dot{H}^{\frac{1}{2}}}\|\nabla\overline{u}(t')\|^2_{L^2(\mathbf{R}^2)} dt'. \tag{5.3.16}$$

Plugging (5.3.15) and (5.3.16) into (5.3.14) yields finally

$$\int_0^t \left|(G_k(\delta_k^\varepsilon, \overline{u})|\delta_k^\varepsilon)_{\dot{H}^{\frac{1}{2}}}\right| dt' \leq \frac{\nu}{8}\int_0^t \|\nabla\delta_k^\varepsilon(t')\|^2_{\dot{H}^{\frac{1}{2}}} dt'$$
$$+ \frac{C}{\nu}\int_0^t \|\delta_k^\varepsilon(t')\|^2_{\dot{H}^{\frac{1}{2}}}\left(\frac{C}{\nu^2}\|\overline{u}(t')\|^4_{\dot{H}^{\frac{1}{2}}} + \|\nabla\overline{u}(t')\|^2_{L^2}\right) dt',$$

and the result follows from the fact that

$$\int_{\mathbf{R}^+} \left(\|\overline{u}(t)\|^4_{\dot{H}^{\frac{1}{2}}} + \|\nabla\overline{u}(t)\|^2_{L^2}\right) dt \leq \frac{C}{\nu}\|\overline{u}_0\|^2_{L^2}\left(\frac{1}{\nu^3}\|\overline{u}_0\|^2_{L^2} + 1\right). \tag{5.3.17}$$

Next let us estimate the term containing $G_k(v^\varepsilon, \overline{u})$. We have

$$\begin{aligned}\left|(G_k(v^\varepsilon, \overline{u})|\delta_k^\varepsilon)_{\dot H^{\frac12}}\right| &\leq \|v^\varepsilon \cdot \nabla \overline{u} + \overline{u} \cdot \nabla v^\varepsilon\|_{L^2} \|\nabla \delta_k^\varepsilon\|_{L^2} \\ &\leq \Big(\|v^\varepsilon\|_{L^{\infty,2}_{x_h,x_3}} \|\nabla \overline{u}\|_{L^2(\mathbf{R}^2)} + R\|v^\varepsilon\|_{L^{\infty,2}_{x_h,x_3}} \|\overline{u}\|_{L^2(\mathbf{R}^2)}\Big) \|\nabla \delta_k^\varepsilon\|_{L^2},\end{aligned}$$

so

$$\begin{aligned}\left|(G_k(v^\varepsilon, \overline{u})|\delta_k^\varepsilon)_{\dot H^{\frac12}}\right| &\leq \|\nabla \delta_k^\varepsilon\|_{L^2}^2 \Big(\|\nabla \overline{u}\|_{L^2} + \|v^\varepsilon\|_{L^{\infty,2}_{x_h,x_3}}\Big) \\ &\quad + C\|v^\varepsilon\|^2_{L^{\infty,2}_{x_h,x_3}} \|\nabla \overline{u}\|_{L^2} + C_R \|v^\varepsilon\|_{L^{\infty,2}_{x_h,x_3}} \|\overline{u}\|^2_{L^2}.\end{aligned}$$

Finally

$$\begin{aligned}\int_0^t \left|(G_k(v^\varepsilon, \overline{u})|\delta_k^\varepsilon)_{\dot H^{\frac12}}\right| dt' &\leq \frac{\nu}{8} \int_0^t \|\nabla \delta_k^\varepsilon(t')\|^2_{\dot H^{\frac12}} dt' \\ &+ \frac{C}{\nu} \int_0^t \|\delta_k^\varepsilon(t')\|^2_{\dot H^{\frac12}} \Big(\|\nabla \overline{u}(t')\|^2_{L^2} + \|v^\varepsilon(t')\|^2_{L^{\infty,2}_{x_h,x_3}}\Big) dt' \\ &+ \int_0^t \|v^\varepsilon(t')\|_{L^{\infty,2}_{x_h,x_3}} \Big(C_R \|\overline{u}(t')\|^2_{L^2} + C\|v^\varepsilon(t')\|_{L^{\infty,2}_{x_h,x_3}} \|\nabla \overline{u}(t')\|_{L^2}\Big) dt'.\end{aligned}$$

But using the a priori bounds on $\overline{u}$ along with the Strichartz estimates on v^ε we have

$$\begin{aligned}\int_0^t \|v^\varepsilon(t')\|_{L^{\infty,2}_{x_h,x_3}} &\Big(C_R \|\overline{u}(t')\|^2_{L^2} + C\|v^\varepsilon(t')\|_{L^{\infty,2}_{x_h,x_3}} \|\nabla \overline{u}(t')\|_{L^2}\Big) dt' \\ &\leq C_{r,R} \|w_0\|_{\dot H^{\frac12}} \|\overline{u}_0\|_{L^2} \Big(\varepsilon^{\frac14} \|\overline{u}_0\|_{L^2} + \varepsilon^{\frac18} \|w_0\|_{\dot H^{\frac12}}\Big).\end{aligned}$$

So finally, we get

$$\begin{aligned}\int_0^t \left|(G_k(v^\varepsilon, \overline{u})|\delta_k^\varepsilon)_{\dot H^{\frac12}}\right| dt' &\leq \frac{\nu}{8} \int_0^t \|\nabla \delta_k^\varepsilon(t')\|^2_{\dot H^{\frac12}} dt' \\ &+ \frac{C}{\nu} \int_0^t \|\delta_k^\varepsilon(t')\|^2_{\dot H^{\frac12}} \Big(\|\nabla \overline{u}(t')\|^2_{L^2} + \|v^\varepsilon(t')\|^2_{L^{\infty,2}_{x_h,x_3}}\Big) dt' \\ &+ C_{r,R} \varepsilon^{\frac18} \|w_0\|_{\dot H^{\frac12}} \|\overline{u}_0\|_{L^2} \Big(\varepsilon^{\frac18} \|\overline{u}_0\|_{L^2} + 1\Big). \qquad (5.3.18)\end{aligned}$$

Let us notice that

$$\int_{\mathbf{R}^+} \Big(\|\nabla \overline{u}(t')\|^2_{L^2} + \|v^\varepsilon(t')\|^2_{L^{\infty,2}_{x_h,x_3}}\Big) dt' \leq \frac{C}{\nu} \|\overline{u}_0\|^2_{L^2} + C_{r,R} \varepsilon^{\frac14} \|w_0\|^2_{\dot H^{\frac12}}, \qquad (5.3.19)$$

and that for ε small enough,

$$C_{r,R} \varepsilon^{\frac18} \|w_0\|_{\dot H^{\frac12}} \|\overline{u}_0\|_{L^2} \Big(\varepsilon^{\frac18} \|\overline{u}_0\|_{L^2} + 1\Big) \leq \frac{\eta^2}{2}.$$

The third term is therefore correctly estimated.

The last term is very easy to estimate: we simply write

$$\left|(F_k(v^\varepsilon)|\delta_k^\varepsilon)_{\dot H^{\frac12}}\right| \leq C\|v^\varepsilon\|_{L^\infty}\|\nabla v^\varepsilon\|_{L^2}\|\nabla \delta_k^\varepsilon\|_{L^2}$$

to find finally

$$\int_0^t \left|(F_k(v^\varepsilon)|\delta_k^\varepsilon)_{\dot H^{\frac12}}\right| \, dt' \leq \frac{\nu}{8}\int_0^t \|\nabla\delta_k^\varepsilon(t')\|^2_{\dot H^{\frac12}}\, dt'$$
$$+\frac{C}{\nu}\int_0^t \|\delta_k^\varepsilon(t')\|^2_{\dot H^{\frac12}}\|\nabla v^\varepsilon(t')\|^2_{\dot H^{\frac12}}\, dt' + C\int_0^t \|v^\varepsilon(t')\|^2_{L^\infty}\|v^\varepsilon(t')\|_{\dot H^{\frac12}}\, dt'.$$

Since

$$\int_0^t \|v^\varepsilon(t')\|^2_{L^\infty}\|v^\varepsilon(t')\|_{\dot H^{\frac12}}\, dt' \leq \frac{C_{r,R}}{\nu}\varepsilon^{\frac14}\|w_0\|^2_{\dot H^{\frac12}}$$

we find

$$\int_0^t \left|(F_k(v^\varepsilon)|\delta_k^\varepsilon)_{\dot H^{\frac12}}\right| \, dt' \leq \frac{\nu}{8}\int_0^t \|\nabla\delta_k^\varepsilon(t')\|^2_{\dot H^{\frac12}}\, dt'$$
$$+\frac{C}{\nu}\int_0^t \|\delta_k^\varepsilon(t')\|^2_{\dot H^{\frac12}}\|\nabla v^\varepsilon(t')\|^2_{\dot H^{\frac12}}\, dt' + \frac{C_{r,R}}{\nu}\varepsilon^{\frac14}\|w_0\|^2_{\dot H^{\frac12}}. \tag{5.3.20}$$

Finally we use the fact that

$$\int_{\mathbf{R}^+} \|\nabla v^\varepsilon(t')\|^2_{\dot H^{\frac12}}\, dt' \leq \frac{C}{\nu}\|w_0\|^2_{\dot H^{\frac12}}, \tag{5.3.21}$$

and that for ε small enough,

$$\frac{C_{r,R}}{\nu}\varepsilon^{\frac14}\|w_0\|^2_{\dot H^{\frac12}} \leq \frac{\eta^2}{2}.$$

Putting together estimates (5.3.12), (5.3.17), (5.3.18) and (5.3.20) yields the result. We notice in particular that

$$f^\varepsilon(t) = C_\nu\left(\|\nabla v^\varepsilon(t)\|^4_{L^2} + \|v^\varepsilon(t)\|^2_{L^{\infty,2}_{x_h,x_3}} + \|\overline{u}(t)\|^4_{\dot H^{\frac12}(\mathbf{R}^2)} + \|\nabla\overline{u}(t)\|^2_{L^2(\mathbf{R}^2)}\right)$$

so as computed in (5.3.13), (5.3.17), (5.3.19), (5.3.21) we have

$$\int_0^t |f^\varepsilon(t')|\, dt' \leq C_\nu\Big(\|w_0\|^2_{\dot H^{\frac12}} + \|w_0\|^4_{\dot H^{\frac12}} + \|\overline{u}_0\|^2_{L^2}$$
$$+\|\overline{u}_0\|^4_{L^2} + C_{r,R}\varepsilon^{\frac14}\|w_0\|^2_{\dot H^{\frac12}}\Big).$$

Proposition 5.2 is proved. □

6

The periodic case

6.1 Setting of the problem, and statement of the main result

This chapter deals with the rotating-fluid equations (NSC) in a purely periodic setting. Let a_1, a_2 and a_3 be three positive real numbers and define the periodic box

$$\mathbf{T}^3 \stackrel{\text{def}}{=} \prod_{j=1}^{3} \mathbf{R}_{/a_j \mathbf{Z}}$$

unlike the previous chapter where $\mathbf{T}^3$ denoted the unit periodic box in $\mathbf{R}^3$. In this chapter the size of the box will have some importance in the analysis.

The system we shall study is the following.

$$(\mathrm{NSC}^{\varepsilon}_{\mathbf{T}}) \begin{cases} \partial_t u^\varepsilon - \nu\Delta u^\varepsilon + \mathbf{P}(u^\varepsilon \cdot \nabla u^\varepsilon) + \dfrac{1}{\varepsilon}\mathbf{P}(e^3 \wedge u^\varepsilon) = 0 \text{ in } \mathbf{T}^3 \\ u^\varepsilon{}_{|t=0} = u_0^\varepsilon \text{ with } \operatorname{div} u_0^\varepsilon = 0, \end{cases}$$

where $\mathbf{P}$ is the Leray projector onto the space of divergence-free vector fields. Let us recall its definition in Fourier variables. As we are in a periodic setting, the Fourier variables are discrete variables $n = (n_1, n_2, n_3) \in \mathbf{Z}^3$ and we will denote throughout this chapter

$$\widetilde{n} = (\widetilde{n}_1, \widetilde{n}_2, \widetilde{n}_3) \quad \text{with } \widetilde{n}_j \stackrel{\text{def}}{=} \frac{n_j}{a_j}, \; j \in \{1, 2, 3\}.$$

Then the Fourier transform of any function h is

$$\forall\, n \in \mathbf{Z}^3, \quad \widehat{h}(n) = \mathcal{F}h(n) \stackrel{\text{def}}{=} \int_{\mathbf{T}^3} e^{-2i\pi\widetilde{n}\cdot x} h(x)\, dx, \tag{6.1.1}$$

and $\widetilde{n} \cdot x$ is the scalar product of $(\widetilde{n}_1, \widetilde{n}_2, \widetilde{n}_3)$ by (x_1, x_2, x_3). The expression of $\mathbf{P}$ in Fourier variables is the following:

$$\widehat{\mathbf{P}}(n) = \mathrm{Id} - \frac{1}{|\widetilde{n}|^2} \begin{pmatrix} \widetilde{n}_1^2 & \widetilde{n}_1\widetilde{n}_2 & \widetilde{n}_1\widetilde{n}_3 \\ \widetilde{n}_2\widetilde{n}_1 & \widetilde{n}_2^2 & \widetilde{n}_2\widetilde{n}_3 \\ \widetilde{n}_3\widetilde{n}_1 & \widetilde{n}_3\widetilde{n}_2 & \widetilde{n}_3^2 \end{pmatrix},$$

where Id denotes the identity matrix in Fourier space.

Let us also recall the definition of Sobolev spaces on $\mathbf{T}^3$. We shall say that a function h is in $H^s(\mathbf{T}^3)$ if

$$\|h\|_{H^s(\mathbf{T}^3)} \stackrel{\text{def}}{=} \left(\sum_{n\in\mathbf{Z}^3}(1+|\widetilde{n}|^2)^s|\widehat{h}(n)|^2\right)^{\frac{1}{2}} < +\infty.$$

Similarly h is in the homogeneous Sobolev space $\dot{H}^s(\mathbf{T}^3)$ if

$$\|h\|_{\dot{H}^s(\mathbf{T}^3)} \stackrel{\text{def}}{=} \left(\sum_{n\in\mathbf{Z}^3}|\widetilde{n}|^{2s}|\widehat{h}(n)|^2\right)^{\frac{1}{2}} < +\infty.$$

A vector field is in $H^s(\mathbf{T}^3)$ (respectively in $\dot{H}^s(\mathbf{T}^3)$) if each of its components is. All the vector fields considered in this chapter will be supposed to be mean-free, and one can notice that for such vector fields, homogeneous and inhomogeneous spaces coincide on $\mathbf{T}^3$.

Our goal is to study the behavior of the solutions of $(\mathrm{NSC}^{\varepsilon}_{\mathbf{T}})$ as the Rossby number ε goes to zero. As in the $\mathbf{R}^3$ case studied in the previous chapter, we will be concerned with the existence and the convergence of both weak and strong solutions.

Let us start by discussing the case of weak solutions. We have the following definition.

Definition 6.1 *We shall say that u is a weak solution of* $(\mathrm{NSC}^{\varepsilon}_{\mathbf{T}})$ *with initial data u_0 in $\mathcal{H}(\mathbf{T}^3)$ if and only if u belongs to the space*

$$C(\mathbf{R}^+;\mathcal{V}')\cap L^\infty_{\text{loc}}(\mathbf{R}^+;\mathcal{H})\cap L^2_{\text{loc}}(\mathbf{R}^+;\mathcal{V}),$$

and for any function Ψ in $C^1(\mathbf{R}^+;\mathcal{V}_\sigma)$,

$$\int_{\mathbf{T}^3}u(t,x)\cdot\Psi(t,x)\,dx - \int_0^t\int_{\mathbf{T}^3}\left(u\cdot\partial_t\Psi + \int_0^t\int_{\mathbf{T}^3}\nu\nabla u:\nabla\Psi\right)(t',x)\,dxdt'$$
$$+\int_0^t\int_{\mathbf{T}^3}\left(\frac{e^3\wedge u}{\varepsilon}\cdot\Psi - u\otimes u:\nabla\Psi\right)(t',x)\,dxdt' = \int_{\mathbf{T}^3}u_0(x)\cdot\Psi(0,x)\,dx.$$

Let us recall the definition of the Coriolis operator L:

$$Lw \stackrel{\text{def}}{=} \mathbf{P}(e^3\wedge w).$$

As L is skew-symmetric and commutes with derivatives, $(\mathrm{NSC}^{\varepsilon}_{\mathbf{T}})$ satisfies the same energy estimates as the three-dimensional Navier–Stokes equations in all H^s spaces. The existence of global Leray solutions for u_0 in $\mathcal{H}$, and of stable, local-in-time solutions for $u_0 \in H^{\frac{1}{2}}$, is therefore obtained exactly as in the case of the Navier–Stokes equations. We recommend to the reader, as an exercise, to rewrite the proof of the following theorem (see Chapters 2 and 3).

Theorem 6.1 *Let u_0 be a vector field in $\mathcal{H}$, and let $\varepsilon > 0$ be given. There exists a global weak solution u^ε to $(\mathrm{NSC}^\varepsilon_{\mathbf{T}})$, satisfying for all $t \geq 0$*

$$\frac{1}{2}\int_{\mathbf{T}^3} |u^\varepsilon(t,x)|^2\,dx + \nu\int_0^t\int_{\mathbf{T}^3} |\nabla u^\varepsilon(t',x)|^2\,dxdt' \leq \frac{1}{2}\|u_0\|^2_{L^2}.$$

Moreover, if u_0 belongs to $H^{\frac{1}{2}}(\mathbf{T}^3)$, there exists a positive time T independent of ε such that u^ε is unique on $[0,T]$ and the family (u^ε) is bounded in $C([0,T]; H^{\frac{1}{2}}(\mathbf{T}^3)) \cap L^2([0,T]; H^{\frac{3}{2}}(\mathbf{T}^3))$. Finally there is a constant $c > 0$, independent of ε, such that if $\|u_0\|_{H^{\frac{1}{2}}(\mathbf{T}^3)} \leq c\nu$, then $T = +\infty$.

In this chapter we are interested in the asymptotic behavior of u^ε as ε goes to zero. The first step of the analysis consists (in Section 6.2) in deriving a limit system for $(\mathrm{NSC}^\varepsilon_{\mathbf{T}})$, which will enable us to state and prove a convergence theorem for weak solutions. The main issue of the chapter consists in studying the behavior of strong solutions, and in proving in particular the following global well-posedness theorem.

Theorem 6.2 (Global well-posedness) *For every triplet of positive real numbers (a_1, a_2, a_3), the following result holds. Let u_0^ε converge to a divergence-free vector field u_0 in $H^{\frac{1}{2}}(\mathbf{T}^3)$. Then for ε small enough (depending on the parameters $(a_j)_{1<j<3}$ and on u_0), there is a unique global solution to the system $(\mathrm{NSC}^\varepsilon_{\mathbf{T}})$ in the space $C^0_b(\mathbf{R}^+; H^{\frac{1}{2}}(\mathbf{T}^3)) \cap L^2(\mathbf{R}^+; H^{\frac{3}{2}}(\mathbf{T}^3))$.*

Before entering into the structure of the proof, let us compare this theorem with Theorem 5.7, page 108. Those two theorems state essentially the same result: for any initial data, if the rotation parameter ε is small enough, the Navier–Stokes–Coriolis system is globally well-posed. In other words, the rotation term has a stabilizing effect. As we saw in the previous chapter, in the case of the whole space $\mathbf{R}^3$ this global well-posedness for small enough ε is due to the fact that the Rossby waves go to infinity immediately; this is a dispersive effect. In the case of the torus, there is of course no dispersive effect. The global well-posedness comes in a totally different way: it is a consequence of the analysis of resonances of Rossby waves in the non-linear term $v \cdot \nabla v$.

The proof of Theorem 6.2 relies on the construction of families of approximated solutions. Let us state the key lemma, where we have used the following notation (see page 78):

$$\|u\|^2_{\frac{1}{2}} \stackrel{\text{def}}{=} \sup_{t\geq 0}\left(\|u(t)\|^2_{H^{\frac{1}{2}}} + 2\nu\int_0^t \|\nabla u(t')\|^2_{H^{\frac{1}{2}}}\,dt'\right) < +\infty.$$

Lemma 6.1 *Let u_0 be in $H^{\frac{1}{2}}$. For any positive real number η, a family $(u^{\varepsilon,\eta}_{\text{app}})$ exists such that*

$$\limsup_{\eta\to 0}\limsup_{\varepsilon\to 0}\|u^{\varepsilon,\eta}_{\text{app}}\|_{\frac{1}{2}} \stackrel{def}{=} U_0 < \infty.$$

Moreover, the families $(u^{\varepsilon,\eta}_{\text{app}})$ are approximate solutions of

$$(\text{NSC}_\varepsilon)\begin{cases}\partial_t u^\varepsilon - \nu\Delta u^\varepsilon + \mathbf{P}(u^\varepsilon\cdot\nabla u^\varepsilon) + \dfrac{1}{\varepsilon}\mathbf{P}(e^3\wedge u^\varepsilon) = 0 \text{ in } \mathbf{T}^3\\ u^\varepsilon{}_{|t=0} = u^\varepsilon_0 \text{ with } \operatorname{div} u^\varepsilon_0 = 0\end{cases}$$

in the sense that $u^{\varepsilon,\eta}_{\text{app}}$ satisfies

$$\begin{cases}\partial_t u^{\varepsilon,\eta}_{\text{app}} - \nu\Delta u^{\varepsilon,\eta}_{\text{app}} + \mathbf{P}(u^{\varepsilon,\eta}_{\text{app}}\cdot\nabla u^{\varepsilon,\eta}_{\text{app}}) + \dfrac{1}{\varepsilon}\mathbf{P}(e^3\wedge u^{\varepsilon,\eta}_{\text{app}}) = R^{\varepsilon,\eta} \text{ in } \mathbf{T}^3\\ \lim\limits_{\eta\to 0}\lim\limits_{\varepsilon\to 0}\|u^{\varepsilon,\eta}_{\text{app}|t=0} - u_0\|_{H^{\frac{1}{2}}} = 0,\end{cases}$$

with $\lim\limits_{\eta\to 0}\lim\limits_{\varepsilon\to 0}\|R^{\varepsilon,\eta}\|_{L^2(\mathbf{R}^+;H^{-\frac{1}{2}})} = 0.$

The goal of this chapter is the construction of the families $(u^{\varepsilon,\eta}_{\text{app}})$. For the time being, let us prove that the above lemma implies Theorem 6.2.

Proof of Theorem 6.2 This is based on Theorem 3.6, page 78. Let us observe that, as the Coriolis operator is skew-symmetric in all Sobolev spaces, all the theorems of Section 3.5 are still valid because their proofs rely on (H^s) energy estimates. With the notation of Theorem 3.6, let us fix η_0 such that, for the associated $(u^{\varepsilon,\eta_0}_{\text{app}})$, we have

$$\limsup_{\varepsilon\to 0}\|R^{\varepsilon,\eta_0}_{\text{app}}\|_{L^2(\mathbf{R}^+;H^{-\frac{1}{2}})} \leq \frac{c}{4}\nu^2\exp\left(-2\frac{C}{\nu^4}U_0^4\right).$$

Let us choose ε_0 such that, for all $\varepsilon\leq\varepsilon_0$,

$$\|u^{\varepsilon,\eta_0}_{\text{app}|t=0} - u_0\|^2_{H^{\frac{1}{2}}} + \frac{4}{\nu}\|R^{\varepsilon,\eta_0}_{\text{app}}\|^2_{L^2(\mathbf{R}^+;H^{-\frac{1}{2}})} \leq \frac{c}{\nu}\exp\left(-2\frac{C}{\nu^4}U_0^4\right)$$

and

$$\|u^{\varepsilon,\eta_0}_{\text{app}}\|_{\frac{1}{2}} \leq 2U_0.$$

Thus, for any $\varepsilon\leq\varepsilon_0$, we have

$$\|u^{\varepsilon,\eta_0}_{\text{app}|t=0} - u_0\|^2_{H^{\frac{1}{2}}} + \frac{4}{\nu}\|R^{\varepsilon,\eta_0}_{\text{app}}\|^2_{L^2(\mathbf{R}^+;H^{-\frac{1}{2}})} \leq \frac{c}{\nu}\exp\left(-\frac{C}{\nu^4}\|u^{\varepsilon,\eta_0}_{\text{app}}\|^4_{\frac{1}{2}}\right).$$

Theorem 3.6, page 78, implies that the solution u^ε is global and the global well-posedness Theorem 6.2 is proved. □

Remark In fact, we can prove a little bit more. For any positive real number δ, let us choose η_0, such that, for any $\eta \leq \eta_0$, a positive ε_η exists such that

$$\forall \varepsilon \leq \varepsilon_\eta\,, \quad \left(\|u_{\text{app}}^{\varepsilon,\eta}(0) - u_0\|^2_{H^{\frac{1}{2}}} + \frac{4}{\nu}\|R^{\varepsilon,\eta}\|_{L^2(\mathbf{R}^+;H^{-\frac{1}{2}})} \right)^{\frac{1}{2}}$$
$$\leq \min\left\{ \frac{\delta}{4}, \frac{c}{4}\nu^2 \exp\left(-2\frac{C}{\nu^4}U_0^4\right) \right\}.$$

Then, inequality (3.6.1), page 80, implies that

$$\forall \varepsilon \leq \varepsilon_\eta\,, \quad \|u_{\text{app}}^{\varepsilon,\eta} - u^\varepsilon\|_{\frac{1}{2}} \leq \delta.$$

Structure of the proof of Lemma 6.1 Let us explain how we will prove Lemma 6.1. Fast time oscillations prevent any result of strong convergence to a fixed function. In order to bypass this difficulty, we are going to introduce a procedure of filtration of the time oscillations. This will lead us to the concept of limit system. The purpose of Section 6.2 is to define the filtering operator, the limit system and to establish that the weak closure of u^ε is included in the set of weak solutions of the limit system.

In Section 6.3 we prove that the non-linear terms in the limit system have a special structure, very close to the structure of the non-linear term in the two-dimensional Navier–Stokes equations, which makes it possible in Section 6.4 to prove the global well-posedness of the limit system, as well as its stability.

Section 6.5 is then devoted to the construction of the families $(u_{\text{app}}^{\varepsilon,\eta})$.

6.2 Derivation of the limit system in the energy space

In this section we shall derive a limit system to $(\mathrm{NSC}_{\mathbf{T}}^\varepsilon)$, when the initial data are in $\mathcal{H}(\mathbf{T}^3)$. Since, according to Theorem 6.1, there is a bounded family of solutions $(u^\varepsilon)_{\varepsilon>0}$ associated with that data, one can easily extract a subsequence and find a weak limit to $(u^\varepsilon)_{\varepsilon>0}$. Unfortunately finding the equation satisfied by that weak limit is no easy matter, as one cannot prove the time equicontinuity of $(u^\varepsilon)_{\varepsilon>0}$ (a quick look at the equation shows that $\partial_t u^\varepsilon$ is not uniformly bounded in ε). So one needs some refined analysis to understand the asymptotic behavior of $(u^\varepsilon)_{\varepsilon>0}$.

Let $\mathcal{L}$ be the evolution group associated with the Coriolis operator L. The vector field $\mathcal{L}(t)w_0$ is the solution at time t of the equation

$$\partial_t w + Lw = 0, \quad w_{|t=0} = w_0.$$

As L is skew-symmetric, the operator $\mathcal{L}(t)$ is unitary for all times t, in all Sobolev spaces $H^s(\mathbf{T}^3)$. In particular if we define the "filtered solution" associated with u^ε, then by Theorem 6.1 the family

$$\widetilde{u}^\varepsilon \stackrel{\text{def}}{=} \mathcal{L}\left(-\frac{t}{\varepsilon}\right) u^\varepsilon$$

is uniformly bounded in the space $L^\infty(\mathbf{R}^+; L^2(\mathbf{T}^3)) \cap L^2(\mathbf{R}^+; H^1(\mathbf{T}^3))$. It satisfies the following system:

$$(\widetilde{\mathrm{NSC}}^\varepsilon_{\mathbf{T}}) \begin{cases} \partial_t \widetilde{u}^\varepsilon - \mathcal{Q}^\varepsilon(\widetilde{u}^\varepsilon, \widetilde{u}^\varepsilon) - \nu\Delta\widetilde{u}^\varepsilon = 0 \\ \widetilde{u}^\varepsilon_{|t=0} = u_0^\varepsilon, \end{cases}$$

noticing that $\mathcal{L}(t/\varepsilon)$ is equal to the identity when $t = 0$. We have used the fact that the operator L commutes with all derivation operators, and we have noted

$$\mathcal{Q}^\varepsilon(a,b) \stackrel{\mathrm{def}}{=} -\frac{1}{2}\left(\mathcal{L}\left(-\frac{t}{\varepsilon}\right)\mathbf{P}\left(\mathcal{L}\left(\frac{t}{\varepsilon}\right)a\cdot\nabla\mathcal{L}\left(\frac{t}{\varepsilon}\right)b\right)\right.$$
$$\left.+\mathcal{L}\left(-\frac{t}{\varepsilon}\right)\mathbf{P}\left(\mathcal{L}\left(\frac{t}{\varepsilon}\right)b\cdot\nabla\mathcal{L}\left(\frac{t}{\varepsilon}\right)a\right)\right). \qquad (6.2.1)$$

The point in introducing the filtered vector field $\widetilde{u}^\varepsilon$ is that one can find a limit system to $(\widetilde{\mathrm{NSC}}^\varepsilon_{\mathbf{T}})$ (contrary to the case of $(\mathrm{NSC}^\varepsilon_{\mathbf{T}})$): if u_0 is in $L^2(\mathbf{T}^3)$, it is not difficult to see that, contrary to the original system, the family $(\partial_t\widetilde{u}^\varepsilon)_{\varepsilon>0}$ is bounded, for instance in the space $L^{\frac{4}{3}}([0,T]; H^{-1}(\mathbf{T}^3))$ for all $T > 0$. Indeed that clearly holds for $\Delta\widetilde{u}^\varepsilon$, and also for $\mathcal{Q}^\varepsilon(\widetilde{u}^\varepsilon, \widetilde{u}^\varepsilon)$ due to the following easy sequence of estimates: since $\mathcal{L}(\cdot)$ is an isometry on all Sobolev spaces we have

$$\|\mathcal{Q}^\varepsilon(\widetilde{u}^\varepsilon, \widetilde{u}^\varepsilon)\|_{L^{\frac{4}{3}}([0,T];H^{-1}(\mathbf{T}^3))} = \left\|\mathcal{L}\left(\frac{t}{\varepsilon}\right)\widetilde{u}^\varepsilon\cdot\nabla\mathcal{L}\left(\frac{t}{\varepsilon}\right)\widetilde{u}^\varepsilon\right\|_{L^{\frac{4}{3}}([0,T];H^{-1}(\mathbf{T}^3))}$$

hence the dual Sobolev embeddings proved in Corollary 1.1, page 25, yield

$$\|\mathcal{Q}^\varepsilon(\widetilde{u}^\varepsilon, \widetilde{u}^\varepsilon)\|_{L^{\frac{4}{3}}([0,T];H^{-1}(\mathbf{T}^3))} \leq C\left\|\mathcal{L}\left(\frac{t}{\varepsilon}\right)\widetilde{u}^\varepsilon\cdot\nabla\mathcal{L}\left(\frac{t}{\varepsilon}\right)\widetilde{u}^\varepsilon\right\|_{L^{\frac{4}{3}}([0,T];L^{\frac{6}{5}}(\mathbf{T}^3))}.$$

By a Hölder inequality we get

$$\|\mathcal{Q}^\varepsilon(\widetilde{u}^\varepsilon, \widetilde{u}^\varepsilon)\|_{L^{\frac{4}{3}}([0,T];H^{-1}(\mathbf{T}^3))} \leq C\left\|\mathcal{L}\left(\frac{t}{\varepsilon}\right)\widetilde{u}^\varepsilon\right\|_{L^4([0,T];L^3(\mathbf{T}^3))}$$
$$\times\left\|\nabla\mathcal{L}\left(\frac{t}{\varepsilon}\right)\widetilde{u}^\varepsilon\right\|_{L^2([0,T];L^2(\mathbf{T}^3))}.$$

So by the Sobolev embedding proved in Theorem 1.2, page 23, along with the Gagliardo–Nirenberg inequality (1.9), page 25, we infer

$$\|\mathcal{Q}^\varepsilon(\widetilde{u}^\varepsilon, \widetilde{u}^\varepsilon)\|_{L^{\frac{4}{3}}([0,T];H^{-1}(\mathbf{T}^3))} \leq C_T\left\|\mathcal{L}\left(\frac{t}{\varepsilon}\right)\widetilde{u}^\varepsilon\right\|^{\frac{1}{2}}_{L^\infty([0,T];L^2(\mathbf{T}^3))}$$
$$\times\left\|\mathcal{L}\left(\frac{t}{\varepsilon}\right)\widetilde{u}^\varepsilon\right\|^{\frac{1}{2}}_{L^2([0,T];H^1(\mathbf{T}^3))}\left\|\nabla\mathcal{L}\left(\frac{t}{\varepsilon}\right)\widetilde{u}^\varepsilon\right\|_{L^2([0,T];L^2(\mathbf{T}^3))}$$

hence

$$\|\mathcal{Q}^\varepsilon(\widetilde{u}^\varepsilon, \widetilde{u}^\varepsilon)\|_{L^{\frac{4}{3}}([0,T];H^{-1}(\mathbf{T}^3))} \leq C(T) \qquad (6.2.2)$$

where we have used again the fact that $\mathcal{L}(\cdot)$ is an isometry on $H^s(\mathbf{T}^3)$, along with the a priori bound provided by Theorem 6.1. That yields the boundedness of $(\partial_t \widetilde{u}^\varepsilon)_{\varepsilon>0}$.

The same compactness argument as that yielding the Leray theorem in Section 2.2.2 enables us, up to the extraction of a subsequence, to obtain a weak limit to the sequence $\widetilde{u}^\varepsilon$, called u (we leave the precise argument to the reader). The linear terms $\partial_t \widetilde{u}^\varepsilon$ and $\Delta \widetilde{u}^\varepsilon$ converge weakly towards $\partial_t u$ and Δu, respectively, in $\mathcal{D}'((0,T)\times \mathbf{T}^3)$, so the point is to find the weak limit of the quadratic form $\mathcal{Q}^\varepsilon(\widetilde{u}^\varepsilon, \widetilde{u}^\varepsilon)$. Let us study that term more precisely. For any point $x = (x_1, x_2, x_3)$ in $\mathbf{T}^3$, we shall define as in the previous chapters the horizontal coordinates $x_h \stackrel{\text{def}}{=} (x_1, x_2)$, and similarly we shall denote $\nabla^h \stackrel{\text{def}}{=} (\partial_1, \partial_2)$, $\operatorname{div}_h \stackrel{\text{def}}{=} \nabla^h \cdot$, and $\Delta_h \stackrel{\text{def}}{=} \partial_1^2 + \partial_2^2$. For any vector field $u = (u^1, u^2, u^3)$, we shall define $u^h \stackrel{\text{def}}{=} (u^1, u^2)$, and $\overline{u}$ will be the quantity

$$\overline{u}(x_h) \stackrel{\text{def}}{=} \frac{1}{|a_3|}\int_0^{a_3} u(x_h, x_3)\, dx_3. \tag{6.2.3}$$

Note that if u is divergence-free, then so is $\overline{u}^h$, due to the following easy computation:

$$\operatorname{div}_h \overline{u}^h = \frac{1}{|a_3|}\int_0^{a_3} \operatorname{div}_h u^h(x_h, x_3)\, dx_3 = -\frac{1}{|a_3|}\int_0^{a_3} \partial_3 u^3(x_h, x_3)\, dx_3 = 0.$$

Finally we will decompose u into

$$u = \overline{u} + u_{\text{osc}}, \tag{6.2.4}$$

where the notation u_{osc} stands for the "oscillating part" of u; that denomination will become clearer as we proceed in the study of $(\mathrm{NSC}^c_{\mathbf{T}})$. Now in order to derive formally the limit of $\mathcal{Q}^\varepsilon$, let us compute more explicitly the operators L and $\mathcal{L}$. We have to solve the equation

$$\partial_t w + Lw = 0,$$

where the matrix L is the product of the horizontal rotation by angle $\pi/2$, denoted by $R^h_{\pi/2}$, with the Leray projector $\mathbf{P}$. Writing L_n for the result of the product $\widehat{\mathbf{P}}(n) R^h_{\pi/2}$, a simple computation shows that

$$L_n = \frac{1}{|\widetilde{n}|^2}\begin{pmatrix} \widetilde{n}_1\widetilde{n}_2 & \widetilde{n}_2^2 + \widetilde{n}_3^2 & 0 \\ -\widetilde{n}_1^2 - \widetilde{n}_3^2 & -\widetilde{n}_1\widetilde{n}_2 & 0 \\ \widetilde{n}_2\widetilde{n}_3 & -\widetilde{n}_1\widetilde{n}_3 & 0 \end{pmatrix},$$

and the eigenvalues of L_n are 0, $i\widetilde{n}_3/|\widetilde{n}|$, and $-i\widetilde{n}_3/|\widetilde{n}|$. We will call $e^0(n)$, $e^+(n)$ and $e^-(n)$ the corresponding eigenvectors, which are given by

$$e^0(n) = {}^t(0,0,1) \quad \text{and}$$

$$e^\pm(n) = \frac{1}{\sqrt{2}|\widetilde{n}||\widetilde{n}_h|}\,{}^t\big(\widetilde{n}_1\widetilde{n}_3 \mp i\widetilde{n}_2|\widetilde{n}|,\ \widetilde{n}_2\widetilde{n}_3 \pm i\widetilde{n}_1|\widetilde{n}|, -|\widetilde{n}_h|^2\big)$$

when $n_h \neq 0$, and $e^0(n) = {}^t(0,0,1)$, $e^{\pm}(n) = \frac{1}{\sqrt{2}}{}^t(1, \pm i, 0)$ when $n_h = 0$. Divergence-free elements of the kernel of L are therefore vector fields which do not depend on the third variable; we recover the well-known Taylor–Proudman theorem recalled in Part I, page 3.

Now we are ready to find the limit of the quadratic form $\mathcal{Q}^\varepsilon$. In the following, we will define $\sigma \stackrel{\text{def}}{=} (\sigma_1, \sigma_2, \sigma_3) \in \{+,-\}^3$, any triplet of pluses or minuses, and for any vector field h, its projection (in Fourier variables) along those vector fields will be denoted

$$\forall n \in \mathbf{Z}^3, \ \forall j \in \{1,2,3\}, \quad h^{\sigma_j}(n) \stackrel{\text{def}}{=} (\mathcal{F}h(n) \cdot e^{\sigma_j}(n))\, e^{\sigma_j}(n).$$

Proposition 6.1 *Let $\mathcal{Q}^\varepsilon$ be the quadratic form defined in (6.2.1), and let a and b be two smooth vector fields on $\mathbf{T}^3$. Then one can define*

$$\mathcal{Q}(a,b) \stackrel{\text{def}}{=} \lim_{\varepsilon \to 0} \mathcal{Q}^\varepsilon(a,b) \quad in \quad \mathcal{D}'(\mathbf{R}^+ \times \mathbf{T}^3),$$

and we have

$$\mathcal{F}\mathcal{Q}(a,b)(n) = -\sum_{\substack{\sigma \in \{+,-\}^3 \\ k \in \mathcal{K}_n^\sigma}} [a^{\sigma_1}(k) \cdot (\widetilde{n} - \widetilde{k})]\, [b^{\sigma_2}(n-k) \cdot e^{\sigma_3}(n)] e^{\sigma_3}(n),$$

where $\mathcal{K}_n^\sigma$ is the "resonant set" defined, for any n in $\mathbf{Z}^3$ and any σ in $\{+,-\}^3$, as

$$\mathcal{K}_n^\sigma \stackrel{\text{def}}{=} \left\{ k \in \mathbf{Z}^3 \ / \ \sigma_1 \frac{\widetilde{k}_3}{|\widetilde{k}|} + \sigma_2 \frac{\widetilde{n}_3 - \widetilde{k}_3}{|\widetilde{n} - \widetilde{k}|} - \sigma_3 \frac{\widetilde{n}_3}{|\widetilde{n}|} = 0 \right\}. \tag{6.2.5}$$

Proof We shall write the proof for $a = b$ for simplicity. We can write

$$-\mathcal{F}\mathcal{Q}^\varepsilon(a,a)(n)(t) = \sum_{\substack{(k,m) \in \mathbf{Z}^6, \sigma \in \{+,-\}^3 \\ k+m=n}} e^{-i\frac{t}{\varepsilon}\left(\sigma_1 \frac{\widetilde{k}_3}{|\widetilde{k}|} + \sigma_2 \frac{\widetilde{m}_3}{|\widetilde{m}|} - \sigma_3 \frac{\widetilde{n}_3}{|\widetilde{n}|}\right)}$$
$$\times [a^{\sigma_1}(k) \cdot \widetilde{m}]\, [a^{\sigma_2}(m) \cdot e^{\sigma_3}(n)] e^{\sigma_3}(n).$$

To find the limit of that expression in the sense of distributions as ε goes to zero, one integrates it against a smooth function $\varphi(t)$. That can be seen as the Fourier transform of φ at the point $\frac{1}{\varepsilon}\left(\sigma_1 \frac{\widetilde{k}_3}{|\widetilde{k}|} + \sigma_2 \frac{\widetilde{m}_3}{|\widetilde{m}|} - \sigma_3 \frac{\widetilde{n}_3}{|\widetilde{n}|}\right)$, which clearly goes to zero as ε goes to zero, if $\sigma_1 \frac{\widetilde{k}_3}{|\widetilde{k}|} + \sigma_2 \frac{\widetilde{m}_3}{|\widetilde{m}|} - \sigma_3 \frac{\widetilde{n}_3}{|\widetilde{n}|}$ is not zero. That is also known as the non-stationary phase theorem. In particular, defining, for any $(n, \sigma) \in \mathbf{Z}^3 \backslash \{0\} \times \{+,-\}^3$,

$$\omega_n^\sigma(k) \stackrel{\text{def}}{=} \sigma_1 \frac{\widetilde{k}_3}{|\widetilde{k}|} + \sigma_2 \frac{\widetilde{n}_3 - \widetilde{k}_3}{|\widetilde{n} - \widetilde{k}|} - \sigma_3 \frac{\widetilde{n}_3}{|\widetilde{n}|}, \tag{6.2.6}$$

we get

$$-\mathcal{F}\mathcal{Q}(a,a)(n) = \sum_{\sigma\in\{+,-\}^3} \sum_{k,\omega_n^\sigma(k)=0} [a^{\sigma_1}(k)\cdot(\widetilde{n}-\widetilde{k})][a^{\sigma_2}(n-k)\cdot e^{\sigma_3}(n)]e^{\sigma_3}(n),$$

and Proposition 6.1 is proved. □

So the limit system is the following:

$$(\text{NSCL}) \begin{cases} \partial_t u - \nu\Delta u - \mathcal{Q}(u,u) = 0 \\ u_{|t=0} = u_0. \end{cases}$$

and we have proved the following theorem.

Theorem 6.3 *Let u_0 be a vector field in $\mathcal{H}$, and let $(u^\varepsilon)_{\varepsilon>0}$ be a family of weak solutions to* $(\text{NSC}_\mathbf{T}^\varepsilon)$*, constructed in Theorem 6.1. Then as ε goes to zero, the weak closure of $(\mathcal{L}\left(-\frac{t}{\varepsilon}\right)u^\varepsilon)_{\varepsilon>0}$ is included in the set of weak solutions of* (NSCL).

In the next section, we shall concentrate on the quadratic form $\mathcal{Q}$, and we will prove it has particular properties which make it very similar to the two-dimensional product arising in the two-dimensional incompressible Navier–Stokes equations.

6.3 Properties of the limit quadratic form $\mathcal{Q}$

The key point of this chapter is that the limit quadratic form $\mathcal{Q}$ has a very special structure, which makes it close to the usual bilinear term in the two-dimensional Navier–Stokes equations. The properties stated in the following proposition will enable us in the next section to prove the global well-posedness of the limit system.

Proposition 6.2 *The quadratic form $\mathcal{Q}$ given in Proposition 6.1 satisfies the following properties.*

(1) For any smooth divergence-free vector field h, we have, using notation (6.2.3) and (6.2.4),

$$-\frac{1}{|a_3|}\int_0^{a_3} \mathcal{Q}(h,h)\,dx_3 = \mathbf{P}(\overline{h}\cdot\nabla\overline{h}).$$

(2) If u, v and w are three divergence-free vector fields, then the following two properties hold, with the notation (6.2.3) and (6.2.4):

$$\forall s\geq 0,\quad \left(\mathcal{Q}(\overline{u},v_{\text{osc}})\middle|(-\Delta)^s v_{\text{osc}}\right)_{L^2(\mathbf{T}^3)} = 0, \tag{6.3.1}$$

and

$$\begin{aligned}\left|\left(\mathcal{Q}(u_{\text{osc}},v_{\text{osc}})\middle|w_{\text{osc}}\right)_{H^{\frac12}(\mathbf{T}^3)}\right| \leq C\big(&\|u_{\text{osc}}\|_{H^{\frac12}(\mathbf{T}^3)}\|v_{\text{osc}}\|_{H^1(\mathbf{T}^3)} \\ &+ \|v_{\text{osc}}\|_{H^{\frac12}(\mathbf{T}^3)}\|u_{\text{osc}}\|_{H^1(\mathbf{T}^3)}\big) \\ &\times \|w_{\text{osc}}\|_{H^{\frac32}(\mathbf{T}^3)}.\end{aligned} \tag{6.3.2}$$

Remark The second result of Proposition 6.2 is a typical two-dimensional product rule, although the setting here is three-dimensional. Indeed, one would rather expect to obtain an estimate of the type

$$\left| \left(\mathcal{Q}(u_{\rm osc}, v_{\rm osc}) | w_{\rm osc}\right)_{H^{\frac12}(\mathbf{T}^3)} \right| \leq C\|u_{\rm osc}\|_{H^1(\mathbf{T}^3)}\|v_{\rm osc}\|_{H^1(\mathbf{T}^3)}\|w_{\rm osc}\|_{H^{\frac32}(\mathbf{T}^3)}.$$

We have indeed

$$\begin{aligned}\left|\left(\mathcal{Q}(u_{\rm osc}, v_{\rm osc}) | w_{\rm osc}\right)_{H^{\frac12}(\mathbf{T}^3)}\right| &\leq \|w_{\rm osc}\|_{H^{\frac32}(\mathbf{T}^3)} \\ &\quad\times \left(\|u_{\rm osc}\cdot\nabla v_{\rm osc}\|_{H^{-\frac12}(\mathbf{T}^3)} + \|v_{\rm osc}\cdot\nabla u_{\rm osc}\|_{H^{-\frac12}(\mathbf{T}^3)}\right),\end{aligned}$$

for which Hölder and Sobolev embeddings yield (in a similar way to (6.2.2))

$$\begin{aligned}\left|\left(\mathcal{Q}(u_{\rm osc}, v_{\rm osc}) | w_{\rm osc}\right)_{H^{\frac12}(\mathbf{T}^3)}\right| &\leq C\|w_{\rm osc}\|_{H^{\frac32}(\mathbf{T}^3)} \\ &\quad\times \left(\|u_{\rm osc}\|_{L^6(\mathbf{T}^3)}\|\nabla v_{\rm osc}\|_{L^2(\mathbf{T}^3)} + \|v_{\rm osc}\|_{L^6(\mathbf{T}^3)}\|\nabla u_{\rm osc}\|_{L^2(\mathbf{T}^3)}\right)\end{aligned}$$

and the result follows by the Sobolev embedding $H^1(\mathbf{T}^3) \hookrightarrow L^6(\mathbf{T}^3)$. Estimate (6.3.2) therefore means one gains half a derivative when one takes into account the special structure of the quadratic form $\mathcal{Q}$ compared with the usual product.

Proof of Proposition 6.2 There are two points to be proved here. Let us start with the first one, which is the simplest.

First statement Let us start with the vertical average

$$\frac{1}{|a_3|}\int_0^{a_3} \mathcal{Q}(h,h)\,dx_3.$$

We have

$$\frac{1}{|a_3|}\int_0^{a_3} \mathcal{Q}(h,h)\,dx_3 = \mathcal{F}^{-1}\left(\mathbf{1}_{\{n_3=0\}}\mathcal{F}\mathcal{Q}(h,h)(n)\right)$$

where $\mathbf{1}_{\mathcal{X}}$ denotes the characteristic function of any frequency set $\mathcal{X}$. Recalling the expression for $\mathcal{Q}$ given in Proposition 6.1, we infer that all we need to compute is the form of the resonant set $\mathcal{K}_n^\sigma$ when n_3 is equal to zero. We have

$$\mathcal{K}^\sigma_{n|n_3=0} = \left\{ k\in\mathbf{Z}^3 \ / \ \sigma_1\frac{\widetilde{k}_3}{|\widetilde{k}|} = \sigma_2\frac{\widetilde{k}_3}{|\widetilde{n}-\widetilde{k}|}, \text{ with } n_3=0\right\}.$$

It follows that either $k_3=0$ or $\sigma_1=\sigma_2$. Now if $k_3=0$, we have

$$\mathbf{1}_{\{k_3=0\}}\widehat{h}(k) = \widehat{\overline{h}}(k)$$

so that will account for the expected limit since

$$
\begin{aligned}
&-\frac{1}{|a_3|}\int_0^{a_3} \mathcal{Q}(h,h)\,dx_3 \\
&= \mathbf{P}(\overline{h}\cdot\nabla\overline{h}) + \mathcal{F}^{-1}\sum_{\substack{\sigma\in\{+,-\}^3\\ k_3\neq 0, n_3=0}}\ \sum_{k\in\mathcal{K}_n^\sigma}[h^{\sigma_1}(k)\cdot(\widetilde{n}-\widetilde{k})][h^{\sigma_2}(n-k)\cdot e^{\sigma_3}(n)]e^{\sigma_3}(n).
\end{aligned}
$$

Let us therefore concentrate now on the case $k_3\neq 0$. In that case $\sigma_1=\sigma_2$ and $|\widetilde{k}|=|\widetilde{m}|$. In order to prove the result we are going to go back to the definition of $\mathcal{Q}(u,u)$ as the weak limit of $\mathcal{Q}^\varepsilon(u,u)$. Recalling that

$$
(h\cdot\nabla)h = -h\wedge\operatorname{curl} h + \nabla\frac{|h|^2}{2},
$$

as $\mathbf{P}\left(\nabla\frac{|h|^2}{2}\right)=0$, we therefore want to compute the limit of

$$
\sum_{\substack{(k,m)\in\mathbf{Z}^6,\sigma\in\{+,-\}^3\\ k+m=n}} 1_{\{n_3=0\}}e^{-i\frac{t}{\varepsilon}\left(\sigma_1\frac{\widetilde{k}_3}{|\widetilde{k}|}+\sigma_2\frac{\widetilde{m}_3}{|\widetilde{m}|}\right)}h^{\sigma_1}(k)\wedge(\widetilde{m}\wedge h^{\sigma_2}(m))\cdot e^{\sigma_3}(n)e^{\sigma_3}(n).
$$

Let us separate the complex time exponential into sines and cosines, defining

$$
c^\varepsilon(k)=\cos\left(\sigma_1\frac{\widetilde{k}_3}{|\widetilde{k}|}\frac{t}{\varepsilon}\right)\quad\text{and}\quad s^\varepsilon(k)=\sin\left(\sigma_1\frac{\widetilde{k}_3}{|\widetilde{k}|}\frac{t}{\varepsilon}\right).
$$

Noticing that

$$
\widetilde{m}\wedge(\widetilde{m}\wedge h^{\sigma_2}(m)) = -|\widetilde{m}|^2h^{\sigma_2}(m)
$$

and using the fact that $ih^{\sigma_1}(k)=(\widetilde{k}/|\widetilde{k}|)\wedge h^{\sigma_1}(k)$, we have the following terms in the sum, when n and σ are fixed (recall that $\sigma_1=\sigma_2$):

$$
\begin{aligned}
&c^\varepsilon(k)c^\varepsilon(m)\,h^{\sigma_1}(k)\wedge(\widetilde{m}\wedge h^{\sigma_1}(m)), \\
&-\frac{|\widetilde{m}|}{|\widetilde{k}|}s^\varepsilon(k)s^\varepsilon(m)\,(\widetilde{k}\wedge h^{\sigma_1}(k))\wedge h^{\sigma_1}(m), \\
&-|\widetilde{m}|c^\varepsilon(k)s^\varepsilon(m)\,h^{\sigma_1}(k)\wedge h^{\sigma_2}(m), \\
&\frac{1}{|\widetilde{k}|}s^\varepsilon(k)c^\varepsilon(m)\,(\widetilde{k}\wedge h^{\sigma_1}(k))\wedge(\widetilde{m}\wedge h^{\sigma_1}(m)).
\end{aligned}
$$

We shall only deal with the left-hand column (which is made up of the even terms in t, whereas the right-hand side is made up of the odd terms). Since as noted above $|\widetilde{m}|=|\widetilde{k}|$ and $k_3=-m_3$, the terms of the left-hand side lead to two contributions:

$$
(c^\varepsilon)^2(m)\,h^{\sigma_1}(k)\wedge(\widetilde{m}\wedge h^{\sigma_1}(m))\quad\text{and}\quad(s^\varepsilon)^2(m)(\widetilde{k}\wedge h^{\sigma_1}(k))\wedge h^{\sigma_1}(m).
$$

Then interchanging k and m by symmetry gives finally the contribution

$$-\cos\left(2\sigma_1\frac{\widetilde{m}_3 t}{|\widetilde{m}|\varepsilon}\right) h^{\sigma_1}(k)\wedge(\widetilde{m}\wedge h^{\sigma_1}(m)).$$

The phase in that term is not zero; it follows that the quadratic form, restricted to $n_3 = 0$ and $k_3 \neq 0$, converges weakly to zero, and that proves the first statement of Proposition 6.2.

Second statement Now let us prove the second statement, which is a little more subtle. To simplify the notation we shall take all the a_i to be equal to one in this proof. We first consider the case of

$$(\mathcal{Q}(\overline{u}, v_{\rm osc}) \mid (-\Delta)^s v_{\rm osc})_{L^2(\mathbf{T}^3)}\,.$$

Let us write the proof of (6.3.1) for $s = 0$ to start with. We have

$$\big(\mathcal{Q}(\overline{u}, v_{\rm osc})\big|v_{\rm osc}\big)_{L^2(\mathbf{T}^3)} = -\sum_{\substack{k_3=0,n_3\neq 0\\ k\in K_n^\sigma,\sigma\in\{+,-\}^3}} [\overline{u}^{\sigma_1}(k)\cdot n][v_{\rm osc}^{\sigma_2}(n-k)\cdot v_{\rm osc}^{\sigma_3 *}(n)],$$

where h^* denotes the complex conjugate of any vector field h. Then we notice that necessarily $\sigma_2 = \sigma_3$ and $|n| = |k-n|$, and we exchange n and $k-n$ in the above summation. Since the vector fields are real-valued we have $v_{\rm osc}^{\sigma_3}(n) = v_{\rm osc}^{\sigma_3 *}(-n)$, so we get

$$\big(\mathcal{Q}(\overline{u}, v_{\rm osc})\big|v_{\rm osc}\big)_{L^2(\mathbf{T}^3)} = -\frac{1}{2}\sum_{\sigma,n_3\neq 0}\ \sum_{\substack{k_3=0\\ k\in K_n^\sigma}}\Big(n\cdot\overline{u}^{\sigma_1}(k) + (k-n)\cdot\overline{u}^{\sigma_1}(k)\Big)$$
$$\times\, v_{\rm osc}^{\sigma_2}(k-n)\cdot v_{\rm osc}^{\sigma_3 *}(n).$$

Then the divergence-free condition on $\overline{u}$ yields the result.

To prove (6.3.1) with $s \neq 0$, we just have to notice that the operator $(-\Delta)^s$ corresponds in Fourier variables to multiplying by $|n|^{2s}$. But when k is in K_n^σ, then in particular $n_3 - k_3 = n_3$ and $|n_h - k_h| = |n_h|$; that means that the same argument as in the case $s = 0$ holds, as one can exchange $k - n$ and n in the summation. So the result (6.3.1) is found.

Now let us consider the statement concerning $\big(\mathcal{Q}(u_{\rm osc}, v_{\rm osc}) \mid w_{\rm osc}\big)_{H^{\frac12}(\mathbf{T}^3)}$, which is more complicated, and will require a preliminary step.

Lemma 6.2 *For any $n \in \mathbf{Z}^3\setminus\{0\}$, let $K(n)$ be a subset of $\mathbf{Z}^3$ such that*

$$k\in K(n) \Rightarrow n-k\in K(n) \quad \textit{and} \quad n\in K(k).$$

For any $j \in \mathbf{N}$, define

$$\mathcal{K}_j(n) \stackrel{\rm def}{=} \left\{k\in\mathbf{Z}^3\setminus\{0\}\,/\,2^j \leq |k| < 2^{j+1} \textit{ and } k\in K(n)\right\}.$$

If $C_0 > 0$ and $\gamma > 0$ are two constants (with C_0 possibly depending on $(a_j)_{1\leq j\leq 3}$ if $K(n)$ does) such that

$$\sup_{n\in\mathbf{Z}^3\setminus\{0\}} \operatorname{card}\mathcal{K}_j(n) \leq C_0 2^{j\gamma}, \tag{6.3.3}$$

then there exists a constant C which may depend on $(a_j)_{1\leq j\leq 3}$ such that for all smooth functions a, b and c, and for all real numbers $\alpha < \gamma/2$ and $\beta < \gamma/2$ such that $\alpha + \beta > 0$, we have

$$\sum_{\substack{(k,n)\in\mathbf{Z}^6\\ k\in K(n)}} |\widehat{a}(k)|\,|\widehat{b}(n-k)|\,|\widehat{c}(n)| \leq C\|a\|_{H^\alpha(\mathbf{T}^3)}\|b\|_{H^\beta(\mathbf{T}^3)}\|c\|_{H^{\gamma/2-\alpha-\beta}(\mathbf{T}^3)}.$$

Remark One of course always has $\gamma \leq 3$, and the case $\gamma = 3$ corresponds to nothing more than three-dimensional product rules. Indeed

$$\sum_{(k,n)\in\mathbf{Z}^6} \widehat{a}(k)\,\widehat{b}(n-k)\,\widehat{c^*}(n) = \operatorname{Re}\left(ab|c\right)_{L^2(\mathbf{T}^3)},$$

where c^* denotes the complex conjugate of c, and product rules in $\mathbf{T}^3$ yield (see for instance [46])

$$\|ab\|_{H^{\alpha+\beta-\frac{3}{2}}(\mathbf{T}^3)} \leq C\|a\|_{H^\alpha(\mathbf{T}^3)}\|b\|_{H^\beta(\mathbf{T}^3)},$$

as long as α and β are smaller than $\frac{3}{2}$ and $\alpha+\beta > 0$. It will turn out in our case, where

$$K(n) \stackrel{\text{def}}{=} \bigcup_{\sigma\in\{+,-\}^3} K_n^\sigma$$

and K_n^σ is the resonant set defined in (6.2.5), that $\gamma = 2$. In that case Lemma 6.2 means that two-dimensional product rules hold when Fourier variables are restricted to $K(n)$. In other words, there is a gain of half a derivative compared with a standard product.

Proof of Lemma 6.2 The idea of the proof is based on Littlewood–Paley decomposition and some paraproduct. One can decompose the sum into three parts, writing

$$\sum_{\substack{(k,n)\in\mathbf{Z}^6\\ k\in K(n)}} |\widehat{a}(k)|\,|\widehat{b}(n-k)|\,|\widehat{c}(n)| = \sum_{(j,\ell,q)\in\mathbf{N}^3} \sum_{\substack{n-k\in\mathcal{K}_\ell(n)\\ k\in\mathcal{K}_j(n),n\in\mathcal{K}_q(k)}} |\widehat{a}(k)|\,|\widehat{b}(n-k)|\,|\widehat{c}(n)|,$$

due to the symmetry properties satisfied by the set $K(n)$. Then one considers separately the following three cases (this is the usual paraproduct algorithm):

1. $j \leq \ell - 2$, which implies that $\ell - 1 \leq q < \ell + 2$
2. $\ell \leq j - 2$, which implies that $j - 1 \leq q < j + 2$
3. $\ell - 1 \leq j \leq \ell + 1$, which implies that $q \leq j + 2$.

Let us start by considering Case 1. Using the Cauchy–Schwarz inequality, we infer that the quantity

$$\sum_{\substack{j\leq \ell-2\\ \ell-1\leq q<\ell+2}} \sum_{\substack{k\in\mathcal{K}_j(n), n\in\mathcal{K}_q(k)\\ n-k\in\mathcal{K}_\ell(n)}} |\widehat{a}(k)|\,|\widehat{b}(n-k)|\,|\widehat{c}(n)|$$

is less than or equal to

$$C_{\alpha,\beta}\sum_{\substack{j\leq \ell-2\\ \ell-1\leq q<\ell+2}} \Bigl(\sum_{2^{\ell-1}\leq|n|<2^{\ell+2}} |\widehat{c}(n)|^2\Bigr)^{\frac12} \Bigl(\sup_n \sum_{k\in\mathcal{K}_j(n)} |k|^{-2\alpha}\Bigr)^{\frac12}$$
$$\times\Biggl(\sum_{\substack{2^{\ell-1}\leq|n|<2^{\ell+2}\\ k\in\mathcal{K}_j(n)}} |k|^{2\alpha}|\widehat{a}(k)|^2|\widehat{b}(n-k)|^2|n-k|^{2\beta}\Biggr)^{\frac12} 2^{-\ell\beta},$$

where we have used the fact that $|n-k|\sim 2^\ell$ if $n-k$ is in $\mathcal{K}_\ell(n)$. But by assumption on card $\mathcal{K}_j(n)$, we have

$$\Bigl(\sum_{k\in\mathcal{K}_j(n)} |k|^{-2\alpha}\Bigr)^{\frac12} \leq C_{\alpha,\beta}2^{j(\gamma/2-\alpha)},$$

since on $\mathcal{K}_j(n)$, we have $|k|\sim 2^j$. We can also write

$$2^{-\ell\beta}\Bigl(\sum_{2^{\ell-1}\leq|n|<2^{\ell+2}} |\widehat{c}(n)|^2\Bigr)^{\frac12}$$
$$\leq C_{\alpha,\beta}\Bigl(\sum_{2^{\ell-1}\leq|n|<2^{\ell+2}} |n|^{\gamma-2\alpha-2\beta}|\widehat{c}(n)|^2\Bigr)^{\frac12} 2^{-\ell(\gamma/2-\alpha)},$$

so we get finally that the quantity

$$\sum_{\substack{j\leq \ell-2\\ \ell-1\leq q<\ell+2}} \sum_{\substack{k\in\mathcal{K}_j(n), n\in\mathcal{K}_q(k)\\ n-k\in\mathcal{K}_\ell(n)}} |\widehat{a}(k)|\,|\widehat{b}(n-k)|\,|\widehat{c}(n)|$$

is less than or equal to

$$C_{\alpha,\beta}\sum_{j\leq\ell-2}\Bigl(\sum_{k\in\mathcal{K}_j(n)} |\widehat{a}(k)|^2|k|^{2\alpha}\Bigr)^{\frac12} 2^{(j-\ell)(\gamma/2-\alpha)}\|b\|_{H^\beta(\mathbf{T}^3)}\|c\|_{H^{\gamma/2-\alpha-\beta}(\mathbf{T}^3)}.$$

Using the assumption that $\alpha < \gamma/2$, the result follows by Young's inequality. Note that the replacement of, say,

$$\sum_{k\in\mathbf{Z}^3} |\widehat{a}(k)|^2|k|^{2\alpha} \quad \text{by} \quad \|a\|^2_{H^\alpha(\mathbf{T}^3)}$$

is valid up to multiplication by a constant depending on $(a_j)_{1\leq j\leq 3}$. Cases 2 and 3 are obtained symmetrically: in Case 2, one exchanges the roles of a and b, and in Case 3 the roles of a and c. That yields the conditions $\beta < \gamma/2$ and $\gamma/2-\alpha-\beta < \gamma/2$, respectively. The proof of Lemma 6.2 is complete. □

Now we can finish the proof of the second statement of Proposition 6.2. We can write the quantity

$$|(Q(u_{\rm osc}, v_{\rm osc})|w_{\rm osc})_{H^{\frac12}(\mathbf{T}^3)}|$$

is less than or equal to

$$C\sum_{n\in\mathbf{Z}^3}\sum_{k\in K(n)} \big(|\widehat{u}_{\rm osc}(k)|\,|n-k|\,|\widehat{v}_{\rm osc}(n-k)| \\ + |\widehat{v}_{\rm osc}(k)|\,|n-k|\,|\widehat{u}_{\rm osc}(n-k)|\big)\,|n|\,|\widehat{w}_{\rm osc}(n)|,$$

so let us give an estimate of the type (6.3.3) on the set $K(n)$, defined as

$$K(n) \stackrel{\rm def}{=} \bigcup_{\sigma\in\{+,-\}^3} K_n^\sigma.$$

If one defines

$$\alpha \stackrel{\rm def}{=} k_3|\widetilde{m}||\widetilde{n}|,\quad \beta \stackrel{\rm def}{=} m_3|\widetilde{n}||\widetilde{k}| \quad \text{and} \quad \gamma \stackrel{\rm def}{=} n_3|\widetilde{k}||\widetilde{m}|,$$

with $k+m=n$, then

$$\prod\nolimits_{\sigma\in\{+,-\}^3}\left(\sigma_1\frac{\widetilde{k}_3}{|\widetilde{k}|} + \sigma_2\frac{\widetilde{n}_3-\widetilde{k}_3}{|\widetilde{n}-\widetilde{k}|} - \sigma_3\frac{\widetilde{n}_3}{|\widetilde{n}|}\right) \\ = -\left(\frac{(\alpha^2+\beta^2-\gamma^2)^2 - 4\alpha^2\beta^2}{a_3^4|\widetilde{k}|^4\,|\widetilde{m}|^4\,|\widetilde{n}|^4}\right)^2.$$

To simplify this expression let us choose the a_i's to be equal to one; their exact value is of no importance in the calculation. Then an easy computation shows that the set $K(n)$ is made of the integers $k\in\mathbf{Z}^3$ such that

$$(k_3^2|k-n|^2|n|^2 + (n_3-k_3)^2|k|^2|n|^2 - n_3^2|k-n|^2|k|^2)^2 \\ = 4k_3^2(n_3-k_3)^2|k|^2|n-k|^2|n|^4.$$

That expression can be seen as a polynomial of degree 8 in k_3, where the coefficient of k_3^8 does not vanish. It follows that if n and (k_1, k_2) are fixed, there are at most eight roots to that polynomial in k_3. So we can write

$$\operatorname{card} \mathcal{K}_j(n) \leq 2^{j+1} \sum_{\substack{2^j \leq |k| < 2^{j+1} \\ k \in K(n)}} |k|^{-1}$$

$$\leq 2^{j+1} \left(1 + 8 \sum_{0 < |k_h| < 2^{j+1}} |k_h|^{-1} \right).$$

Finally, assumption (6.3.3) holds with $\gamma = 2$. Now it remains to apply Lemma 6.2 twice: once with $\alpha = \frac{1}{2}$, $\beta = 0$ and $\widehat{a}(k) = \widehat{u}_{\text{osc}}(k), \widehat{b}(n-k) = |n-k|\, \widehat{v}_{\text{osc}}(n-k)$ and $\widehat{c}(n) = |n|\, \widehat{w}_{\text{osc}}(n)$ and once with the same values of α and β and exchanging u_{osc} and v_{osc}. The result follows directly:

$$\begin{aligned} |(\mathcal{Q}_{\text{osc}}(u_{\text{osc}}, v_{\text{osc}}) | w_{\text{osc}})_{H^{\frac{1}{2}}(\mathbf{T}^3)}| \leq C(&\|u_{\text{osc}}\|_{H^{\frac{1}{2}}(\mathbf{T}^3)} \|\nabla v_{\text{osc}}\|_{L^2(\mathbf{T}^3)} \\ &+ \|v_{\text{osc}}\|_{H^{\frac{1}{2}}(\mathbf{T}^3)} \|\nabla u_{\text{osc}}\|_{L^2(\mathbf{T}^3)}) \|\nabla w_{\text{osc}}\|_{H^{\frac{1}{2}}(\mathbf{T}^3)}, \end{aligned}$$

and Proposition 6.2 is proved. □

Finally we have shown that the limit system (NSCL) splits into two parts: in the following, we will say that $u = \overline{u} + u_{\text{osc}}$ satisfies the limit system (NS2D$-$S$_{\text{osc}}$) associated with data $\overline{u}_0 + u_{0,\text{osc}}$ and forcing term $\overline{f} + f_{\text{osc}}$ if it satisfies the two-dimensional Navier–Stokes equation

$$(\text{NS2D}) \begin{cases} \partial_t \overline{u} - \nu \Delta_h \overline{u} + \mathbf{P}_h(\overline{u}^h \cdot \nabla^h \overline{u}) = \overline{f} \\ \overline{u}_{|t=0} = \overline{u}_0, \end{cases}$$

where $\mathbf{P}_h$ denotes the two-dimensional Leray projector onto two-dimensional divergence-free vector fields, coupled with the system

$$(\text{S}_{\text{osc}}) \begin{cases} \partial_t u_{\text{osc}} - \nu \Delta u_{\text{osc}} - \mathcal{Q}(2\overline{u} + u_{\text{osc}}, u_{\text{osc}}) = f_{\text{osc}} \\ u_{\text{osc}|t=0} = u_{0,\text{osc}}. \end{cases}$$

Of course here $\overline{u}_0$ and $\overline{f}$ are two-dimensional divergence-free vector fields, and $u_{0,\text{osc}}$ and f_{osc} have zero vertical average and are divergence-free.

6.4 Global existence and stability for the limit system

The aim of this section is to study the global well-posedness and the stability of the limit system (NS2D $-$ S$_{\text{osc}}$) derived in Section 6.3. In the rest of this part, to simplify notation we will note (for $d = 2$ or 3)

$$E_{\frac{1}{2}}(\mathbf{T}^d) \stackrel{\text{def}}{=} C_b^0(\mathbf{R}^+; H^{\frac{1}{2}}(\mathbf{T}^d)) \cap L^2(\mathbf{R}^+; H^{\frac{3}{2}}(\mathbf{T}^d))$$

and

$$\|b\|_{\frac{1}{2}} \stackrel{\text{def}}{=} \Big(\|b\|^2_{L^\infty(\mathbf{R}^+;H^{\frac{1}{2}}(\mathbf{T}^d))} + \nu \|b\|^2_{L^2(\mathbf{R}^+;H^{\frac{3}{2}}(\mathbf{T}^d))} \Big)^{\frac{1}{2}}.$$

Let us prove the following results.

Proposition 6.3 (Global well-posedness) *Let $\overline{u}_0$ and $u_{0,\mathrm{osc}}$ be two divergence-free vector fields, respectively, in $H^{\frac{1}{2}}(\mathbf{T}^2)$ and in $H^{\frac{1}{2}}(\mathbf{T}^3)$ such that $u_{0,\mathrm{osc}}(x_h,\cdot)$ is mean-free on $\mathbf{T}^1$, and define $u_0 = \overline{u}_0 + u_{0,\mathrm{osc}}$. Let us consider also an external force $f = \overline{f} + f_{\mathrm{osc}}$, where $\overline{f}$ is in $L^2(\mathbf{R}^+; H^{-\frac{1}{2}}(\mathbf{T}^2))$ and f_{osc} in $L^2(\mathbf{R}^+; H^{-\frac{1}{2}}(\mathbf{T}^3))$ and mean-free on $\mathbf{T}^1_{x_3}$. Then there exists a unique global solution u to the system $(\mathrm{NS2D} - \mathrm{S}_{\mathrm{osc}})$, and*

$$u = \overline{u} + u_{\mathrm{osc}} \quad \text{with } \overline{u} \in E_{\frac{1}{2}}(\mathbf{T}^2) \text{ and } u_{\mathrm{osc}} \in E_{\frac{1}{2}}(\mathbf{T}^3).$$

Moreover, the solution u goes to zero as time goes to infinity:

$$\lim_{t\to+\infty} \|u(t,\cdot)\|_{H^{\frac{1}{2}}} = 0. \tag{6.4.1}$$

Proposition 6.4 (Stability) *The application*

$$(u_0, f) \in \mathcal{E} \mapsto u \in \Big(E_{\frac{1}{2}}(\mathbf{T}^2) + E_{\frac{1}{2}}(\mathbf{T}^3) \Big),$$

where

$$\mathcal{E} \stackrel{\text{def}}{=} \Big(H^{\frac{1}{2}}(\mathbf{T}^2) + H^{\frac{1}{2}}(\mathbf{T}^3) \Big) \times \Big(L^2(\mathbf{R}^+; H^{-1}(\mathbf{T}^2)) + L^2(\mathbf{R}^+; H^{-\frac{1}{2}}(\mathbf{T}^3)) \Big)$$

mapping the initial data $u_0 = \overline{u}_0 + u_{0,\mathrm{osc}}$ and external force $f = \overline{f} + f_{\mathrm{osc}}$ to the solution $u = \overline{u} + u_{\mathrm{osc}}$ of $(\mathrm{NS2D} - \mathrm{S}_{\mathrm{osc}})$ given in Proposition 6.3 is Lipschitz on bounded subsets of $\mathcal{E}$.

Proof of Proposition 6.3 The fact that there is a unique global solution $\overline{u}$ to (NS2D) in the space $E_{\frac{1}{2}}$ is nothing but Theorem 3.7, page 80.

Now let us consider the system $(\mathrm{S}_{\mathrm{osc}})$; it has a priori the structure of the three-dimensional Navier–Stokes equations, so it is much less obvious that it can be solved uniquely, globally in time. The key to the proof of that result is that the three-dimensional interactions in $\mathcal{Q}(u_{\mathrm{osc}}, u_{\mathrm{osc}})$ are sufficiently few to enable one to write two-dimensional-type estimates, and hence to conclude. That was studied extensively in the previous section, so we shall be referring here to Proposition 6.2.

Let us start by noticing that due to the skew-symmetry of $\mathcal{Q}$ shown in (6.3.1) (with $s = 0$), we have as for Leray solutions of the three-dimensional Navier–Stokes equations the energy estimate

$$\|u_{\mathrm{osc}}(t)\|^2_{L^2(\mathbf{T}^3)} + \nu \int_0^t \|u_{\mathrm{osc}}(t')\|^2_{H^1(\mathbf{T}^3)}\, dt'$$
$$\leq \|u_{0,\mathrm{osc}}\|^2_{L^2(\mathbf{T}^3)} + \frac{C}{\nu} \int_0^t \|f_{\mathrm{osc}}(t')\|^2_{H^{-1}(\mathbf{T}^3)}\, dt'. \tag{6.4.2}$$

Moreover due to (6.3.1), one can apply standard arguments of the theory of three-dimensional Navier–Stokes equations to solve (S_{osc}) locally in time (see Theorem 3.5); one obtains a unique solution

$$u_{\rm osc} \in C^0([0,T]; H^{\frac{1}{2}}(\mathbf{T}^3)) \cap L^2([0,T]; H^{\frac{3}{2}}(\mathbf{T}^3)),$$

for some time $T > 0$. Furthermore, if that time is not infinite, then according to Corollary 3.10, page 79, there is a unique maximal time $T^* \in]0, +\infty[$ with

$$\lim_{\substack{T\to T^* \\ T<T^*}} \|u_{\rm osc}\|_{L^2([0,T];H^{\frac{3}{2}}(\mathbf{T}^3))} = +\infty. \tag{6.4.3}$$

In order to prove the proposition, one therefore just has to check that the norm of $u_{\rm osc}$ in $L^2([0,T]; H^{\frac{3}{2}}(\mathbf{T}^3))$ can be bounded uniformly in T, which will automatically extend the solution globally in time, due to (6.4.3). In order to do so, we are going to write an energy estimate on the system (S_{osc}) in the space $H^{\frac{1}{2}}(\mathbf{T}^3)$. The computations will be very similar to the two-dimensional case treated above. We have

$$\begin{aligned}
&\frac{1}{2}\frac{d}{dt}\|u_{\rm osc}(t)\|^2_{H^{\frac{1}{2}}(\mathbf{T}^3)} + \nu\|u_{\rm osc}(t)\|^2_{H^{\frac{3}{2}}(\mathbf{T}^3)} \\
&\quad = (f_{\rm osc}|u_{\rm osc})_{H^{\frac{1}{2}}(\mathbf{T}^3)} + 2\,(\mathcal{Q}(\overline{u}, u_{\rm osc})|u_{\rm osc})_{H^{\frac{1}{2}}(\mathbf{T}^3)} \\
&\qquad + (\mathcal{Q}(u_{\rm osc}, u_{\rm osc})|u_{\rm osc})_{H^{\frac{1}{2}}(\mathbf{T}^3)}\,.
\end{aligned} \tag{6.4.4}$$

We then get from Proposition 6.2 the following estimate:

$$\begin{aligned}
&\frac{1}{2}\frac{d}{dt}\|u_{\rm osc}(t)\|^2_{H^{\frac{1}{2}}(\mathbf{T}^3)} + \nu\|u_{\rm osc}(t)\|^2_{H^{\frac{3}{2}}(\mathbf{T}^3)} \\
&\quad \leq \|f_{\rm osc}(t)\|_{H^{-\frac{1}{2}}(\mathbf{T}^3)}\|u_{\rm osc}(t)\|_{H^{\frac{3}{2}}(\mathbf{T}^3)} \\
&\qquad + C\,\|u_{\rm osc}(t)\|_{H^{\frac{1}{2}}(\mathbf{T}^3)}\|u_{\rm osc}(t)\|_{H^1(\mathbf{T}^3)}\|u_{\rm osc}(t)\|_{H^{\frac{3}{2}}(\mathbf{T}^3)},
\end{aligned}$$

where the term containing $\overline{u}$ has disappeared due to (6.3.1). We infer that

$$\begin{aligned}
&\frac{d}{dt}\|u_{\rm osc}(t)\|^2_{H^{\frac{1}{2}}(\mathbf{T}^3)} + \nu\|u_{\rm osc}(t)\|^2_{H^{\frac{3}{2}}(\mathbf{T}^3)} \\
&\quad \leq \frac{C}{\nu}\big(\|u_{\rm osc}(t)\|^2_{H^{\frac{1}{2}}(\mathbf{T}^3)}\|u_{\rm osc}(t)\|^2_{H^1(\mathbf{T}^3)} + \|f_{\rm osc}(t)\|^2_{H^{-\frac{1}{2}}(\mathbf{T}^3)}\big).
\end{aligned}$$

Then (6.4.2), associated with a Gronwall estimate, yields finally

$$\begin{aligned}
&\|u_{\rm osc}(t)\|^2_{H^{\frac{1}{2}}(\mathbf{T}^3)} + \nu\int_0^t \|u_{\rm osc}(s)\|^2_{H^{\frac{3}{2}}(\mathbf{T}^3)}\,ds \\
&\quad \leq \left(\|u_{0,\rm osc}\|^2_{H^{\frac{1}{2}}(\mathbf{T}^3)} + \frac{C}{\nu}\int_0^t \|f_{\rm osc}(s)\|^2_{H^{-\frac{1}{2}}(\mathbf{T}^3)}\,ds\right) \\
&\qquad \times \exp\left(\frac{C}{\nu^2}\left(\|u_{0,\rm osc}\|^2_{L^2(\mathbf{T}^3)} + \frac{C}{\nu}\|f_{\rm osc}\|^2_{L^2(\mathbf{R}^+;H^{-1}(\mathbf{T}^3))}\right)\right),
\end{aligned}$$

and the global existence is proved. Now to end the proof of Proposition 6.3, we just need to prove the asymptotic result (6.4.1). That result is quite straightforward: we just need to recall that the solution u is in the energy space

$$L^\infty(\mathbf{R}^+; L^2(\mathbf{T}^3)) \cap L^2(\mathbf{R}^+; H^1(\mathbf{T}^3)),$$

so in particular u is in $L^4(\mathbf{R}^+; H^{\frac{1}{2}}(\mathbf{T}^3))$ and there is a time T_0 such that $\|u(T_0,\cdot)\|_{H^{\frac{1}{2}}} \leq c\nu$, where $c\nu$ is the smallness constant of Theorem 3.5, page 73. So for a large enough time we are in the small-data situation and (6.4.1) follows from Theorem 3.5.

The proof of Proposition 6.3 is now complete. □

Proof of Proposition 6.4 This proposition follows directly from the following lemma.

Lemma 6.3 *Let u_1 and u_2 be two solutions of* (NS2D − $\mathrm{S_{osc}}$), *associated with data $u_{0,1}$ and $u_{0,2}$ and forcing terms f_1 and f_2, respectively, satisfying the assumptions of Proposition 6.3. Then $w \stackrel{def}{=} u_1 - u_2$ satisfies the following estimates, with $g \stackrel{\mathrm{def}}{=} f_1 - f_2$:*

$$\|\overline{w}(t)\|_{\frac{1}{2}}^2 \leq \left(\|\overline{w}_0\|_{H^{\frac{1}{2}}(\mathbf{T}^2)}^2 + \frac{C}{\nu}\int_0^t \|\overline{g}(t')\|_{H^{-\frac{1}{2}}(\mathbf{T}^3)}^2 \, dt' \right) G_1, \tag{6.4.5}$$

$$\begin{aligned} \|w_{\mathrm{osc}}(t)\|_{\frac{1}{2}}^2 \leq \Big(& \|w_{0,\mathrm{osc}}\|_{H^{\frac{1}{2}}(\mathbf{T}^3)}^2 + \|\overline{w}_0\|_{L^2(\mathbf{T}^2)}^2 \\ & + \frac{C}{\nu}\int_0^t \|g_{\mathrm{osc}}(t')\|_{H^{-\frac{1}{2}}(\mathbf{T}^3)}^2 \, dt' \Big) H_1 \end{aligned} \tag{6.4.6}$$

where G_1 is a function of $\|\overline{u}_{0,i}\|_{L^2(\mathbf{T}^2)}$ and $\|\overline{f}_i\|_{L^2(\mathbf{R}^+;H^{-1}(\mathbf{T}^2))}$ for i equal to 1 or 2, and H_1 is a function of $\|\overline{u}_{0,i}\|_{L^2(\mathbf{T}^2)}$ and $\|\overline{f}_i\|_{L^2(\mathbf{R}^+;H^{-1}(\mathbf{T}^2))}$ as well as of $\|u_{\mathrm{osc},i}\|_{H^{\frac{1}{2}}(\mathbf{T}^3)}$ and $\|f_{\mathrm{osc},i}\|_{L^2(\mathbf{R}^+;H^{-\frac{1}{2}}(\mathbf{T}^3))}$ for $i \in \{1,2\}$.

Proof We shall start with the estimates on $\overline{w}$. As $\overline{u}_1$ and $\overline{u}_2$ satisfy the two-dimensional Navier–Stokes equations in $\mathbf{T}^2$, we can apply the stability inequality of Theorem 3.2, page 56, to find that

$$\begin{aligned} & \|\overline{w}(t)\|_{L^2(\mathbf{T}^2)}^2 + \nu\int_0^t \|\nabla^h \overline{w}(s)\|_{L^2(\mathbf{T}^2)}^2 \, ds \\ & \qquad \leq \left(\|\overline{w}_0\|_{L^2(\mathbf{T}^2)}^2 + \frac{1}{\nu}\int_0^t \|\overline{g}(t')\|_{H^{-1}(\mathbf{T}^2)}^2 \, dt' \right) G_1, \end{aligned}$$

where $G_1 \stackrel{\mathrm{def}}{=} \exp\left(\dfrac{CE^2(t)}{\nu^4}\right)$, and $E(t)$ denotes

$$\min\left\{ \|\overline{u}_{0,1}\|_{L^2}^2 + \frac{1}{\nu}\int_0^t \|\overline{f}_1(t')\|_{H^{-1}}^2 \, dt' \,,\, \|\overline{u}_{0,2}\|_{L^2}^2 + \frac{1}{\nu}\int_0^t \|\overline{f}_2(t')\|_{H^{-1}}^2 \, dt' \right\}.$$

The result for the L^2 norm is proved.

Now let us prove the $H^{\frac{1}{2}}$ estimate (6.4.5). Since $\overline{w}$ satisfies the equation

$$\partial_t \overline{w} - \nu\Delta_h \overline{w} + \mathbf{P}(\overline{w}\cdot\nabla^h\overline{w}) = -\mathbf{P}(\overline{w}^h\cdot\nabla^h\overline{u}_2 + \overline{u}_2\cdot\nabla^h\overline{w}^h) + \overline{g}, \qquad (6.4.7)$$

we can apply Theorem 3.7, page 80, to $\overline{w}$ considering as external force

$$f = -\mathbf{P}(\overline{w}^h\cdot\nabla^h\overline{u}_2 + \overline{u}_2\cdot\nabla^h\overline{w}^h) + \overline{g}.$$

We have therefore, by Theorem 3.7,

$$\begin{aligned}\|\overline{w}(t)\|^2_{H^{\frac{1}{2}}} + \nu\int_0^t \|\nabla\overline{w}(t')\|^2_{H^{\frac{1}{2}}}dt' &\leq \exp\left(\frac{C}{\nu}\int_0^t\|\nabla\overline{w}(t')\|^2_{L^2}dt'\right)\\ \times\left(\|\overline{w}_0\|^2_{H^{\frac{1}{2}}} + \frac{C}{\nu}\int_0^t\|f(t')\|^2_{H^{-\frac{1}{2}}}\right.&\left.\exp\left(-\frac{C}{\nu}\int_{t'}^t\|\nabla\overline{w}(t'')\|^2_{L^2}dt''\right)dt'\right),\end{aligned}$$

and we need to control f in $H^{-\frac{1}{2}}(\mathbf{T}^2)$. By definition $\overline{g}$ is an element of $H^{-\frac{1}{2}}(\mathbf{T}^2)$ so let us now study $\overline{w}^h\cdot\nabla^h\overline{u}_2 + \overline{u}_2\cdot\nabla^h\overline{w}^h$. On the one hand, we can write, by the dual Sobolev embeddings of Corollary 1.1, page 25,

$$\|\overline{w}^h\cdot\nabla^h\overline{u}_2\|_{H^{-\frac{1}{2}}} \leq C\|\overline{w}^h\cdot\nabla^h\overline{u}_2\|_{L^{\frac{4}{3}}}$$

so a Hölder estimate yields, using the embedding of $L^4(\mathbf{T}^2)$ into $H^{\frac{1}{2}}(\mathbf{T}^2)$,

$$\|\overline{w}^h\cdot\nabla^h\overline{u}_2\|_{H^{-\frac{1}{2}}} \leq C\|w\|_{H^{\frac{1}{2}}}\|\nabla^h\overline{u}_2\|_{L^2}.$$

It follows that

$$\begin{aligned}&\int_0^t\|\overline{w}^h(t')\cdot\nabla^h\overline{u}_2(t')\|^2_{H^{-\frac{1}{2}}}\exp\left(-\frac{C}{\nu}\int_{t'}^t\|\nabla w(t'')\|^2_{L^2}dt''\right)dt'\\ &\quad\leq C\int_0^t\|w(t')\|^2_{H^{\frac{1}{2}}}\|\nabla^h\overline{u}_2(t')\|^2_{L^2}\exp\left(-\frac{C}{\nu}\int_{t'}^t\|\nabla w(t'')\|^2_{L^2}dt''\right)dt'.\end{aligned}$$

On the other hand we have, by the same arguments,

$$\|\overline{u}_2\cdot\nabla^h\overline{w}^h\|_{H^{-\frac{1}{2}}} \leq C\|\nabla^h\overline{w}^h\|_{L^2}\|\overline{u}_2\|_{H^{\frac{1}{2}}},$$

so an interpolation inequality yields

$$\|\overline{u}_2\cdot\nabla^h\overline{w}^h\|_{H^{-\frac{1}{2}}} \leq C\|\overline{w}^h\|^{\frac{1}{2}}_{H^{\frac{1}{2}}}\|\nabla^h\overline{w}^h\|^{\frac{1}{2}}_{H^{\frac{1}{2}}}\|\overline{u}_2\|_{H^{\frac{1}{2}}}.$$

Now we notice that for any function F,

$$\begin{aligned}\exp\left(\frac{C}{\nu}\int_0^t\|\nabla\overline{w}(t')\|^2_{L^2}\,dt'\right)&\int_0^t F(t')\exp\left(-\frac{C}{\nu}\int_{t'}^t\|\nabla w(t'')\|^2_{L^2}\,dt''\right)dt'\\ &\leq\int_0^t F(t')\exp\left(\frac{C}{\nu}\int_0^{t'}\|\nabla w(t'')\|^2_{L^2}\,dt''\right)dt',\end{aligned}$$

and we have that the quantity

$$W_0(t) \stackrel{\text{def}}{=} \int_0^t \|\overline{u}_2(t') \cdot \nabla^h \overline{w}^h(t')\|_{H^{-\frac{1}{2}}} \exp\left(\frac{C}{\nu}\int_0^{t'} \|\nabla w(t'')\|_{L^2}^2\, dt''\right) dt'$$

is less than and equal to

$$C\int_0^t \|\overline{w}^h(t')\|_{H^{\frac{1}{2}}} \|\nabla^h \overline{w}^h(t')\|_{H^{\frac{1}{2}}} \|\overline{u}_2(t')\|_{H^{\frac{1}{2}}}^2 \times \exp\left(\frac{C}{\nu}\int_0^{t'} \|\nabla w(t'')\|_{L^2}^2 dt''\right) dt'.$$

We infer that

$$W_0(t) \leq \frac{\nu}{2}\int_0^t \|\nabla^h \overline{w}^h(t')\|_{H^{\frac{1}{2}}}^2\, dt' + \frac{C}{\nu}\int_0^t \|\overline{w}^h(t')\|_{H^{\frac{1}{2}}}^2 \|\overline{u}_2(t')\|_{H^{\frac{1}{2}}}^4 \exp\left(\frac{2C}{\nu}\int_0^{t'} \|\nabla w(t'')\|_{L^2}^2 dt''\right) dt',$$

and finally putting all the estimates together we have obtained that

$$\|\overline{w}(t)\|_{H^{\frac{1}{2}}}^2 + \frac{\nu}{2}\int_0^t \|\nabla \overline{w}(t')\|_{H^{\frac{1}{2}}}^2\, dt' \leq \left(\sum_{j=1}^3 W_j(t)\right) \exp\left(\frac{C}{\nu}\int_0^t \|\nabla \overline{w}(t')\|_{L^2}^2\, dt'\right)$$

with

$$W_1(t) \stackrel{\text{def}}{=} \|\overline{w}_0\|_{H^{\frac{1}{2}}}^2 + \int_0^t \|\overline{g}(t')\|_{H^{-\frac{1}{2}}}^2\, dt'$$

$$W_2(t) \stackrel{\text{def}}{=} C\int_0^t \|w(t')\|_{H^{\frac{1}{2}}}^2 \|\nabla^h \overline{u}_2(t')\|_{L^2}^2 dt'$$

$$W_3(t) \stackrel{\text{def}}{=} \frac{C}{\nu}\int_0^t \|\overline{w}^h(t')\|_{H^{\frac{1}{2}}}^2 \|\overline{u}_2(t')\|_{H^{\frac{1}{2}}}^4 dt'.$$

Then a Gronwall lemma, associated with the energy estimate satisfied by $\overline{u}_2$, yields the expected result.

Finally we are left with the proof of estimate (6.4.6). The function w_{osc} satisfies the following equation:

$$\partial_t w_{\text{osc}} - \nu\Delta w_{\text{osc}} - \mathcal{Q}(w_{\text{osc}} + 2\overline{w}, u_{1,\text{osc}}) - \mathcal{Q}(u_{2,\text{osc}} + 2\overline{u}_2, w_{\text{osc}}) = g_{\text{osc}}$$

recalling that

$$g_{\text{osc}} \stackrel{\text{def}}{=} f_{\text{osc},1} - f_{\text{osc},2}.$$

An energy estimate in $H^{\frac{1}{2}}$ yields

$$\frac{1}{2}\frac{d}{dt}\|w_{\rm osc}\|^2_{H^{\frac{1}{2}}(\mathbf{T}^3)} + \nu\|\nabla w_{\rm osc}\|^2_{H^{\frac{1}{2}}(\mathbf{T}^3)}$$
$$\leq \left|(\mathcal{Q}(w_{\rm osc}, u_{1,\rm osc} + u_{2,\rm osc})|w_{\rm osc})_{H^{\frac{1}{2}}(\mathbf{T}^3)}\right| + \left|(g_{\rm osc}|w_{\rm osc})_{H^{\frac{1}{2}}(\mathbf{T}^3)}\right|$$
$$2\left|(\mathcal{Q}(\overline{w}, u_{1,\rm osc})|w_{\rm osc})_{H^{\frac{1}{2}}(\mathbf{T}^3)}\right| + 2\left|(\mathcal{Q}(\overline{u}_2, w_{\rm osc})|w_{\rm osc})_{H^{\frac{1}{2}}(\mathbf{T}^3)}\right|. \qquad (6.4.8)$$

The last term on the right-hand side is zero due to (6.3.1), so let us estimate the others. For the first one, we use (6.3.2) which yields that the quantity

$$\mathcal{Q}_1 \stackrel{\rm def}{=} \left|(\mathcal{Q}(w_{\rm osc} u_{1,\rm osc} + u_{2,\rm osc})|w_{\rm osc})_{H^{\frac{1}{2}}(\mathbf{T}^3)}\right|$$

is less than or equal to

$$C\|w_{\rm osc}\|_{H^{\frac{1}{2}}(\mathbf{T}^3)}\left(\|u_{1,\rm osc}\|_{H^1(\mathbf{T}^3)} + \|u_{2,\rm osc}\|_{H^1(\mathbf{T}^3)}\right)\|\nabla w_{\rm osc}\|_{H^{\frac{1}{2}}(\mathbf{T}^3)}$$
$$+ C\|w_{\rm osc}\|_{H^1(\mathbf{T}^3)}\|\nabla w_{\rm osc}\|_{H^{\frac{1}{2}}(\mathbf{T}^3)}\left(\|u_{1,\rm osc}\|_{H^{\frac{1}{2}}(\mathbf{T}^3)} + \|u_{2,\rm osc}\|_{H^{\frac{1}{2}}(\mathbf{T}^3)}\right).$$

This gives

$$\mathcal{Q}_1 \leq \frac{\nu}{4}\|\nabla w_{\rm osc}\|^2_{H^{\frac{1}{2}}(\mathbf{T}^3)}$$
$$+ \frac{C}{\nu}\|w_{\rm osc}\|^2_{H^{\frac{1}{2}}(\mathbf{T}^3)}\left(\|u_{1,\rm osc}\|^2_{H^1(\mathbf{T}^3)} + \|u_{2,\rm osc}\|^2_{H^1(\mathbf{T}^3)}\right)$$
$$+ \frac{C}{\nu^3}\|w_{\rm osc}\|^2_{H^{\frac{1}{2}}(\mathbf{T}^3)}\left(\|u_{1,\rm osc}\|^4_{H^{\frac{1}{2}}(\mathbf{T}^3)} + \|u_{2,\rm osc}\|^4_{H^{\frac{1}{2}}(\mathbf{T}^3)}\right). \qquad (6.4.9)$$

Now let us estimate $(\mathcal{Q}(\overline{w}, u_{1,\rm osc})|w_{\rm osc})_{H^{\frac{1}{2}}(\mathbf{T}^3)}$. We have

$$\left|(\mathcal{Q}(\overline{w}, u_{1,\rm osc})|w_{\rm osc})_{H^{\frac{1}{2}}(\mathbf{T}^3)}\right| \leq A_1 + A_2$$

with

$$A_1 \stackrel{\rm def}{=} \left|(\overline{w}\cdot\nabla u_{1,\rm osc}|w_{\rm osc})_{H^{\frac{1}{2}}(\mathbf{T}^3)}\right|,$$
$$A_2 \stackrel{\rm def}{=} \left|(u_{1,\rm osc}\cdot\nabla\overline{w}|w_{\rm osc})_{H^{\frac{1}{2}}(\mathbf{T}^3)}\right|. \qquad (6.4.10)$$

The term A_1 is estimated as follows.

$$A_1 \leq \int_{\mathbf{T}^1}\|\overline{w}\|_{L^4(\mathbf{T}^2)}\|\nabla u_{1,\rm osc}(\cdot, x_3)\|_{L^2(\mathbf{T}^2)}\|\nabla w_{\rm osc}(\cdot, x_3)\|_{L^4(\mathbf{T}^2)}\,dx_3$$
$$\leq \|\overline{w}\|_{L^4(\mathbf{T}^2)}\|\nabla u_{1,\rm osc}\|_{L^2(\mathbf{T}^3)}\|\nabla w_{\rm osc}\|_{L^2(\mathbf{T}^1;L^4(\mathbf{T}^2))}$$

by a Cauchy–Schwartz inequality in the third variable. Now we claim that for any function h,

$$\|h\|_{L^2(\mathbf{T}^1;L^4(\mathbf{T}^2))} \leq C\|h\|_{H^{\frac{1}{2}}(\mathbf{T}^3)}.$$

Indeed Sobolev embeddings yield

$$\|h\|^2_{L^2(\mathbf{T}^1;L^4(\mathbf{T}^2))} \leq C \int_{\mathbf{T}^1} \|h(\cdot,x_3)\|^2_{H^{\frac12}(\mathbf{T}^2)}\, dx_3.$$

But

$$\begin{aligned}\int_{\mathbf{T}^1} \|h(\cdot,x_3)\|^2_{H^{\frac12}(\mathbf{T}^2)}\, dx_3 &= \int_{\mathbf{T}^1} \sum_{n^h\in\mathbf{Z}^2} (1+|n^h|^2)^{\frac12} |\mathcal{F}h(n^h,x_3)|^2\, dx_3\\ &= (2\pi)^{-1} \sum_{n\in\mathbf{Z}^3} (1+|n^h|^2)^{\frac12} |\widehat{h}(n)|^2\\ &\leq (2\pi)^{-1} \|h\|^2_{H^{\frac12}(\mathbf{T}^3)}\end{aligned}$$

so the claim follows, and we obtain

$$A_1 \leq C\|\overline{w}\|_{H^{\frac12}(\mathbf{T}^2)} \|\nabla u_{1,\mathrm{osc}}\|_{L^2(\mathbf{T}^3)} \|\nabla w_{\mathrm{osc}}\|_{H^{\frac12}(\mathbf{T}^3)}. \tag{6.4.11}$$

The estimate of (A_2) is along the same lines. We have indeed

$$A_2 \leq \int_{\mathbf{T}^1} \|u_{1,\mathrm{osc}}(\cdot,x_3)\|_{L^4(\mathbf{T}^2)} \|\nabla^h \overline{w}\|_{L^2(\mathbf{T}^2)} \|\nabla w_{\mathrm{osc}}(\cdot,x_3)\|_{L^4(\mathbf{T}^2)}\, dx_3.$$

So as above, we claim

$$A_2 \leq C\|\nabla^h\overline{w}\|_{L^2(\mathbf{T}^2)} \|\nabla w_{\mathrm{osc}}\|_{H^{\frac12}(\mathbf{T}^3)} \|u_{1,\mathrm{osc}}\|_{H^{\frac12}(\mathbf{T}^3)}. \tag{6.4.12}$$

Plugging (6.4.11) and (6.4.12) into (6.4.10) yields

$$\begin{aligned}\left|(\mathcal{Q}(\overline{w},u_{1,\mathrm{osc}})|w_{\mathrm{osc}})_{H^{\frac12}(\mathbf{T}^3)}\right| &\leq \frac{\nu}{4}\|\nabla w_{\mathrm{osc}}\|^2_{H^{\frac12}(\mathbf{T}^3)}\\ &\quad + \frac{C}{\nu}\|\overline{w}\|^2_{H^{\frac12}(\mathbf{T}^2)} \|\nabla u_{1,\mathrm{osc}}\|^2_{L^2(\mathbf{T}^3)}\\ &\quad + \frac{C}{\nu}\|\nabla^h\overline{w}\|^2_{L^2(\mathbf{T}^2)} \|u_{1,\mathrm{osc}}\|^2_{H^{\frac12}(\mathbf{T}^3)}.\end{aligned} \tag{6.4.13}$$

Going back to estimate (6.4.8), and putting (6.4.9) and (6.4.13) together we get

$$\frac{d}{dt}\|w_{\mathrm{osc}}\|^2_{H^{\frac12}(\mathbf{T}^3)} + \nu\|\nabla w_{\mathrm{osc}}\|^2_{H^{\frac12}(\mathbf{T}^3)} \leq \sum_{j=1}^{3} \mathcal{W}_j$$

with

$$\mathcal{W}_1 \stackrel{\mathrm{def}}{=} \frac{C}{\nu}\left(\|\overline{w}\|^2_{H^{\frac12}(\mathbf{T}^2)} \|\nabla u_{1,\mathrm{osc}}\|^2_{L^2(\mathbf{T}^3)} + \|g_{\mathrm{osc}}\|^2_{H^{-\frac12}}\right)$$

$$\begin{aligned}\mathcal{W}_2 \stackrel{\mathrm{def}}{=} \frac{C}{\nu}\Big(&\|\nabla^h\overline{w}\|^2_{L^2(\mathbf{T}^2)} \|u_{1,\mathrm{osc}}\|^2_{H^{\frac12}(\mathbf{T}^3)}\\ &+\|w_{\mathrm{osc}}\|^2_{H^{\frac12}(\mathbf{T}^3)} \left(\|u_{1,\mathrm{osc}}\|^2_{H^1(\mathbf{T}^3)} + \|u_{2,\mathrm{osc}}\|^2_{H^1(\mathbf{T}^3)}\right)\Big) \quad \text{and}\end{aligned}$$

$$\mathcal{W}_3 \stackrel{\mathrm{def}}{=} \frac{C}{\nu^3}\|w_{\mathrm{osc}}\|^2_{H^{\frac12}(\mathbf{T}^3)} \left(\|u_{1,\mathrm{osc}}\|^4_{H^{\frac12}(\mathbf{T}^3)} + \|u_{2,\mathrm{osc}}\|^4_{H^{\frac12}(\mathbf{T}^3)}\right).$$

The result follows by integration in time and a Gronwall lemma, using the estimates proved previously on $u_{1,\mathrm{osc}}$, $u_{2,\mathrm{osc}}$ and $\overline{w}$. Lemma 6.3 is proved, and the stability result of Proposition 6.4 follows directly. □

6.5 Construction of an approximate solution

In this section we shall prove Lemma 6.1 using the results of the previous paragraphs. The goal of this section is therefore to construct an approximation of u^ε, called $u_{\mathrm{app}}^{\varepsilon,\eta}$, which exists globally in time. In fact we will do more, as in the construction of $u_{\mathrm{app}}^{\varepsilon,\eta}$ we will also show that $u_{\mathrm{app}}^{\varepsilon,\eta}$ is a (strong) approximation of the limit solution u built in the previous section. That will enable us to infer the following theorem, which is a more precise version of the weak convergence Theorem 6.3.

Theorem 6.4 *Let u_0 be in $H^{\frac{1}{2}}$ and let u be the unique, global solution constructed in Proposition 6.3 (with $f = 0$). For any positive real number η, a family $(u_{\mathrm{app}}^{\varepsilon,\eta})_{\varepsilon>0}$ exists such that*

$$\lim_{\eta\to 0}\limsup_{\varepsilon\to 0}\left\| u_{\mathrm{app}}^{\varepsilon,\eta} - \mathcal{L}\left(\frac{t}{\varepsilon}\right)u\right\|_{\frac{1}{2}} = 0.$$

Moreover, the family $(u_{\mathrm{app}}^{\varepsilon,\eta})$ satisfies

$$\begin{cases} \partial_t u_{\mathrm{app}}^{\varepsilon,\eta} - \nu\Delta u_{\mathrm{app}}^{\varepsilon,\eta} + \mathbf{P}(u_{\mathrm{app}}^{\varepsilon,\eta}\cdot\nabla u_{\mathrm{app}}^{\varepsilon,\eta}) + \dfrac{1}{\varepsilon}\mathbf{P}(e^3\wedge u_{\mathrm{app}}^{\varepsilon,\eta}) = R^{\varepsilon,\eta} \\ \lim\limits_{\eta\to 0}\limsup\limits_{\varepsilon\to 0}\|u_{\mathrm{app}|t=0}^{\varepsilon,\eta} - u_0\|_{H^{\frac{1}{2}}} = 0, \end{cases} \tag{6.5.1}$$

with $\lim_{\eta\to 0}\limsup_{\varepsilon\to 0}\|R^{\varepsilon,\eta}\|_{L^2(\mathbf{R}^+;H^{-\frac{1}{2}})} = 0.$

Remark Theorem 6.4 clearly proves Lemma 6.1, page 120, simply by a triangular inequality since

$$\|u_{\mathrm{app}}^{\varepsilon,\eta}\|_{\frac{1}{2}} \leq \left\|u_{\mathrm{app}}^{\varepsilon,\eta} - \mathcal{L}\left(\frac{t}{\varepsilon}\right)u\right\|_{\frac{1}{2}} + \left\|\mathcal{L}\left(\frac{t}{\varepsilon}\right)u\right\|_{\frac{1}{2}} \leq C < +\infty.$$

Proof of Theorem 6.4 Let η be an arbitrary positive number. We define, for any positive integer N,

$$u_N = \mathbf{P}_N u \stackrel{\mathrm{def}}{=} \mathcal{F}^{-1}\left(\mathbf{1}_{|n|\leq N}\widehat{u}(n)\right),$$

and obviously there is $N_\eta > 0$ such that

$$\left\|\mathcal{L}\left(\frac{t}{\varepsilon}\right)(u_{N_\eta} - u)\right\|_{\frac{1}{2}} \leq \rho^{\varepsilon,\eta},$$

where $\rho^{\varepsilon,\eta}$ denotes from now on any non-negative quantity such that

$$\lim_{\eta\to 0}\limsup_{\varepsilon\to 0}\rho^{\varepsilon,\eta} = 0.$$

We will also denote generically by $R^{\varepsilon,\eta}$ any vector field satisfying

$$\|R^{\varepsilon,\eta}\|_{L^2(\mathbf{R}^+;H^{-\frac{1}{2}})} = \rho^{\varepsilon,\eta}.$$

From now on we can concentrate on u_{N_η}, and our goal is to approximate $\mathcal{L}(t/\varepsilon)u_{N_\eta}$ in such a way as to satisfy system (6.5.1). So let us write

$$u_{\text{app}}^{\varepsilon,\eta} = \mathcal{L}\left(\frac{t}{\varepsilon}\right) u_{N_\eta} + \varepsilon u_1^{\varepsilon,\eta}$$

where $u_1^{\varepsilon,\eta}$ is a smooth, divergence-free vector field to be determined. To simplify we also define the notation

$$u_0^{\varepsilon,\eta} \stackrel{\text{def}}{=} \mathcal{L}\left(\frac{t}{\varepsilon}\right) u_{N_\eta},$$

as well as the operator

$$L_\nu^\varepsilon w \stackrel{\text{def}}{=} \partial_t w - \nu\Delta w + \frac{1}{\varepsilon} e^3 \wedge w.$$

Then we have

$$L_\nu^\varepsilon u_{\text{app}}^{\varepsilon,\eta} + u_{\text{app}}^{\varepsilon,\eta} \cdot \nabla u_{\text{app}}^{\varepsilon,\eta} = L_\nu^\varepsilon u_0^{\varepsilon,\eta} + \varepsilon L_\nu^\varepsilon u_1^{\varepsilon,\eta} + u_{\text{app}}^{\varepsilon,\eta} \cdot \nabla u_{\text{app}}^{\varepsilon,\eta}, \tag{6.5.2}$$

and the only point left to prove is that there is a smooth, divergence-free vector field $u_1^{\varepsilon,\eta}$ such that the right-hand side of (6.5.2) is a remainder term.

We notice that by definition of $u_0^{\varepsilon,\eta}$,

$$\begin{aligned}
\mathbf{P} L_\nu^\varepsilon u_0^{\varepsilon,\eta} &= \mathbf{P}(\partial_t u_0^{\varepsilon,\eta} - \nu\Delta u_0^{\varepsilon,\eta} + \frac{1}{\varepsilon} e^3 \wedge u_0^{\varepsilon,\eta}) \\
&= \mathcal{L}\left(\frac{t}{\varepsilon}\right)(\partial_t u_{N_\eta} - \nu\Delta u_{N_\eta}) + \frac{1}{\varepsilon}\partial_\tau \mathcal{L}\left(\frac{t}{\varepsilon}\right) u_{N_\eta} + \frac{1}{\varepsilon}\mathbf{P}\left(e^3 \wedge \mathcal{L}\left(\frac{t}{\varepsilon}\right) u_{N_\eta}\right) \\
&= \mathcal{L}\left(\frac{t}{\varepsilon}\right) \mathbf{P}_{N_\eta} \mathcal{Q}(u,u),
\end{aligned}$$

since by definition of $\mathcal{L}$, $\partial_\tau \mathcal{L}(t/\varepsilon)u_{N_\eta} + \mathbf{P}\left(e^3 \wedge \mathcal{L}(t/\varepsilon)u_{N_\eta}\right) = 0$. But it is easy to prove that

$$\left\|\mathcal{L}\left(\frac{t}{\varepsilon}\right) \mathbf{P}_{N_\eta} \mathcal{Q}(u,u) - \mathcal{L}\left(\frac{t}{\varepsilon}\right) \mathcal{Q}(u_{N_\eta}, u_{N_\eta})\right\|_{L^2(\mathbf{R}^+;H^{-\frac{1}{2}}(\mathbf{T}^3))} \leq \rho^{\varepsilon,\eta}. \tag{6.5.3}$$

Indeed we have, on the one hand, by Lebesgue's theorem and using the fact that u is in $E_{\frac{1}{2}}$,

$$\|\mathbf{P}_{N_\eta} \mathcal{Q}(u,u) - \mathcal{Q}(u,u)\|_{L^2(\mathbf{R}^+;H^{-\frac{1}{2}})} \leq \rho^{\varepsilon,\eta}.$$

On the other hand, we can write by the usual Sobolev embeddings

$$\begin{aligned}\|\mathcal{Q}(u,u) - \mathcal{Q}(u_{N_\eta}, u_{N_\eta})\|_{L^2(\mathbf{R}^+;H^{-\frac12})} \\ \leq C\|u\|_{L^\infty(\mathbf{R}^+;H^{\frac12})}\|\nabla(u_{N_\eta} - u)\|_{L^2(\mathbf{R}^+;H^{\frac12})},\end{aligned}$$

where we have used the fact that

$$\|u_{N_\eta}\|_{L^\infty(\mathbf{R}^+;H^{\frac12})} \leq \|u\|_{L^\infty(\mathbf{R}^+;H^{\frac12})}.$$

The result (6.5.3) therefore simply follows again from Lebesgue's theorem. We infer that

$$\begin{aligned}\mathbf{P}L_\nu^\varepsilon u_{\mathrm{app}}^{\varepsilon,\eta} + \mathbf{P}(u_{\mathrm{app}}^{\varepsilon,\eta} \cdot \nabla u_{\mathrm{app}}^{\varepsilon,\eta}) = R^{\varepsilon,\eta} + \mathcal{L}\left(\frac{t}{\varepsilon}\right)\mathcal{Q}(u_{N_\eta}, u_{N_\eta}) \\ + \varepsilon \mathbf{P}L_\nu^\varepsilon u_1^{\varepsilon,\eta} + \mathbf{P}(u_{\mathrm{app}}^{\varepsilon,\eta} \cdot \nabla u_{\mathrm{app}}^{\varepsilon,\eta}).\end{aligned}$$

Now we write, by definition of $\mathcal{Q}^\varepsilon$,

$$\mathbf{P}(u_{\mathrm{app}}^{\varepsilon,\eta} \cdot \nabla u_{\mathrm{app}}^{\varepsilon,\eta}) = -\mathcal{L}\left(\frac{t}{\varepsilon}\right)\mathcal{Q}^\varepsilon\left(\mathcal{L}\left(-\frac{t}{\varepsilon}\right)u_{\mathrm{app}}^{\varepsilon,\eta}, \mathcal{L}\left(-\frac{t}{\varepsilon}\right)u_{\mathrm{app}}^{\varepsilon,\eta}\right),$$

hence since $\mathcal{L}(-t/\varepsilon)u_0^{\varepsilon,\eta} = u_{N_\eta}$, we get

$$\mathbf{P}(u_{\mathrm{app}}^{\varepsilon,\eta} \cdot \nabla u_{\mathrm{app}}^{\varepsilon,\eta}) = -\mathcal{L}\left(\frac{t}{\varepsilon}\right)\mathcal{Q}^\varepsilon(u_{N_\eta}, u_{N_\eta}) + F^{\varepsilon,\eta},$$

where

$$F^{\varepsilon,\eta} \stackrel{\mathrm{def}}{=} -\varepsilon\mathcal{L}\left(\frac{t}{\varepsilon}\right)\mathcal{Q}^\varepsilon\left(u_{N_\eta}, \mathcal{L}\left(-\frac{t}{\varepsilon}\right)u_1^{\varepsilon,\eta}\right) - \varepsilon^2\mathcal{Q}^\varepsilon\left(\mathcal{L}\left(-\frac{t}{\varepsilon}\right)u_1^{\varepsilon,\eta}, \mathcal{L}\left(-\frac{t}{\varepsilon}\right)u_1^{\varepsilon,\eta}\right).$$

Going back to the equation on $u_{\mathrm{app}}^{\varepsilon,\eta}$ we find that

$$\begin{aligned}\mathbf{P}L_\nu^\varepsilon u_{\mathrm{app}}^{\varepsilon,\eta} + \mathbf{P}(u_{\mathrm{app}}^{\varepsilon,\eta} \cdot \nabla u_{\mathrm{app}}^{\varepsilon,\eta}) = R^{\varepsilon,\eta} + \mathcal{L}\left(\frac{t}{\varepsilon}\right)\mathcal{Q}(u_{N_\eta}, u_{N_\eta}) \\ - \mathcal{L}\left(\frac{t}{\varepsilon}\right)\mathcal{Q}^\varepsilon(u_{N_\eta}, u_{N_\eta}) + F^{\varepsilon,\eta} + \varepsilon\mathbf{P}L_\nu^\varepsilon u_1^{\varepsilon,\eta}.\end{aligned}$$

Let us postpone for a while the proof of the following lemma.

Lemma 6.4 *Let $\eta > 0$ be given. There is a family of divergence-free vector fields $u_1^{\varepsilon,\eta}$, bounded in $(L^\infty \cap L^1)(\mathbf{R}^+; H^s(\mathbf{T}^3))$ for all $s \geq 0$, such that*

$$\mathcal{L}\left(\frac{t}{\varepsilon}\right)(\mathcal{Q}^\varepsilon - \mathcal{Q})(u_{N_\eta}, u_{N_\eta}) = \varepsilon\mathbf{P}L_\nu^\varepsilon u_1^{\varepsilon,\eta} + R^{\varepsilon,\eta}.$$

Lemma 6.4 implies that

$$\mathbf{P}L^{\varepsilon}_{\nu}u^{\varepsilon,\eta}_{\text{app}} + \mathbf{P}(u^{\varepsilon,\eta}_{\text{app}} \cdot \nabla u^{\varepsilon,\eta}_{\text{app}}) = R^{\varepsilon,\eta} + F^{\varepsilon}_{N}$$

and the only point left to check is that $F^{\varepsilon,\eta}$ is a remainder term. But that is obvious due to the smoothness of $u^{\varepsilon,\eta}_1$ which implies that $F^{\varepsilon,\eta}$ is $O(\varepsilon)$ in $L^2(\mathbf{R}^+; H^{-\frac{1}{2}})$ for instance, for all η. So the theorem is proved, up to the proof of Lemma 6.4. □

Proof of Lemma 6.4 We start by noticing that by definition,

$$(\mathcal{Q}^{\varepsilon} - \mathcal{Q})(u_{N_\eta}, u_{N_\eta}) = -\mathcal{F}^{-1} \sum_{\substack{k \notin \mathcal{K}^{\sigma}_n \\ \sigma \in \{+,-\}^3}} e^{-i\frac{t}{\varepsilon}\omega^{\sigma}_n(k)} \mathbf{1}_{|k|\leq N_\eta} \mathbf{1}_{|n-k|\leq N_\eta} [u^{\sigma_1}(k) \cdot (n-k)]$$
$$\times [u^{\sigma_2}(n-k) \cdot e^{\sigma_3}(n)] e^{\sigma_3}(n),$$

recalling that

$$\omega^{\sigma}_n(k) = \sigma_1 \frac{\widetilde{k}_3}{|\widetilde{k}|} + \sigma_2 \frac{\widetilde{n}_3 - \widetilde{k}_3}{|\widetilde{n} - \widetilde{k}|} - \sigma_3 \frac{\widetilde{n}_3}{|\widetilde{n}|}$$

and that $k \notin \mathcal{K}^{\sigma}_n$ means that $\omega^{\sigma}_n(k)$ is not zero, so $(\mathcal{Q}^{\varepsilon} - \mathcal{Q})(u_{N_\eta}, u_{N_\eta})$ is an oscillating term in time. Moreover the frequency truncation implies that $|\omega^{\sigma}_n(k)|$ is bounded from below, by a constant depending on η. That enables us to define

$$\widetilde{u}^{\varepsilon,\eta}_1 \stackrel{\text{def}}{=} \mathcal{F}^{-1} \sum_{\substack{k \notin \mathcal{K}^{\sigma}_n \\ \sigma \in \{+,-\}^3}} \frac{e^{-i\frac{t}{\varepsilon}\omega^{\sigma}_n(k)}}{i\omega^{\sigma}_n(k)} \mathbf{1}_{|k|\leq N_\eta} \mathbf{1}_{|n-k|\leq N_\eta} [u^{\sigma_1}(k) \cdot (n-k)]$$
$$\times [u^{\sigma_2}(n-k) \cdot e^{\sigma_3}(n)] e^{\sigma_3}(n),$$

and $u^{\varepsilon,\eta}_1 \stackrel{\text{def}}{=} \mathcal{L}\left(\frac{t}{\varepsilon}\right) \widetilde{u}^{\varepsilon,\eta}_1$. Then

$$\varepsilon \partial_t \widetilde{u}^{\varepsilon,\eta}_1 = (\mathcal{Q} - \mathcal{Q}^{\varepsilon})(u_{N_\eta}, u_{N_\eta}) + \varepsilon R^{\varepsilon,t}_{\eta},$$

where $R^{\varepsilon,t}_{\eta}$ is the inverse Fourier transform of

$$2 \sum_{\substack{k \notin \mathcal{K}^{\sigma}_n \\ \sigma \in \{+,-\}^3}} \frac{e^{-i\frac{t}{\varepsilon}\omega^{\sigma}_n(k)}}{i\omega^{\sigma}_n(k)} \mathbf{1}_{|k|\leq N_\eta} \mathbf{1}_{|n-k|\leq N_\eta} [\partial_t u^{\sigma_1}(k) \cdot (n-k)][u^{\sigma_2}(n-k) \cdot e^{\sigma_3}(n)] e^{\sigma_3}(n).$$

Notice that $\varepsilon \widetilde{u}^{\varepsilon,\eta}_1$ is defined as the primitive in time of the oscillating term $\mathcal{Q}^{\varepsilon} - \mathcal{Q}$, and it is precisely the time oscillations that imply that $\widetilde{u}^{\varepsilon,\eta}_1$ is uniformly bounded in ε.

We therefore have

$$\varepsilon \partial_t u^{\varepsilon,\eta}_1 = \varepsilon \mathcal{L}\left(\frac{t}{\varepsilon}\right) \partial_t \widetilde{u}^{\varepsilon,\eta}_1 + \partial_\tau \mathcal{L}\left(\frac{t}{\varepsilon}\right) \widetilde{u}^{\varepsilon,\eta}_1$$
$$= \mathcal{L}\left(\frac{t}{\varepsilon}\right) (\mathcal{Q} - \mathcal{Q}^{\varepsilon})(u_{N_\eta}, u_{N_\eta}) + \varepsilon \mathcal{L}\left(\frac{t}{\varepsilon}\right) R^{\varepsilon,t}_{\eta} - \mathbf{P}\left(\mathcal{L}\left(\frac{t}{\varepsilon}\right) \widetilde{u}^{\varepsilon,\eta}_1 \wedge e^3\right),$$

recalling that $\partial_\tau \mathcal{L} + \mathbf{P} R_{\frac{\pi}{2}} \mathcal{L} = 0$ where $R_{\frac{\pi}{2}}$ is the rotation by angle $\pi/2$. Finally we have

$$\varepsilon \partial_t u_1^{\varepsilon,\eta} + \mathbf{P}\left(u_1^{\varepsilon,\eta} \wedge e^3\right) = \mathcal{L}\left(\frac{t}{\varepsilon}\right)(\mathcal{Q} - \mathcal{Q}^\varepsilon)(u_{N_\eta}, u_{N_\eta}) + \varepsilon \mathcal{L}\left(\frac{t}{\varepsilon}\right) R_\eta^{\varepsilon,t}.$$

Since $u_1^{\varepsilon,\eta}$ is arbitrarily smooth (for a fixed η) and so is $R_\eta^{\varepsilon,t}$, we find finally that

$$\varepsilon \mathbf{P} L_\nu^\varepsilon u_1^{\varepsilon,\eta} = R^{\varepsilon,\eta} + \mathcal{L}\left(\frac{t}{\varepsilon}\right)(\mathcal{Q} - \mathcal{Q}^\varepsilon)(u_{N_\eta}, u_{N_\eta}),$$

and Lemma 6.4 is proved. □

6.6 Study of the limit system with anisotropic viscosity

In Chapter 7, we shall be dealing with a fluid evolving between two fixed horizontal plates at $x_3 = 0$ and $x_3 = 1$, with Dirichlet boundary conditions on the plates, and periodic horizontal boundary conditions. In that case we will see that the limit system is quite similar to the purely periodic case, apart from the fact that there is no vertical viscosity. There is also an additional, positive operator $\mathcal{E}$. In other words the limit system in that case is

$$(\mathrm{NSE}_{\nu,\mathcal{E}}) \begin{cases} \partial_t u - \nu \Delta_h u + \mathcal{E} u + \mathcal{Q}(u,u) = 0 \\ \operatorname{div} u = 0 \\ u_{|t=0} = u_0. \end{cases}$$

The well-posedness result will be useful in the next chapter. Before stating and proving that result, we need some additional notation. We define

$$H^{0,1} \stackrel{\text{def}}{=} \{ f \in L^2(\Omega),\ \partial_3 f \in L^2(\Omega) \}$$

and

$$\|f\|_{H^{0,1}}^2 = \|f\|_{L^2}^2 + \|\partial_3 f\|_{L^2}^2.$$

Moreover we denote by $\mathcal{H}_0$ the space of divergence-free vector fields with a vanishing horizontal mean.

Finally we shall need the notion of admissible classes of periodic boxes.

Definition 6.2 *We say that $\mathbf{T}^3$ satisfies condition $(\mathcal{A})$ if one of the following conditions is satisfied:*

- *the following implication holds:*

$$(\mathcal{R}) \qquad k \in \mathcal{K}_n^\sigma \Rightarrow k_3 n_3 = 0;$$

- *condition $(\mathcal{R})$ does not hold and $a_i/a_j \in \mathbf{Q}$ for all $(i,j) \in \{1,2,3\}^2$;*
- *condition $(\mathcal{R})$ does not hold and there are $i \neq j$, $i \neq k$, $j \neq k$ such that $a_i^2/a_j^2 \in \mathbf{Q}$ and a_j^2/a_k^2 is not algebraic of degree 4.*

Remark It can be proved that almost all periodic boxes satisfy condition $(\mathcal{R})$ (see [47], [48]). Condition $(\mathcal{A})$ is introduced because it is under that assumption that we will be able to prove that the horizontal mean of the solution of the limit system is preserved; a vanishing horizontal mean will be needed to construct the boundary layers in Section 7.4 page 176.

Proposition 6.5 (Global well-posedness) *Let u_0 be in $\mathcal{H} \cap H^{0,1}$, and suppose that the operator $\mathcal{E}$ is continuous and non-negative on $\mathcal{H} \cap H^{0,1}$. There is a unique solution u in $C^0(\mathbf{R}^+, \mathcal{H} \cap H^{0,1})$ to $(\mathrm{NSE}_{\nu,\mathcal{E}})$ such that $\nabla^h u$ is in $L^2_{\mathrm{loc}}(\mathbf{R}^+; H^{0,1})$. Moreover, defining*

$$u = \overline{u} + \widetilde{u} \quad \textit{with } \partial_3 \overline{u} = 0 \textit{ and } \int_0^1 \widetilde{u}\, dx_3 = 0,$$

we have the following estimates:

$$\|\overline{u}(t)\|^2_{L^2(\mathbf{T}^2)} + 2\int_0^t \|\nabla^h \overline{u}(t')\|^2_{L^2(\mathbf{T}^2)}\, dt' \le \|\overline{u}_0\|^2_{L^2(\mathbf{T}^2)}, \tag{6.6.1}$$

and

$$\|\widetilde{u}(t)\|^2_{H^{0,1}} + \int_0^t \|\nabla^h \widetilde{u}(t')\|^2_{H^{0,1}}\, dt'$$
$$\le \|\widetilde{u}_0\|^2_{H^{0,1}} \times \exp\left(C_\nu t^{\frac{1}{2}} \|\widetilde{u}_0\|_{L^2}(1 + t^{\frac{1}{2}}\|\widetilde{u}_0\|_{L^2})\right).$$

Finally we have $\mathcal{Q}(u,u) \in L^2_{\mathrm{loc}}(\mathbf{R}^+, H^{-1,0})$, and if $\mathbf{T}^3$ satisfies condition $(\mathcal{A})$ then

$$\forall t \ge 0, \quad \int_{\mathbf{T}^2} \mathcal{Q}(u,u)(t, x_h, x_3)\, dx_h = 0. \tag{6.6.2}$$

In particular the horizontal mean of the solution is preserved.

Remark Note that the global well-posedness result is no a priori trivial fact, as the structure of that system is that of the three-dimensional Navier–Stokes equations, supplemented with vanishing viscosity in the vertical variable. Similarly to the above, the reason why global well-posedness holds relies on the very special structure of the non-linear term $\mathcal{Q}$.

Proof of Proposition 6.5 Let us start by proving the uniqueness of the solutions. Suppose u and v are two solutions associated to the same initial data u_0, and define $w = u - v$. Then

$$\partial_t w - \nu \Delta_h w + \mathcal{E} w = \mathcal{Q}(w,w) + \mathcal{Q}(u,w) + \mathcal{Q}(w,u)$$

and an energy estimate in L^2 implies that

$$\|w(t)\|^2_{L^2} + 2\nu \int_0^t \|\nabla^h w(t')\|^2_{L^2}\, dt'$$
$$\le \|w_0\|^2_{L^2} + \int_0^t |((\mathcal{Q}(w,w)|w)_{L^2} + (\mathcal{Q}(w,u)|w)_{L^2} + (\mathcal{Q}(u,w)|w)_{L^2})\,(t')|\, dt'$$

and the symmetry properties of $\mathcal{Q}$ enable us to write that

$$\|w(t)\|_{L^2}^2 + 2\nu \int_0^t \|\nabla^h w(t')\|_{L^2}^2\, dt' \le \|w_0\|_{L^2}^2 + 2\int_0^t |(\mathcal{Q}(w,w)|u)_{L^2}|\, dt'.$$

In order to estimate the last term, it is enough to prove the following lemma.

Lemma 6.5 *Let δ be a divergence-free vector field vanishing on the boundary. Then for any vector field b,*

$$\int_0^T (\delta(t)\cdot\nabla\delta(t)|b(t))_{L^2}\, dt \le \int_0^T \|\delta(t)\|_{L^2}\|\nabla^h\delta(t)\|_{L^2} \times \big(\|\nabla^h b(t)\|_{H^{0,1}} + \|\partial_3 b(t)\|_{L^2}\|\partial_3\nabla^h b(t)\|_{L^2}\big)\, dt.$$

Proof The divergence-free condition implies that

$$\int_0^t (\delta(t')\cdot\nabla\delta(t')|b(t'))_{L^2}\, dt' = \int_0^t (\delta(t')\otimes\delta(t')|\nabla b(t'))_{L^2}\, dt' = A_h + A_v,$$

where

$$A_h = \sum_{\substack{k\in\{1,2\}\\ j\in\{1,2,3\}}} \int_0^t \delta^k(t')\delta^j(t')\partial_k b^j(t')\, dt'.$$

Let us start by estimating A_h. We have

$$|A_h| \le \sum_{\substack{k\in\{1,2\}\\ j\in\{1,2,3\}}} \int_0^t\int_0^1 \|\delta^k(t',\cdot,x_3)\|_{L_h^4}\|\delta^j(t',\cdot,x_3)\|_{L_h^4}\|\partial_k b^j(t',\cdot,x_3)\|_{L_h^2}\, dx_3 dt'$$

$$\le \int_0^t\int_0^1 \|\delta(t',\cdot,x_3)\|_{L_h^2}\|\nabla^h\delta(t',\cdot,x_3)\|_{L_h^2}\, dx_3 \|\partial_k b^j(t')\|_{L^\infty([0,1];L_h^2)}\, dt'$$

by the Gagliardo–Nirenberg inequality. The result follows from the continuous embedding of $H^1([0,1])$ into $L^\infty([0,1])$.

Now let us estimate A_v. We have

$$|A_v| \le \int_0^t\int_0^1 \|\delta^3(t',\cdot,x_3)\|_{L_h^2}\|\delta(t',\cdot,x_3)\|_{L_h^4}\|\partial_3 b(t',\cdot,x_3)\|_{L_h^4}\, dx_3 dt'$$

$$\le \int_0^t\int_0^1 \|\delta^3(t',\cdot,x_3)\|_{L_h^2}\|\delta(t',\cdot,x_3)\|_{L_h^2}^{\frac12}\|\nabla^h\delta(t',\cdot,x_3)\|_{L_h^2}^{\frac12} \times \|\partial_3 b(t',\cdot,x_3)\|_{L_h^2}^{\frac12}\|\partial_3\nabla^h b(t',\cdot,x_3)\|_{L_h^2}^{\frac12}\, dx_3 dt'.$$

But

$$\begin{aligned}\|\delta^3(\cdot,x_3)\|_{L^2_h}^2 &= \int_0^{x_3} \frac{d}{dy_3}\|\delta^3(\cdot,y_3)\|_{L^2_h}^2\,dy_3\\ &= 2\int_0^{x_3}\int_{\mathbf{T}^2} \partial_3\delta^3(x_h,y_3)\delta^3(x_h,y_3)\,dx_h dy_3\\ &\leq 2\int_0^{x_3} \|\nabla^h\delta(\cdot,y_3)\|_{L^2_h}\|\delta(\cdot,y_3)\|_{L^2_h}\,dy_3,\end{aligned}$$

where we have used the divergence-free condition on δ. Finally

$$|A_v| \leq C\int_0^t \|\nabla^h\delta(t')\|_{L^2}\|\delta(t')\|_{L^2}\|\partial_3 b(t')\|_{L^2}^{\frac12}\|\partial_3\nabla^h b(t')\|_{L^2}^{\frac12}\,dt',$$

which proves the lemma. □

Remark It is easy to adapt the proof (and the result) of Lemma 6.5 to the quadratic form $\mathcal{Q}$; it is enough to use the fact that $\mathcal{L}$ is unitary together with the definition of $\mathcal{Q}$. We infer

$$\begin{aligned}\int_0^t |(\mathcal{Q}(w,w)|u)_{L^2}|\,dt' \leq C\int_0^t &\|w(t)\|_{L^2}\|\nabla^h w(t')\|_{L^2}\\ &\times\left(\|\nabla^h u(t')\|_{H^{0,1}} + \|\partial_3 u(t')\|_{L^2}\|\partial_3\nabla^h u(t')\|_{L^2}\right)dt'.\end{aligned}$$

We find that

$$\begin{aligned}&\|w(t)\|_{L^2}^2 + 2\nu\int_0^t \|\nabla^h w(t)\|_{L^2}^2\\ &\quad\leq C\int_0^t \|w(t')\|_{L^2}^2 \times \left(\|\nabla^h u(t')\|_{H^{0,1}}^2 + \|\partial_3 u(t')\|_{L^2}^2\|\partial_3\nabla^h u(t')\|_{L^2}^2\right)dt'.\end{aligned}$$

and the result follows from Gronwall's lemma and the regularity properties of u.

Now let us prove the global existence result. That result is not trivial for several reasons. First, the system $(\mathrm{NSE}_{\nu,\varepsilon})$ is similar to a three-dimensional Navier–Stokes equation, for which such a result is not known. However, as we saw in Section 6.3, the bilinear term has in fact a special structure which makes it close to a two-dimensional equation (the product rules it satisfies are of two-dimensional type, with the loss of only one derivative instead of 3/2). The additional difficulty here is that the Laplacian is only two-dimensional, so there is a priori no smoothing effect in the third direction. However using the fact that the velocity field is divergence-free, we can recover one vertical derivative on the third component of the velocity field and we shall see that this fact will be enough to conclude. So let us prove Proposition 6.5. We recall that, with notation (6.2.3) and (6.2.4), the vector field $\overline{u}$ satisfies a two-dimensional (damped) Navier–Stokes equation so the existence of $\overline{u}$ satisfying (6.6.1) is obvious. Now let us concentrate on the vector field $u_{\rm osc} = u - \overline{u}$, which satisfies the following system:

$$\begin{cases} \partial_t u_{\rm osc} - \nu\Delta_h u_{\rm osc} + \mathcal{E} u_{\rm osc} - \mathcal{Q}(u_{\rm osc}, u_{\rm osc} + 2\overline{u}) = 0 \\ \operatorname{div} u_{\rm osc} = 0 \\ u_{{\rm osc}|\partial\Omega} = 0 \\ u_{|t=0} = u_{{\rm osc},0}. \end{cases}$$

In order to avoid additional notation, we shall write energy estimates on $u_{\rm osc}$ and $\partial_3 u_{\rm osc}$ directly, and not on regularizations; we shall omit the final, now classical step consisting of taking the limit of the sequence of regularized solutions satisfying the energy estimate. The properties recalled in Proposition 6.2 imply the following estimate (the operator $\mathcal{E}$ is non-negative on $\mathcal{H} \cap H^{0,1}$ so we shall no longer consider it here):

$$\|u_{\rm osc}(t)\|_{L^2}^2 + 2\nu \int_0^t \|\nabla^h u_{\rm osc}(t')\|_{L^2}^2\, dt' \leq \|u_{{\rm osc},0}\|_{L^2}^2. \tag{6.6.3}$$

Now we want to write an energy estimate on $\partial_3 u_{\rm osc}$. We have

$$\begin{aligned} \frac{1}{2}\frac{d}{dt}\|\partial_3 u_{\rm osc}(t)\|_{L^2}^2 + \nu\|\nabla^h \partial_3 u_{\rm osc}(t)\|_{L^2}^2 \leq & \left| (\partial_3 \mathcal{Q}(u_{\rm osc}, u_{\rm osc})|\partial_3 u_{\rm osc})_{L^2} \right| \\ & + 2\left| (\partial_3 \mathcal{Q}(\overline{u}, u_{\rm osc})|\partial_3 u_{\rm osc})_{L^2} \right|. \end{aligned} \tag{6.6.4}$$

By Proposition 6.2 we have $(\partial_3 \mathcal{Q}(\overline{u}, u_{\rm osc})|\partial_3 u_{\rm osc})_{L^2} = 0$, so let us estimate the first term on the right-hand side. We cannot simply use the methods of the above sections since we are missing the regularization effect in the third variable, so we need to use the fact that $u_{\rm osc}$ is divergence-free in order to gain that missing derivative. The method is as follows. We define

$$\mathcal{Q}(u_{\rm osc}, u_{\rm osc}) = \mathcal{Q}_h(u_{\rm osc}, u_{\rm osc}) + \mathcal{Q}_v(u_{\rm osc}, u_{\rm osc})$$

where for any vector field a and b we have defined

$$\mathcal{Q}_h(a,b) \stackrel{\rm def}{=} \lim_{\varepsilon\to 0} \mathcal{L}\left(-\frac{t}{\varepsilon}\right)\left(\left(\mathcal{L}\left(\frac{t}{\varepsilon}\right)a\right)^h \cdot \nabla^h \mathcal{L}\left(\frac{t}{\varepsilon}\right) b\right) \tag{6.6.5}$$

and

$$\mathcal{Q}_v(a,b) \stackrel{\rm def}{=} \lim_{\varepsilon\to 0} \mathcal{L}\left(-\frac{t}{\varepsilon}\right)\left(\left(\mathcal{L}\left(\frac{t}{\varepsilon}\right)a\right)^3 \partial_3 \mathcal{L}\left(\frac{t}{\varepsilon}\right) b\right). \tag{6.6.6}$$

Then we can write

$$\begin{aligned} \left| (\partial_3 \mathcal{Q}(u_{\rm osc}, u_{\rm osc})|\partial_3 u_{\rm osc})_{L^2} \right| \leq & \left| (\partial_3 \mathcal{Q}_h(u_{\rm osc}, u_{\rm osc})|\partial_3 u_{\rm osc})_{L^2} \right| \\ & + \left| (\partial_3 \mathcal{Q}_v(u_{\rm osc}, u_{\rm osc})|\partial_3 u_{\rm osc})_{L^2} \right| \end{aligned} \tag{6.6.7}$$

and we are going to show that each of those two terms can be estimated in the same way. In the following computations we shall use the notation of the above sections. The computations will be quite similar to Section 6.4, apart from the fact that one needs to take into account the anisotropy of the problem here (since the Laplacian is anisotropic). The first term in (6.6.7), where $\mathcal{Q}_h$ appears, is a finite sum of terms of the form

$$\sum_{\substack{(k,n)\in\mathbf{Z}^6\\ k\in K(n)}} \Big(\widehat{u}^j_{\mathrm{osc}}(k)(k_h-n_h)(k_3-n_3)\widehat{u}^{j'}_{\mathrm{osc}}(n-k)$$

$$+ k_3\widehat{u}^j_{\mathrm{osc}}(k)(k_h-n_h)\widehat{u}^{j'}_{\mathrm{osc}}(n-k)\Big)n_3\widehat{u}^{j''}_{\mathrm{osc}}(n), \tag{6.6.8}$$

where j, j' and j'' are in $\{1,2,3\}$. We have denoted $K_n \stackrel{\mathrm{def}}{=} \cup_\sigma \mathcal{K}^\sigma_n$ with notation (6.2.5). Now we are going to show that $(\partial_3\mathcal{Q}_v(u_{\mathrm{osc}},u_{\mathrm{osc}})|\partial_3 u_{\mathrm{osc}})_{L^2}$ can be written in the same way. We have

$$\Big(\partial_3\mathcal{Q}_v(u_{\mathrm{osc}},u_{\mathrm{osc}})\Big|\partial_3 u_{\mathrm{osc}}\Big)_{L^2}$$

$$= -\lim_{\varepsilon\to 0}\left(\partial_3\left(\left(\mathcal{L}\left(\frac{t}{\varepsilon}\right)u_{\mathrm{osc}}\right)^3\partial_3\mathcal{L}\left(\frac{t}{\varepsilon}\right)u_{\mathrm{osc}}\right)\Big|\partial_3\mathcal{L}\left(\frac{t}{\varepsilon}\right)u_{\mathrm{osc}}\right)_{L^2}$$

and using the fact that $\mathcal{L}\left(\frac{t}{\varepsilon}\right)u_{\mathrm{osc}}$ is divergence-free, we can write on the one hand

$$\left(\partial_3\left(\mathcal{L}\left(\frac{t}{\varepsilon}\right)u_{\mathrm{osc}}\right)^3\partial_3\mathcal{L}\left(\frac{t}{\varepsilon}\right)u_{\mathrm{osc}}\Big|\partial_3\mathcal{L}\left(\frac{t}{\varepsilon}\right)u_{\mathrm{osc}}\right)_{L^2}$$

$$= -\left(\mathrm{div}_h\left(\mathcal{L}\left(\frac{t}{\varepsilon}\right)u_{\mathrm{osc}}\right)^h\partial_3\mathcal{L}\left(\frac{t}{\varepsilon}\right)u_{\mathrm{osc}}\Big|\partial_3\mathcal{L}\left(\frac{t}{\varepsilon}\right)u_{\mathrm{osc}}\right)_{L^2}$$

whereas, on the other hand, an integration by parts implies that

$$\left(\left(\mathcal{L}\left(\frac{t}{\varepsilon}\right)u_{\mathrm{osc}}\right)^3\partial_3^2\mathcal{L}\left(\frac{t}{\varepsilon}\right)u_{\mathrm{osc}}\Big|\partial_3\mathcal{L}\left(\frac{t}{\varepsilon}\right)u_{\mathrm{osc}}\right)_{L^2}$$

$$= -\frac{1}{2}\left(\partial_3\left(\mathcal{L}\left(\frac{t}{\varepsilon}\right)u_{\mathrm{osc}}\right)^3\partial_3\mathcal{L}\left(\frac{t}{\varepsilon}\right)u_{\mathrm{osc}}\Big|\partial_3\mathcal{L}\left(\frac{t}{\varepsilon}\right)u_{\mathrm{osc}}\right)_{L^2}.$$

So finally using once again the fact that $\mathcal{L}\left(\frac{t}{\varepsilon}\right)u_{\mathrm{osc}}$ is divergence-free we get

$$\Big(\partial_3\mathcal{Q}_v(u_{\mathrm{osc}},u_{\mathrm{osc}})\Big|\partial_3 u_{\mathrm{osc}}\Big)_{L^2}$$

$$= \frac{1}{2}\lim_{\varepsilon\to 0}\left(\mathrm{div}_h\left(\mathcal{L}\left(\frac{t}{\varepsilon}\right)u_{\mathrm{osc}}\right)^h\partial_3\mathcal{L}\left(\frac{t}{\varepsilon}\right)u_{\mathrm{osc}}\Big|\partial_3\mathcal{L}\left(\frac{t}{\varepsilon}\right)u_{\mathrm{osc}}\right)_{L^2}.$$

So we are back to the formulation (6.6.8). The following lemma shows how to estimate a sum of the type (6.6.8).

Lemma 6.6 *There is a constant C such that the following estimate holds: for all vector fields a, b, c of the form 7.0.5, we have*

$$\sum_{\substack{(k,n)\in\mathbf{Z}^6\\ k\in K_n}} |\widehat{a}(k)\widehat{b}(n-k)\widehat{c}(n)| \leq C\Big(\sum_{n\in\mathbf{Z}^3}(1+|n_h|^2)^{\frac{1}{2}}\,|\widehat{a}(n)|^2\Big)^{\frac{1}{2}}$$
$$\times\Big(\sum_{n\in\mathbf{Z}^3}|\widehat{b}(n)|^2\Big)^{\frac{1}{2}}\Big(\sum_{n\in\mathbf{Z}^3}(1+|n_h|^2)^{\frac{1}{2}}\,|\widehat{c}(n)|^2\Big)^{\frac{1}{2}}.$$

Remark This result is similar to Lemma 6.2, only the spaces appearing on the right-hand side are anisotropic-type Sobolev spaces. Contrary to the case of Lemma 6.2 and in order to avoid any additional complication, we shall not prove that result in any more generality (with a general "resonant set" and with more general Sobolev norms on the right-hand side).

Before proving Lemma 6.6, let us finish the proof of Proposition 6.5. We use Lemma 6.6 twice, once with $a = u_{\text{osc}}$, $b = \partial_3\nabla^h u_{\text{osc}}$ and $c = \partial_3 u_{\text{osc}}$, and once with $a = \partial_3 u_{\text{osc}}$, $b = \nabla^h u_{\text{osc}}$ and $c = \partial_3 u_{\text{osc}}$. We get

$$\begin{aligned}
&|(\partial_3 Q(u_{\text{osc}}, u_{\text{osc}})|\partial_3 u_{\text{osc}})_{L^2}|\\
&\quad\leq \Big(\|\partial_3 u_{\text{osc}}\|_{L^2} + \|\,|\nabla^h|^{\frac{1}{2}}\partial_3 u_{\text{osc}}\|_{L^2}\Big)^2\|\nabla^h u_{\text{osc}}\|_{L^2}\\
&\qquad+\Big(\|u_{\text{osc}}\|_{L^2} + \|\,|\nabla^h|^{\frac{1}{2}} u_{\text{osc}}\|_{L^2}\Big)\|\nabla^h\partial_3 u_{\text{osc}}\|_{L^2}\\
&\qquad\times\Big(\|\partial_3 u_{\text{osc}}\|_{L^2} + \||\nabla^h|^{\frac{1}{2}}\partial_3 u_{\text{osc}}\|_{L^2}\Big).
\end{aligned}$$

An easy computation using the interpolation inequality

$$\||\nabla^h|^{\frac{1}{2}}u_{\text{osc}}(t)\|_{L^2}^2 \leq \|u_{\text{osc}}(t)\|_{L^2}\|\nabla^h u_{\text{osc}}(t)\|_{L^2}, \tag{6.6.9}$$

yields the following estimate:

$$\begin{aligned}
&|(\partial_3 Q(u_{\text{osc}}, u_{\text{osc}})|\partial_3 u_{\text{osc}})_{L^2}|\\
&\quad\leq \frac{\nu}{4}\|\nabla^h\partial_3 u_{\text{osc}}\|_{L^2}^2 + \|\partial_3 u_{\text{osc}}\|_{L^2}^2\|\nabla^h u_{\text{osc}}\|_{L^2}\\
&\qquad+\frac{C}{\nu}\|\partial_3 u_{\text{osc}}\|_{L^2}^2(\|\nabla^h u_{\text{osc}}\|_{L^2}^2 + \|u_{\text{osc}}\|_{L^2}^2) + \frac{C}{\nu}\|u_{\text{osc}}\|_{L^2}^2\|\partial_3|\nabla^h|^{\frac{1}{2}}u_{\text{osc}}\|_{L^2}^2\\
&\qquad+\frac{C}{\nu}\|\partial_3 u_{\text{osc}}\|_{L^2}^2\||\nabla^h|^{\frac{1}{2}}u_{\text{osc}}\|_{L^2}^2 + \frac{C}{\nu}\|\partial_3|\nabla^h|^{\frac{1}{2}}u_{\text{osc}}\|_{L^2}^2\||\nabla^h|^{\frac{1}{2}}u_{\text{osc}}\|_{L^2}^2.
\end{aligned}$$

Using (6.6.9) once again yields

$$
\begin{aligned}
&|(\partial_3 Q(u_{\rm osc}, u_{\rm osc})|\partial_3 u_{\rm osc})_{L^2}| \\
&\quad \le \frac{\nu}{2}\|\nabla^h \partial_3 u_{\rm osc}\|_{L^2}^2 + \|\partial_3 u_{\rm osc}\|_{L^2}^2 \|\nabla^h u_{\rm osc}\|_{L^2} \\
&\qquad + \frac{C}{\nu}\|\partial_3 u_{\rm osc}\|_{L^2}^2 \left(\|\nabla^h u_{\rm osc}\|_{L^2}^2 + \|u_{\rm osc}\|_{L^2}^2 + \frac{C}{\nu^2}\|\partial_3 u_{\rm osc}\|_{L^2}^2 \|u_{\rm osc}\|_{L^2}^2\right).
\end{aligned}
$$

Going back to (6.6.3). we have finally

$$
\begin{aligned}
&\frac{d}{dt}\|\partial_3 u_{\rm osc}(t)\|_{L^2}^2 + \|\nabla^h \partial_3 u_{\rm osc}(t)\|_{L^2}^2 \\
&\quad \le \|\partial_3 u_{\rm osc}\|_{L^2}^2 \left(\|\nabla^h u_{\rm osc}\|_{L^2} + \frac{C}{\nu}(\|\nabla^h u_{\rm osc}\|_{L^2}^2 + \|u_{\rm osc}\|_{L^2}^2)\right) \\
&\qquad + \frac{C}{\nu^3}\|\partial_3 u_{\rm osc}\|_{L^2}^2 \|u_{\rm osc}\|_{L^2}^2.
\end{aligned}
$$

The result follows from the L^2 energy estimate (6.6.3) on $u_{\rm osc}$ thanks to the Gronwall lemma.

To end the proof of Proposition 6.5, we still need to prove (6.6.2). This result relies on some refined analysis of diophantine equations, and is beyond the scope of this book. We will therefore not prove the result, but refer to [49] where the proof can be found. Proposition 6.5 is proved. □

Proof of Lemma 6.6 This is, as in the case of Proposition 6.2, due to the special structure of the quadratic form Q. We recall to this end that if n and k_h are fixed, then there are at most eight integers k_3 such that k is in the resonant set K_n. So we can write, using the Cauchy–Schwarz inequality.

$$
\sum_{\substack{(k,n)\in \mathbf{Z}^6 \\ k\in K_n}} |\widehat{a}(k)\widehat{b}(n-k)\widehat{c}(n)| \le C \sum_{(k_h,n)\in \mathbf{Z}^5} |\widehat{c}(n)| \left(\sum_{k_3\in \mathbf{Z}} |\widehat{a}(k)|^2 |\widehat{b}(n-k)|^2\right)^{\frac12}
$$

which again by a Cauchy–Schwarz inequality yields

$$
\begin{aligned}
\sum_{\substack{(k,n)\in \mathbf{Z}^6 \\ k\in K_n}} |\widehat{a}(k)\widehat{b}(n-k)\widehat{c}(n)| \le C \sum_{(k_h,n_h)\in \mathbf{Z}^4} &\left(\sum_{n_3\in \mathbf{Z}} |\widehat{c}(n)|^2\right)^{\frac12} \\
&\times \left(\sum_{(k_3,n_3)\in \mathbf{Z}^2} |\widehat{a}(k)|^2 |\widehat{b}(n-k)|^2\right)^{\frac12}.
\end{aligned}
$$

We deduce that

$$\sum_{\substack{(k,n)\in\mathbf{Z}^6\\ k\in K_n}} |\widehat{a}(k)\widehat{b}(n-k)\widehat{c}(n)|$$

$$\leq C \sum_{(k_h,n_h)\in\mathbf{Z}^4} \Big(\sum_{n_3\in\mathbf{Z}} |\widehat{c}(n)|^2\Big)^{\frac12} \Big(\sum_{k_3} |\widehat{a}(k)|^2\Big)^{\frac12} \Big(\sum_{\ell_3} |\widehat{b}(n_h-k_h,\ell_3)|^2\Big)^{\frac12}.$$

Two-dimensional product rules enable us to write, for all two-dimensional vector fields A, B and C

$$\sum_{(k_h,n_h)\in\mathbf{Z}^4} |\widehat{A}(n_h)\widehat{B}(n_h-k_h)\widehat{C}(k_h)|$$

$$\leq C\left(\sum_{n_h\in\mathbf{Z}^2} (1+|n_h|^2)^{\frac12}|\widehat{A}(n_h)|^2\right)^{\frac12}$$

$$\times\left(\sum_{n_h\in\mathbf{Z}^2} |\widehat{B}(n_h)|^2\right)^{\frac12}\left(\sum_{n_h\in\mathbf{Z}^2} (1+|n_h|^2)^{\frac12}|\widehat{C}(n_h)|^2\right)^{\frac12}$$

so the result follows, taking $\widehat{A}(n_h) = \big(\sum_{n_3\in\mathbf{Z}} |\widehat{a}(n)|^2\big)^{\frac12}$ and similarly for B and C. Lemma 6.6 is proved. □

Note that in particular Lemma 6.6 implies the following result.

Proposition 6.6 *The quadratic form $\mathcal{Q}$ satisfies, for any divergence-free vector field a,*

$$\|\mathcal{Q}(a,a)\|_{H^{-1,0}} \leq C\|a\|_{L^2}\|a\|_{H^{1,0}} + \|\partial_3 a\|_{L^2}\|a\|_{L^2}^{\frac12}\|a\|_{H^{1,0}}^{\frac12}.$$

In particular $\mathcal{Q}$ is bilinear continuous from

$$L^\infty([0,T];H^{0,1})\times L^2([0,T];H^{1,0}) \quad \textit{into} \quad L^2([0,T];H^{-1,0}).$$

Proof As in (6.6.5) and (6.6.6) let us decompose

$$\mathcal{Q}(a,a) = \mathcal{Q}_h(a,a) + \mathcal{Q}_v(a,a).$$

Let $b\in H^{1,0}$ be given. The divergence-free condition on a implies that

$$(\mathcal{Q}(a,a)|b) \leq \sum_{k\in K_n^\sigma} |\widehat{a}(k)||\widehat{a}(n-k)||n_h||\widehat{b}(n)| + \sum_{k\in K_n^\sigma} |\widehat{a}(k)||\widehat{a}(n-k)|n_3|\widehat{b}(n)|.$$

Let us start with the first term. We can write

$$|(\mathcal{Q}_h(a,a)|b)| \leq \sum_{k\in K_n^\sigma} |\widehat{a}(k)||\widehat{a}(n-k)||n_h||\widehat{b}(n)|,$$

so Lemma 6.6 implies that

$$|(\mathcal{Q}_h(a,a)|b)| \leq C \left(\sum_{n\in\mathbf{Z}^3} (1+|n_h|^2)^{\frac{1}{2}} |\widehat{a}(n)|^2 \right) \left(\sum_{n\in\mathbf{Z}^3} (1+|n_h|^2) |\widehat{b}(n)|^2 \right)^{\frac{1}{2}}.$$

By the Gagliardo–Nirenberg inequality we get

$$|(\mathcal{Q}(a,a)|b)| \leq C \|a\|_{L^2} \|a\|_{H^{1,0}} \|b\|_{H^{1,0}}.$$

For $\mathcal{Q}_v$ we write

$$\begin{aligned} |(\mathcal{Q}_v(a,a)|b)| \leq & \sum_{k\in K_n^\sigma} k_3 |\widehat{a}(k)| |\widehat{a}(n-k)| |\widehat{b}(n)| \\ & + \sum_{k\in K_n^\sigma} |\widehat{a}(k)| |n_3 - k_3 |\widehat{a}(n-k)| |\widehat{b}(n)|, \end{aligned}$$

and Lemma 6.6 along with the Gagliardo–Nirenberg inequality implies that

$$|(\mathcal{Q}_v(a,a)|b)| \leq C \|\partial_3 a\|_{L^2} \|a\|_{L^2}^{\frac{1}{2}} \|a\|_{H^{1,0}}^{\frac{1}{2}} \|b\|_{L^2}^{\frac{1}{2}} \|b\|_{H^{1,0}}^{\frac{1}{2}},$$

and Proposition 6.6 is proved. □

7

Ekman boundary layers for rotating fluids

In this chapter, we investigate the problem of rapidly rotating viscous fluids between two horizontal plates with Dirichlet boundary conditions. We present the model with so-called "turbulent" viscosity. More precisely, we shall study the limit when ε tends to 0 of the system

$$(\mathrm{NSC}_\varepsilon)\quad \begin{cases} \partial_t u^\varepsilon + \operatorname{div}(u^\varepsilon \otimes u^\varepsilon) - \nu\Delta_h u^\varepsilon - \beta\varepsilon\partial_3^2 u^\varepsilon + \dfrac{e^3 \wedge u^\varepsilon}{\varepsilon} = -\nabla p^\varepsilon \\ \operatorname{div} u^\varepsilon = 0 \\ u^\varepsilon_{|\partial\Omega} = 0 \\ u^\varepsilon_{|t=0} = u_0^\varepsilon, \end{cases}$$

where $\Omega = \Omega_h \times]0,1[$: here Ω_h will be the torus $\mathbf{T}^2$ or the whole plane $\mathbf{R}^2$. We shall use, as in the previous chapters, the following notation: if u is a vector field on Ω we state $u = (u^h, u^3)$. In all that follows, we shall assume that on the boundary $\partial\Omega$, $u_0^\varepsilon \cdot n = u_0^{\varepsilon,3} = 0$, and that $\operatorname{div} u_0^\varepsilon = 0$. The condition $u_0^3 = 0$ on the boundary implies the following fact: for any vector field $u \in \mathcal{H}(\Omega)$, the function $\partial_3 u^3$ is $L^2(]0,1[)$ with respect to the variable x_3 with values in $H^{-1}(\Omega_h)$ due to the divergence-free condition. So by integration, we get

$$u^3(x_h, 1) - u^3(x_h, 0) = -\int_0^1 \operatorname{div}_h u^h(x_h, x_3)dx_3 = 0. \tag{7.0.1}$$

So the vertical mean value of the horizontal part of the vector field is divergence-free as a vector field on Ω_h.

We proved in Chapter 2 that a weak solution u^ε exists such that if

$$E_t^\varepsilon(v) \stackrel{\text{def}}{=} \|v(t)\|_{L^2}^2 + 2\nu\int_0^t \|\nabla v(t')\|_{L^2}^2 dt' + 2\beta\varepsilon\int_0^t \|\partial_3 v(t')\|_{L^2}^2 dt', \tag{7.0.2}$$

then we have

$$E_t^\varepsilon(u^\varepsilon) \leq \|u_0\|_{L^2}^2. \tag{7.0.3}$$

It follows that the family $(u^\varepsilon)_{\varepsilon>0}$ has some weak limit points, for instance in the space $L^2_{\mathrm{loc}}(\mathbf{R}^+ \times \Omega)$, and as seen in the introduction in Part I they do not depend on the vertical variable. Let us recall the arguments. Let $\overline{u}$ be in the weak closure of u^ε. Obviously, the term

$$\partial_t u^\varepsilon + \operatorname{div}(u^\varepsilon \otimes u^\varepsilon) - \nu\Delta_h u^\varepsilon - \beta\varepsilon\partial_3^2 u^\varepsilon$$

is, up to an extraction of a subsequence, convergent in $\mathcal{D}'(\mathbf{R}^+ \times \Omega)$. Thus this implies that

$$e_3 \wedge \overline{u} = -\nabla \overline{p}$$

which can be written

$$\begin{pmatrix} -\overline{u}^2 \\ \overline{u}^1 \\ 0 \end{pmatrix} = \begin{pmatrix} -\partial_1 \overline{p} \\ -\partial_2 \overline{p} \\ -\partial_3 \overline{p} \end{pmatrix},$$

from which we deduce that $\partial_3 \overline{u}^h = 0$ and $\mathrm{div}_h \, \overline{u}^h = 0$. As $\overline{u}$ is a divergence-free vector field on the three-dimensional domain Ω, then $\partial_3 \overline{u}^3 = 0$.

In the following we will also define

$$E_T(v) \stackrel{\mathrm{def}}{=} \sup_{t \in [0,T]} \|v(t)\|^2_{L^2(\Omega)} + 2\nu \int_0^T \|\nabla^h v(t)\|^2_{L^2(\Omega)} dt.$$

The aim of this chapter is to study the asymptotics of the family $(u^\varepsilon)_{\varepsilon > 0}$ as ε goes to zero. The easiest situation occurs when the initial data do not depend on the vertical variable x_3 (which corresponds to the so-called "well-prepared" case). The precise theorem in the well-prepared case is the following. It will be proved in Section 7.3, using a fundamental linear lemma stated and proved in Section 7.1. Let us first introduce space with anisotropic regularity.

Definition 7.1 *We shall denote by $H^{1,0}(\Omega)$ (or by $H^{1,0}$ when no confusion is possible) the space of L^2 functions u on Ω such that $\nabla^h u$ also belongs to $L^2(\Omega)$.*

Theorem 7.1 *Let $(u^\varepsilon)_{\varepsilon > 0}$ be a family of weak solutions of* $(\mathrm{NSC}_\varepsilon)$ *associated with a family of initial data $u_0^\varepsilon \in \mathcal{H}(\Omega)$ such that*

$$\lim_{\varepsilon \to 0} u_0^\varepsilon = (\overline{u}_0^h, 0) \quad in \quad \mathcal{H}(\Omega),$$

where $\overline{u}_0^h$ is a horizontal vector field in $\mathcal{H}(\Omega_h)$. Denoting by $\overline{u}$ the global solution of the two-dimensional Navier–Stokes-type equations on Ω_h

$$(\mathrm{NSE}_{\nu,\beta}) \begin{cases} \partial_t \overline{u} + \mathrm{div}_h(\overline{u} \otimes \overline{u}) - \nu \Delta_h \overline{u} + \sqrt{2\beta}\, \overline{u} = -\nabla^h \overline{p} \\ \mathrm{div}_h \, \overline{u} = 0 \\ \overline{u}_{|t=0} = \overline{u}_0^h, \end{cases}$$

then u^ε goes to $(\overline{u}, 0)$ as ε goes to zero, in the space

$$L^\infty_{\mathrm{loc}}(\mathbf{R}_+; L^2(\Omega)) \cap L^2_{\mathrm{loc}}(\mathbf{R}_+; H^{1,0}).$$

Remark We will omit the proof that there is a unique global solution to the system $(\mathrm{NSE}_{\nu,\beta})$ with initial data $\overline{u}_0^h$ in $\mathcal{H}(\Omega_h)$, as it is exactly the same proof as for the two-dimensional Navier–Stokes equations studied in Part II, Section 2.2 and Chapter 3. In particular the solution $\overline{u}$ belongs to the space

$$C_b^0(\mathbf{R}^+, \mathcal{H}(\Omega_h)) \cap L^2(\mathbf{R}^+, \mathcal{V}_\sigma(\Omega_h)),$$

and for all $t \geq 0$,

$$\frac{1}{2}\|\overline{u}(t)\|_{L^2}^2 + \nu \int_0^t \|\nabla\overline{u}(t')\|_{L^2}^2\, dt' + \sqrt{2\beta} \int_0^t \|\overline{u}(t')\|_{L^2}^2\, dt' = \frac{1}{2}\|\overline{u}_0\|_{L^2}^2. \quad (7.0.4)$$

We shall investigate the more delicate case when the initial data u_0 do actually depend on the vertical variable x_3 (the so-called "ill-prepared" case) in Section 7.4, where we shall compute explicitly approximate solutions for the linear problem. The study of the full non-linear problem depends strongly on the domain Ω. In the case when $\Omega = \mathbf{R}^2 \times]0,1[$, the result is the following, proved in Section 7.6.

Theorem 7.2 *Let u_0 be a vector field in $\mathcal{H}(\mathbf{R}^2 \times]0,1[)$. Let us consider a family $(u^\varepsilon)_{\varepsilon>0}$ of weak solutions of $(\mathrm{NSC}_\varepsilon)$ associated with the initial data u_0. Denoting by $\overline{u}$ the global solution of the two-dimensional Navier–Stokes-type equations $(\mathrm{NSE}_{\nu,\beta})$ with initial data*

$$\overline{u}_{|t=0}(x_h) \stackrel{\text{def}}{=} \int_0^1 u_0^h(x_h, x_3)dx_3,$$

we have, for any positive time T and any compact subset K of $\mathbf{R}^2 \times]0,1[$,

$$\lim_{\varepsilon\to 0} \int_{[0,T]\times K} |u^\varepsilon(t,x) - (\overline{u}(t,x_h), 0)|^2\, dx_h dx_3 = 0.$$

The key point of the proof of this theorem is that the dispersive phenomena studied in Chapter 5 are not affected by the Ekman boundary layer.

When the domain Ω_h is periodic, the result is different, due to the absence of dispersion; in particular the methods of Chapter 6 will be used here to deal with the periodic horizontal boundary conditions. We will therefore be using some notation of the purely periodic case: in the coming theorem, $\mathcal{L}$ is the filtering operator of Chapter 6, and $\mathcal{Q}$ is the quadratic form defined in Proposition 6.1, page 124. Before stating the result let us define the periodic setting: the horizontal torus will be defined as

$$\mathbf{T}^2 \stackrel{\text{def}}{=} \prod_{j=1}^{2} \mathbf{R}_{/a_j}\mathbf{Z},$$

where the a_j's are positive real numbers. We will see that solving the problem in Ω with Dirichlet boundary conditions corresponds to considering vector fields having the following symmetries:

$$u(x_h, x_3) = (u^h(x_h, -x_3), -u^3(x_h, -x_3)).$$

These symmetry properties are clearly preserved by the rotating-fluid equations. Such vector fields will be decomposed in the following way:

$$u(x) = \sum_{k\in\mathbf{N}^3} \begin{pmatrix} u^{k,h} e^{i\pi k_h\cdot x_h} \cos(k_3\pi x_3) \\ -\dfrac{i}{k_3\pi} k_h \cdot u^{k,h} e^{i\pi k_h\cdot x_h} \sin(k_3\pi x_3) \end{pmatrix}. \tag{7.0.5}$$

This form is very similar to the decomposition of periodic vector fields used in Chapter 6, and in particular the Fourier components are

$$(u^{k,1}, u^{k,2}, u^{k,3}) \quad \text{with} \quad u^{k,3} = -\frac{i}{k_3\pi} k_h \cdot u^{k,h}.$$

Comparing the vertical Fourier transform defined in (7.0.5) with the definition of the classical Fourier transform (6.1.1), page 117, shows that with the notation of Chapter 6 we have $a_3 = 2$, so as in Chapter 6 we will define

$$\widetilde{k} \stackrel{\text{def}}{=} \left(\frac{k_1}{a_1}, \frac{k_2}{a_2}, \frac{k_3}{a_3}\right) \quad \text{with} \quad a_3 = 2.$$

The theorem proved in Section 7.7 is the following. We have denoted by $H^{0,1}$ the space of functions $f \in L^2(\Omega)$ such that $\partial_3 f$ is in $L^2(\Omega)$, and $\mathcal{H}_0$ the space of vector fields in $\mathcal{H}$ with vanishing horizontal mean.

Theorem 7.3 *A non-negative, continuous operator $\mathcal{E}$ on $\mathcal{H}_0 \cap H^{0,1}(\Omega)$ exists (it is defined by (7.4.39) below) such that the following results hold. Let us consider initial data u_0 in $\mathcal{H}_0 \cap H^{0,1}$, and let u be the associated solution of the system*

$$(\text{NSE}_{\nu,\mathcal{E}}) \begin{cases} \partial_t u - \nu\Delta_h u + \mathcal{E} u + \mathcal{Q}(u,u) = 0 \\ \operatorname{div} u = 0 \\ u_{|t=0} = u_0 \end{cases}$$

given by Proposition 6.5, page 145. If $\mathbf{T}^2$ satisfies condition $(\mathcal{A})$ in the sense of Definition 6.2, then the following result is true. Let $(u^\varepsilon)_{\varepsilon>0}$ be a family of weak solutions of (NSC_ε) associated with the initial data u_0. Then for all $T > 0$,

$$\lim_{\varepsilon\to 0} E_T\left(u^\varepsilon - \mathcal{L}\left(\frac{t}{\varepsilon}\right) u\right) = 0.$$

Note that the structure of the limiting equation in the periodic case is very different from the case of the whole space. The filtering operator acts like the identity on two-dimensional vector fields, and that is why it does not appear in the statement of Theorem 7.3, as the three-dimensional vector fields disappear due to dispersion. In the same way, the operator $\mathcal{E}$ acts like $\sqrt{2\beta}\,\text{Id}$ on two-dimensional vector fields. In the case of $\mathbf{T}^2 \times]0,1[$, the above theorem generalizes Theorem 7.1. Here three-dimensional vector fields remain in the limit, since u is no longer two dimensional. We recall (see Section 6.6, page 144) that the fact that the periodic box satisfies condition $(\mathcal{A})$ ensures that the horizontal mean of the solution of the limit system is preserved – a vanishing horizontal mean is needed to construct the boundary layers in Section 7.4.

7.1 The well-prepared linear problem

The goal of this section is to construct approximate solutions to the linear system (Stokes–Coriolis equations)

$$(\mathrm{SC}_\varepsilon)\begin{cases}\partial_t v^\varepsilon - \nu\Delta_h v^\varepsilon - \nu_V\partial_3^2 v^\varepsilon + \dfrac{e_3\wedge v^\varepsilon}{\varepsilon} = -\nabla p^\varepsilon\\ \operatorname{div} v^\varepsilon = 0\\ v^\varepsilon_{|t=0} = v_0,\end{cases}$$

when v_0 belongs to $\mathcal{H}(\Omega_h)$.

Let us denote by R the rotation of angle $\pi/2$ in the (x_1, x_2) plane and by L^ε the operator

$$L^\varepsilon w \stackrel{\text{def}}{=} \begin{pmatrix}\partial_t w^h - \nu\Delta_h w^h + \dfrac{Rw^h}{\varepsilon} - \nu_V\partial_3^2 w^h\\ \partial_t w^3 - \nu\Delta_h w^3 - \nu_V\partial_3^2 w^3\end{pmatrix}.$$

The system $(\mathrm{SC}_\varepsilon)$ can be rewritten as

$$(\mathrm{SC}_\varepsilon)\begin{cases}L^\varepsilon v^\varepsilon = -\nabla p^\varepsilon\\ \operatorname{div} v^\varepsilon = 0\\ v^\varepsilon_{|t=0} = v_0.\end{cases}$$

We will assume that ν_V depends on the parameter ε, and that these two parameters go to 0. It is of course possible to investigate this double limit without any further assumption, however the only interesting regime arises when ν_V/ε converges to some positive constant β. To shorten the study we have therefore chosen to restrict ourselves to the case when

$$\nu_V = \beta\varepsilon, \tag{7.1.1}$$

with $\beta > 0$, which is a physically relevant regime.

We immediately note that the Taylor–Proudman theorem (i.e. the limit is independent of the vertical variable) is not compatible with Dirichlet conditions (except in the degenerate case when $v_0 \equiv 0$). As usual in these situations, boundary layers appear near $x_3 = 0$ and $x_3 = 1$.

We want to do the following: a horizontal two-dimensional divergence-free vector field v being given, we want to construct a family of approximate solutions $(v^\varepsilon_{\mathrm{app}})_{\varepsilon>0}$ and an operator $\underline{L}$ such that $L^\varepsilon v^\varepsilon_{\mathrm{app}} = \underline{L}v$ up to small remainder terms (i.e. which will tend to 0 when ε tends to 0.) As we shall see later on, the operator $\underline{L}$ is related to the system $(\mathrm{NSE}_{\nu,\beta})$.

A typical approach is to look for approximate solutions of the form (the Ansatz)

$$\begin{aligned} v^\varepsilon &= v_{0,\mathrm{int}} + v_{0,\mathrm{BL}} + \varepsilon v_{1,\mathrm{int}} + \varepsilon v_{1,\mathrm{BL}} + \cdots,\\ p^\varepsilon &= \frac{1}{\varepsilon}p_{-1,\mathrm{int}} + \frac{1}{\varepsilon}p_{-1,\mathrm{BL}} + p_{0,\mathrm{int}} + p_{0,\mathrm{BL}} + \cdots, \end{aligned} \tag{7.1.2}$$

where "int" stands for "interior", namely smooth functions of (x_h, x_3). and "BL" stands for "boundary layers": smooth functions of the form

$$(x_h, x_3) \mapsto f\left(t, x_h, \frac{x_3}{\delta}\right) + g\left(t, x_h, \frac{1-x_3}{\delta}\right).$$

where $f(x_h, \zeta)$ and $g(x_h, \zeta)$ decrease rapidly at infinity in ζ. In that expression δ, which goes to 0 in the limiting process, denotes the size of the boundary layer. We will prove below that δ is of order $\sqrt{\nu_V \varepsilon} = \varepsilon\sqrt{\beta}$.

Now we plug the Ansatz (7.1.2) into the above system $(\mathrm{SC}_\varepsilon)$ and identify powers of ε, functions of the type "int" and "BL", keeping in mind that the divergence-free condition and the Dirichlet boundary condition must be satisfied.

First, the third component of $(\mathrm{SC}_\varepsilon)$ at order $\varepsilon^{-1}\delta^{-1}$ gives in the boundary layer

$$\partial_3 p_{-1,\mathrm{BL}} = 0.$$

As, by definition, $p_{-1,\mathrm{BL}}$ goes to 0 as ζ goes to infinity, we deduce that

$$p_{-1,\mathrm{BL}} = 0.$$

We recover a classical principle of fluid mechanics which claims that the pressure does not vary in a boundary layer.

Next, horizontal components of $(\mathrm{SC}_\varepsilon)$ to leading order in the boundary layer yield

$$-\varepsilon\beta\partial_3^2 v_{0,\mathrm{BL}}^h + \frac{R v_{0,\mathrm{BL}}^h}{\varepsilon} = 0. \tag{7.1.3}$$

Note that $\partial_3^2 v_{0,\mathrm{BL}}^h$ is of order δ^{-2}, hence the previous equation indicates that $\nu_V \delta^{-2}$ must be of order ε. We thus define

$$\delta = \sqrt{2\nu_V \varepsilon} = \varepsilon\sqrt{2\beta},$$

the coefficient $\sqrt{2}$ being considered for algebraic simplicity. In physics, it is usual to introduce the Ekman number

$$E \stackrel{\mathrm{def}}{=} 2\varepsilon\nu_V = 2\varepsilon^2\beta.$$

The boundary layer size δ is then simply $\sqrt{E}$. Let us search for $v_{0,\mathrm{BL}}^h$ of the form

$$(x_h, x_3) \mapsto M_0\left(\frac{x_3}{\sqrt{E}}\right) v_{0,\mathrm{int}}^h + M_0\left(\frac{1-x_3}{\sqrt{E}}\right) v_{0,\mathrm{int}}^h,$$

where M_0 is a 2×2 matrix with real-valued coefficients. Then, equation (7.1.3) implies that M_0 satisfies the following ordinary differential equation

$$M_0'' = 2RM_0 \quad \text{with} \quad M_0(0) = -\mathrm{Id} \quad \text{and} \quad M_0(+\infty) = 0, \tag{7.1.4}$$

which gives

$$M_0(\zeta) = -e^{-\zeta} R_{-\zeta}, \tag{7.1.5}$$

where R_α denotes rotation by the angle α. So we have

$$v_{0,\mathrm{BL}} = \left(\left(M_0\left(\frac{x_3}{\sqrt{E}}\right) + M_0\left(\frac{1-x_3}{\sqrt{E}}\right)\right) v^h_{0,\mathrm{int}}, 0\right). \tag{7.1.6}$$

Note that

$$L^\varepsilon v_{0,\mathrm{BL}} = (\partial_t v_{0,\mathrm{int}} - \nu\Delta_h v_{0,\mathrm{int}}, 0) \quad \text{and} \tag{7.1.7}$$

$$(v_{0,\mathrm{int}} + v_{0,\mathrm{BL}})_{|\partial\Omega} = \left(M_0\left(\frac{1}{\sqrt{E}}\right) v^h_{0,\mathrm{int}}, 0\right) \tag{7.1.8}$$

which means that the boundary condition is satisfied to leading order, up to exponentially small terms that will be treated later on.

Let us continue with the study of the Ansatz. The third component of $(\mathrm{SC}_\varepsilon)$ in the interior, to order ε^{-1}, gives

$$\partial_3 p_{-1,\mathrm{int}} = 0,$$

hence $p_{-1,\mathrm{int}}$ only depends on the horizontal coordinates. The horizontal components of the same equation lead to

$$R v^h_{0,\mathrm{int}} = -\nabla^h p_{-1,\mathrm{int}}.$$

Therefore

$$\begin{pmatrix} -v^2_{0,\mathrm{int}} \\ v^1_{0,\mathrm{int}} \end{pmatrix} = \begin{pmatrix} \partial_1 p_{-1,\mathrm{int}} \\ \partial_2 p_{-1,\mathrm{int}} \end{pmatrix}.$$

This implies, using the incompressibility of $v_{0,\mathrm{int}}$, that $v^h_{0,\mathrm{int}}$ does not depend on the vertical variable x_3, that $v^3_{0,\mathrm{int}} = 0$, and that

$$\mathrm{div}_h\, v^h_{0,\mathrm{int}} = 0, \tag{7.1.9}$$

which is exactly the Taylor–Proudman theorem.

The next step is to ensure the incompressibility condition in the boundary layer, namely

$$\mathrm{div}_h\, v^h_{0,\mathrm{BL}} + \varepsilon\partial_3 v^3_{1,\mathrm{BL}} = 0. \tag{7.1.10}$$

As $v_{0,\mathrm{BL}}$ is explicitly known as a function of $v^h_{0,\mathrm{int}}$, it is possible to solve (7.1.10) in order to compute $v^3_{1,\mathrm{BL}}$. A simple integration of (7.1.6) gives

$$v^3_{1,\mathrm{BL}} = \sqrt{2\beta}\left(f\left(\frac{x_3}{\sqrt{E}}\right) - f\left(\frac{1-x_3}{\sqrt{E}}\right)\right) \mathrm{curl}_h v^h_{0,\mathrm{int}}, \tag{7.1.11}$$

where

$$f(\zeta) = -\frac{1}{2} e^{-\zeta}(\sin\zeta + \cos\zeta).$$

Now we turn to $v^3_{1,\mathrm{int}}$. The boundary values of $v^3_{1,\mathrm{BL}}$ at $x_3 = 0$ and $x_3 = 1$ provide the boundary condition for $v^3_{1,\mathrm{int}}$, namely

$$v^3_{1,\mathrm{int}} = \frac{1}{2}\sqrt{2\beta}\,\mathrm{curl}_h v^h_{0,\mathrm{int}} \quad \text{at} \quad x_3 = 0 \quad \text{and} \tag{7.1.12}$$

$$v^3_{1,\mathrm{int}} = -\frac{1}{2}\sqrt{2\beta}\,\mathrm{curl}_h v^h_{0,\mathrm{int}} \quad \text{at} \quad x_3 = 1. \tag{7.1.13}$$

We recall that under (7.1.1), $\sqrt{E}/\varepsilon$ is a constant, namely

$$\frac{\sqrt{E}}{\varepsilon} = \sqrt{2\beta}.$$

Let us lift those boundary conditions by a vector field $v_{1,\mathrm{int}}$ over the interior domain. A natural choice is

$$v^3_{1,\mathrm{int}} = \sqrt{2\beta}\left(\frac{1}{2} - x_3\right)\mathrm{curl}_h v^h_{0,\mathrm{int}}, \tag{7.1.14}$$

which, completed with

$$v^h_{1,\mathrm{int}} = \sqrt{2\beta}\,R_{-\frac{\pi}{2}} v^h_{0,\mathrm{int}}, \tag{7.1.15}$$

leads to a divergence-free vector field $v_{1,\mathrm{int}}$. Let us note that, as in (7.1.8), we have exponentially small error terms at the boundary, namely

$$\left(v^3_{1,\mathrm{BL}} + v^3_{1,\mathrm{int}}\right)_{|\partial\Omega} = (-1)^{1-x_3}_{|\partial\Omega}\sqrt{2\beta}\, f\left(\frac{1}{\sqrt{E}}\right)\mathrm{curl}_h v^h_{0,\mathrm{int}}. \tag{7.1.16}$$

Next, (SC_ε) to order $O(1)$ in the interior gives

$$\left\{\begin{array}{l} \partial_t v^h_{0,\mathrm{int}} - \nu\Delta_h v^h_{0,\mathrm{int}} + R v^h_{1,\mathrm{int}} = -\nabla^h p_{0,\mathrm{int}} \\ \partial_t v^3_{0,\mathrm{int}} - \nu\Delta_h v^3_{0,\mathrm{int}} = -\partial_3 p_{0,\mathrm{int}}. \end{array}\right. \tag{7.1.17}$$

Taking the two-dimensional curl of the first two equations gives

$$\partial_t \mathrm{curl}_h v^h_{0,\mathrm{int}} - \nu\Delta_h \mathrm{curl}_h v^h_{0,\mathrm{int}} = -\mathrm{div}_h v^h_{1,\mathrm{int}} = \partial_3 v^3_{1,\mathrm{int}}.$$

The left-hand side being independent of x_3, so is the right-hand side, and integration in x_3 over $]0,1[$ gives

$$\partial_t \mathrm{curl}_h v^h_{0,\mathrm{int}} - \nu\Delta_h \mathrm{curl}_h v^h_{0,\mathrm{int}} = v^3_{1,\mathrm{int}}(x_h, 1) - v^3_{1,\mathrm{int}}(x_h, 0),$$

namely

$$\partial_t \mathrm{curl}_h v^h_{0,\mathrm{int}} - \nu\Delta_h \mathrm{curl}_h v^h_{0,\mathrm{int}} + \sqrt{2\beta}\,\mathrm{curl}_h v^h_{0,\mathrm{int}} = 0. \tag{7.1.18}$$

Note that (7.1.17), combined with (7.1.18), completely determines $v^h_{1,\mathrm{int}}$, up to two-dimensional divergence-free vector fields. To go further we need to enforce an asymptotic expansion for the initial value itself. Here we take u_0 as the initial data, which do not depend on ε. This allows us to go on with the construction, in theory, up to any order. For our purpose, however, it is sufficient to stop the construction at this step, up to small technicalities.

Before dealing with those, let us sum up what we have done:

- We first ensured the boundary condition by introducing $v_{0,\mathrm{BL}}$.
- Then, as $v_{0,\mathrm{BL}}$ violated the divergence-free condition, we introduced $v_{1,\mathrm{BL}}$ in order to ensure the divergence-free condition.
- The introduction of $v_{1,\mathrm{BL}}$ destroyed the boundary condition which we restored by introducing $v_{1,\mathrm{int}}$.

Let us deal with the final technicalities.

- As shown by (7.1.8), the Dirichlet boundary condition is not exactly satisfied.
- The horizontal component of the vector field $v_{1,\mathrm{int}}$ defined in (7.1.15) does not vanish on the boundary.

In order to deal with the second point, let us state

$$\begin{aligned}
v^h_{1,\mathrm{BL}} &\stackrel{\mathrm{def}}{=} -\sqrt{2\beta}\left(\exp\left(\frac{-x_3}{\sqrt{E}}\right)+\exp\left(\frac{-(1-x_3)}{\sqrt{E}}\right)\right)R_{-\frac{\pi}{2}}v^h_{0,\mathrm{int}}\,.\\
v^h_{2,\mathrm{BL}} &\stackrel{\mathrm{def}}{=} 0 \\
v^3_{2,\mathrm{BL}} &\stackrel{\mathrm{def}}{=} -2\beta\left(\exp\left(\frac{-x_3}{\sqrt{E}}\right)+\exp\left(-\frac{(1-x_3)}{\sqrt{E}}\right)\right)\mathrm{curl}_h v^h_{0,\mathrm{int}}.
\end{aligned}\tag{7.1.19}$$

Then, using (7.1.8), (7.1.10), and (7.1.16), we get that

$$\begin{aligned}
&\mathrm{div}\left(v_{0,\mathrm{int}}+v_{0,\mathrm{BL}}+\varepsilon v_{1,\mathrm{int}}+\varepsilon v_{1,\mathrm{BL}}+\varepsilon^2 v_{2,\mathrm{BL}}\right)=0 \quad \text{and}\\
&\left(v_{0,\mathrm{int}}+v_{0,\mathrm{BL}}+\varepsilon v_{1,\mathrm{int}}+\varepsilon v_{1,\mathrm{BL}}+\varepsilon^2 v_{2,\mathrm{BL}}\right)_{|\partial\Omega}=v_R
\end{aligned}$$

where v_R is defined by

$$\begin{aligned}
v_R \stackrel{\mathrm{def}}{=} \Bigg(& M_0\left(\frac{1}{\sqrt{E}}\right)v^h_{0,\mathrm{int}}-\sqrt{E}\exp\left(\frac{-1}{\sqrt{E}}\right)R_{-\frac{\pi}{2}}v^h_{0,\mathrm{int}},\\
&(-1)^{1-x_3}_{|\Omega}\sqrt{E}\,f\left(-\frac{1}{\sqrt{E}}\right)\mathrm{curl}_h v^h_{0,\mathrm{int}}+E\exp\left(-\frac{1}{\sqrt{E}}\right)\mathrm{curl}_h v^h_{0,\mathrm{int}}\Bigg).
\end{aligned}$$

Let us define $v^\varepsilon_{\rm exp}$ by

$$\begin{aligned} v^\varepsilon_{\rm exp} \stackrel{\rm def}{=} & \left(-\cos(2\pi x_3) v^h_R, \frac{\sin(2\pi x_3)}{2\pi} \operatorname{div}_h v^h_R\right) \\ & + \sqrt{E} f\left(\frac{1}{\sqrt{E}}\right)\left(\pi \sin(\pi x_3) R_{-\frac{\pi}{2}} v^h_{0,\rm int}, \cos(\pi x_3) \operatorname{curl}_h v^h_{0,\rm int}\right) \\ & + E \exp\left(-\frac{1}{\sqrt{E}}\right)\left(2\pi \sin(2\pi x_3) R_{-\frac{\pi}{2}} v^h_{0,\rm int}, \cos(2\pi x_3) \operatorname{curl}_h v^h_{0,\rm int}\right). \end{aligned}$$

Then it is obvious that

$$\begin{aligned} & \operatorname{div}\left(v_{0,\rm int} + v_{0,\rm BL} + \varepsilon v_{1,\rm int} + \varepsilon v_{1,\rm BL} + \varepsilon^2 v_{2,\rm BL} + v^\varepsilon_{\rm exp}\right) = 0 \quad \text{and} \\ & \left(v_{0,\rm int} + v_{0,\rm BL} + \varepsilon v_{1,\rm int} + \varepsilon v_{1,\rm BL} + \varepsilon^2 v_{2,\rm BL} + v^\varepsilon_{\rm exp}\right)_{|\partial\Omega} = 0. \end{aligned} \tag{7.1.20}$$

Now we are ready to state the fundamental approximation lemma. We define $H^{s,0}$ to be the space of vector fields f satisfying

$$\|f\|^2_{H^{s,0}} \stackrel{\rm def}{=} \int_0^1 \|f(\cdot, x_3)\|^2_{H^s(\Omega_h)}\, dx_3 < +\infty. \tag{7.1.21}$$

Let us define the limit Stokes–Ekman system

$$(\mathrm{SE}_{\nu,\beta}) \quad \begin{cases} \partial_t \overline{v} - \nu \Delta_h \overline{v} + \sqrt{2\beta}\, \overline{v} = \overline{f} - \nabla^h p \\ \operatorname{div}_h \overline{v} = 0 \\ \overline{v}_{|t=0} = \overline{v}_0 . \end{cases}$$

Lemma 7.1 *Let T be in $\mathbf{R}^+$, let $\overline{v}_0$ be a horizontal vector field in $\mathcal{H}(\Omega_h)$, and let $\overline{f}$ be a horizontal vector field in $L^2([0,T]; \mathcal{V}'_\sigma(\Omega_h))$. We denote by $\overline{v}$ the solution of the system $(\mathrm{SE}_{\nu,\beta})$.*

Then for any positive η, there exists a family $(v^{\varepsilon,\eta}_{\rm app})_{\varepsilon>0}$ of smooth divergence-free vector fields on Ω, vanishing on the boundary, such that

$$\|\mathbf{P} L^\varepsilon v^{\varepsilon,\eta}_{\rm app} - \overline{f}\|_{L^2([0,T]; H^{-1,0})} = \rho^{\varepsilon,\eta}, \quad \textit{and} \tag{7.1.22}$$

$$\|v^{\varepsilon,\eta}_{\rm app} - \overline{v}\|^2_{L^\infty([0,T]; \mathcal{H}(\Omega))} + 2\nu \int_0^T \|\nabla^h (v^{\varepsilon,\eta}_{\rm app} - \overline{v})(t)\|^2_{L^2(\Omega)}\, dt = \rho^{\varepsilon,\eta}, \tag{7.1.23}$$

where, as in the previous chapter, $\rho^{\varepsilon,\eta}$ denotes generically any sequence of non-negative real numbers, possibly depending on T, such that

$$\lim_{\eta\to 0} \limsup_{\varepsilon\to 0} \rho^{\varepsilon,\eta} = 0. \tag{7.1.24}$$

Moreover, $v_{\text{app}}^{\varepsilon,\eta}$ satisfies, for any $t \in [0,T]$, the energy estimate

$$\sup_{t\in[0,T]} E_t^\varepsilon(v_{\text{app}}^{\varepsilon,\eta}) \leq \|\overline{v}_0\|_{L^2}^2 + 2\int_0^t \langle \overline{f}(t'), \overline{v}(t')\rangle\, dt' + \rho^{\varepsilon,\eta}. \tag{7.1.25}$$

Remark In the case when the Fourier transforms of $\overline{f}$ and $\overline{v}_0$ are included in a fixed ball, then the family $v_{\text{app}}^{\varepsilon,\eta}$ can be chosen independently of η, and all the estimates given in Lemma 7.1 hold with $\eta = 0$.

Proof of Lemma 7.1 The proof consists mainly in reading the previous computations. First of all, let us consider a given $\eta > 0$, and let us perform a frequency cut-off, by stating for any $N \in \mathbf{N}$,

$$\overline{v}_{0,N} \stackrel{\text{def}}{=} \mathbf{P}_N \overline{v}_0 \quad \text{and} \quad \overline{f}_N \stackrel{\text{def}}{=} \mathbf{P}_N \overline{f}, \tag{7.1.26}$$

where $\mathbf{P}_N$ is the usual spectral cut-off operator introduced in Chapter 2 (see relation (2.1.6) page 40). Then we denote by $\overline{v}_N$ the solution of $(\text{SE}_{\nu,\beta})$ associated with the initial data $\overline{v}_{0,N}$ and bulk force $\overline{f}_N$. It is obvious that $\overline{v}_{0,N}$ converges towards $\overline{v}_0$ in $\mathcal{H}$ as N goes to infinity, and that (see Proposition 2.3, page 40, for details) for N_η chosen large enough

$$\|\overline{f}_{N_\eta} - \overline{f}\|_{L^2([0,T];\mathcal{V}'_\sigma)} \leq \frac{\eta}{2}. \tag{7.1.27}$$

It follows that for N_η large enough,

$$\forall t \leq T, \quad \|(\overline{v} - \overline{v}_{N_\eta})(t)\|_{\mathcal{H}(\Omega_h)}^2 + \nu \int_0^t \|\nabla^h(\overline{v} - \overline{v}_{N_\eta})(t')\|_{L^2(\Omega_h)}^2\, dt' \leq \eta.$$

Thus inequality (7.1.23) will be proved if we show that

$$\lim_{\varepsilon\to 0}\Bigg(\|v_{\text{app}}^{\varepsilon,\eta} - \overline{v}_{N_\eta}\|_{L^\infty([0,T];\mathcal{H}(\Omega))}^2 + 2\nu \int_0^T \|\nabla^h(v_{\text{app}}^{\varepsilon,\eta} - \overline{v}_{N_\eta})(t)\|_{L^2(\Omega)}^2\, dt \Bigg) = 0. \tag{7.1.28}$$

Now let us define, with the notation introduced at the beginning of this section, (and, in order to avoid excessive heaviness, dropping the index η),

$$v_{\text{app}}^{\varepsilon,\eta} \stackrel{\text{def}}{=} \overline{v}_N + \varepsilon v_{1,\text{int}} + v_{0,\text{BL}} + \varepsilon v_{1,\text{BL}} + \varepsilon^2 v_{2,\text{BL}} + v_{\exp}^\varepsilon$$

where we recall that

$$v_{1,\text{int}} \stackrel{\text{def}}{=} \left(\sqrt{2\beta} R_{-\frac{\pi}{2}} \overline{v}_N, \sqrt{2\beta}\left(\frac{1}{2} - x_3\right) \text{curl}_h \overline{v}_N\right),$$

$$v_{0,\text{BL}} \stackrel{\text{def}}{=} \left(M_0\left(\frac{x_3}{\sqrt{E}}\right)\overline{v}_N + M_0\left(\frac{1-x_3}{\sqrt{E}}\right)\overline{v}_N, 0\right),$$

$$v_{1,\mathrm{BL}} \stackrel{\mathrm{def}}{=} \left(-\sqrt{2\beta}\left(\exp\left(-\frac{x_3}{\sqrt{E}}\right)+\exp\left(-\frac{1-x_3}{\sqrt{E}}\right)\right)R_{-\frac{\pi}{2}}\overline{v}_N,\right.$$
$$\left.\sqrt{2\beta}\left(f\left(\frac{x_3}{\sqrt{E}}\right)-f\left(\frac{1-x_3}{\sqrt{E}}\right)\right)\mathrm{curl}_h\overline{v}_N\right)$$
$$v_{2,\mathrm{BL}} \stackrel{\mathrm{def}}{=} \left(0,-\sqrt{2\beta}\left(\exp(-\frac{x_3}{\sqrt{E}})+\exp(-\frac{1-x_3}{\sqrt{E}})\right)\mathrm{curl}_h\overline{v}_N\right)$$

and finally

$$v^{\varepsilon}_{\mathrm{exp}} \stackrel{\mathrm{def}}{=} \left(-\cos(2\pi x_3)v^h_R, \frac{\sin(2\pi x_3)}{2\pi}\,\mathrm{div}_h\, v^h_R\right)$$
$$+\sqrt{E}\,f\left(\frac{1}{\sqrt{E}}\right)\left(\pi\sin(\pi x_3)R_{-\frac{\pi}{2}}\overline{v}_N, \cos(\pi x_3)\mathrm{curl}_h\overline{v}_N\right)$$
$$+E\exp\left(\frac{1}{\sqrt{E}}\right)\left(2\pi\sin(2\pi x_3)R_{-\frac{\pi}{2}}\overline{v}_N, \cos(2\pi x_3)\mathrm{curl}_h\overline{v}_N\right),$$

with

$$v_R \stackrel{\mathrm{def}}{=} (\overline{v}_N + v_{0,\mathrm{BL}} + \varepsilon v_{1,\mathrm{int}} + \varepsilon v_{1,\mathrm{BL}} + \varepsilon^2 v_{2,\mathrm{BL}})_{|\partial\Omega}.$$

Thanks to (7.1.20) we have

$$\mathrm{div}\, v^{\varepsilon}_{\mathrm{app}} = 0 \quad \text{and} \quad v^{\varepsilon}_{\mathrm{app}|\partial\Omega} = 0,$$

and furthermore $v^{\varepsilon}_{\mathrm{app}}$ is clearly a smooth vector field.

Let us start by estimating $\|L^{\varepsilon}v^{\varepsilon}_{\mathrm{app}} - \overline{f}\|_{L^2([0,T];H^{-1,0})}$. We shall first deal with the boundary layer terms. We have, using (7.1.7),

$$L^{\varepsilon}v_{0.\mathrm{BL}} = \left(M_0\left(\frac{x_3}{\sqrt{E}}\right)+M_0\left(\frac{1-x_3}{\sqrt{E}}\right)\right)(\partial_t\overline{v}_N - \nu\Delta_h\overline{v}_N)$$
$$= \left(M_0\left(\frac{x_3}{\sqrt{E}}\right)+M_0\left(\frac{1-x_3}{\sqrt{E}}\right)\right)(\overline{f}_N - \sqrt{2\beta}\,\overline{v}_N).$$

Hence, by Definition (7.1.21) of the $\|\cdot\|_{H^{-1,0}}$ norm, we get

$$\|L^{\varepsilon}v_{0,\mathrm{BL}}(t)\|_{H^{-1,0}} \leq C\left(\|\overline{f}_N(t)\|_{H^{-1}(\Omega_h)} + \sqrt{2\beta}\,\|\overline{v}_N\|_{H^{-1}(\Omega_h)}\right)E^{\frac{1}{4}}$$
$$\leq C\varepsilon^{\frac{1}{2}}\left(\|\overline{f}_N(t)\|_{H^{-1}(\Omega_h)} + \sqrt{2\beta}\,\|\overline{v}_N\|_{L^2(\Omega_h)}\right).$$

For the higher-order boundary layer terms we simply need to use the fact that

$$\varepsilon\partial_3 g\left(\frac{x_3}{\sqrt{E}}\right) = g'\left(\frac{x_3}{\sqrt{E}}\right),$$

and we leave the details to the reader. We infer that

$$L^{\varepsilon}v^{\varepsilon}_{\mathrm{app}} = L^{\varepsilon}(\overline{v}_N + \varepsilon v_{1,\mathrm{int}}) + R^{\varepsilon,\eta}$$
$$= (\partial_t - \nu\Delta_h)\overline{v}_N + \sqrt{2\beta}\,\overline{v}_N + R^{\varepsilon,\eta},$$

where $R^{\varepsilon,\eta}$ denotes generically a vector field satisfying

$$\lim_{\eta\to 0}\limsup_{\varepsilon\to 0}\|R^{\varepsilon,\eta}\|_{L^2([0,T];H^{-1,0})}=0.$$

Since $(\partial_t-\nu\Delta_h)\overline{v}_N+\sqrt{2\beta}\overline{v}_N=\overline{f}_N$, we conclude, using also (7.1.27), that

$$\|L^\varepsilon v^\varepsilon_{\mathrm{app}}-\overline{f}\|_{L^2([0,T];H^{-1,0})}=\rho^{\varepsilon,\eta}.$$

In order to prove (7.1.28) (hence (7.1.23) as noted above), let us observe that

$$\begin{aligned}\|\nabla^h(v^\varepsilon_{\mathrm{app}}-\overline{v}_N)(t)\|_{L^2}&\le C_\eta\|(v^\varepsilon_{\mathrm{app}}-\overline{v}_N)(t)\|_{L^2}\\&\le C_\eta\left(\sum_{j=0}^{2}\varepsilon^j\|v^\varepsilon_{j,\mathrm{BL}}(t)\|_{L^2}+\varepsilon\|v_{1,\mathrm{int}}(t)\|_{L^2}+\|v^\varepsilon_{\exp}\|_{L^2}\right).\end{aligned}$$

Thus we infer

$$\|\nabla^h(v^\varepsilon_{\mathrm{app}}-\overline{v}_N)(t)\|_{L^2}\le C_\eta\varepsilon^{\frac12}.$$

Now let us prove estimate (7.1.25). By the definition of $v^\varepsilon_{\mathrm{app}}$, we have that the energy

$$\|v^\varepsilon_{\mathrm{app}}(t)\|^2_{L^2}+2\nu\int_0^t\|\nabla^h v^\varepsilon_{\mathrm{app}}(t')\|^2_{L^2}\,dt'+2\beta\varepsilon\int_0^t\|\partial_3 v^\varepsilon_{\mathrm{app}}(t')\|^2_{L^2}\,dt'$$

is less than or equal to

$$\|\overline{v}(t)\|^2_{L^2}+2\nu\int_0^t\|\nabla^h\overline{v}(t')\|^2_{L^2}\,dt'+2\beta\varepsilon\int_0^t\|\partial_3 v_{0,\mathrm{BL}}(t')\|^2_{L^2}\,dt'+C_\eta\varepsilon.$$

An easy computation implies that

$$\begin{aligned}\|\partial_3 v_{0,\mathrm{BL}}(t')\|^2_{L^2}&=\frac{1}{2\beta\varepsilon^2}\left\|\left(M_0'\left(\frac{x_3}{\varepsilon}\right)+M_0'\left(\frac{1-x_3}{\varepsilon}\right)\right)\overline{v}_N\right\|^2_{L^2(\Omega)}+C_\eta\varepsilon\\&=\frac{4}{2\beta\varepsilon^2}\|\overline{v}_N(t)\|^2_{L^2(\Omega_h)}\int_0^1\exp\left(-\frac{2x_3}{\varepsilon\sqrt{2\beta}}\right)dx_3+C_\eta\varepsilon\\&\le\frac{\sqrt{2\beta}}{\beta\varepsilon}\|\overline{v}(t)\|^2_{L^2(\Omega_h)}+C_\eta\varepsilon.\end{aligned}$$

So we get that

$$\begin{aligned}&\|v^\varepsilon_{\mathrm{app}}(t)\|^2_{L^2}+2\nu\int_0^t\|\nabla^h v^\varepsilon_{\mathrm{app}}(t')\|^2_{L^2}\,dt'+2\beta\varepsilon\int_0^t\|\partial_3 v^\varepsilon_{\mathrm{app}}(t')\|^2_{L^2}\,dt'\\&\quad\le\|\overline{v}(t)\|^2_{L^2}+2\nu\int_0^t\|\nabla^h\overline{v}(t')\|^2_{L^2}dt'+2\sqrt{2\beta}\int_0^t\|\overline{v}(t')\|^2_{L^2(\Omega_h)}dt'+C_\eta\varepsilon.\end{aligned}$$

Clearly we have

$$\|\overline{v}(t)\|_{L^2}^2 + 2\nu \int_0^t \|\nabla^h \overline{v}(t')\|_{L^2}^2 dt' + 2\sqrt{2\beta} \int_0^t \|\overline{v}(t')\|_{L^2(\Omega_h)}^2 dt'$$
$$= \|\overline{v}_0\|_{L^2}^2 + 2 \int_0^t \langle \overline{f}(t'), \overline{v}(t')\rangle \, dt',$$

so finally

$$E_t^\varepsilon(v_{\text{app}}^\varepsilon) \leq \|\overline{v}_0\|_{L^2(\Omega_h)}^2 + 2 \int_0^t \langle \overline{f}(t'), \overline{v}(t')\rangle \, dt' + C_\eta \varepsilon,$$

and Lemma 7.1 is proved. □

7.2 Non-linear estimates in the well-prepared case

The two key estimates of this section (Lemmas 7.2 and 7.3) essentially say that Ekman boundary layers do not affect non-linear terms. That will be of great importance in the proof of Theorem 7.1. Actually as we will see in Section 7.5, that fact remains true in the ill-prepared case.

Let us start by proving the following lemma.

Lemma 7.2 *Under the assumptions of Lemma 7.1, the families* $(v_{\text{app}}^{\varepsilon,\eta})_{\varepsilon>0}$ *satisfy*

$$v_{\text{app}}^{\varepsilon,\eta} \cdot \nabla v_{\text{app}}^{\varepsilon,\eta} - \overline{v} \cdot \nabla^h \overline{v} = F^{\varepsilon,\eta} + R^{\varepsilon,\eta},$$

where, with the notation of Lemma 7.1,

$$\|R^{\varepsilon,\eta}\|_{L^2([0,T],H^{-1,0})} = \rho^{\varepsilon,\eta} \quad \textit{and} \quad \forall \eta > 0\,, \ \lim_{\varepsilon \to 0} \|F^{\varepsilon,\eta}\|_{L^2([0,T],L^2(\Omega))} = 0.$$

Proof Let us denote in all the following $L_h^p = L^p(\Omega_h)$. The fact that Q is a continuous map from the space $L^4([0,T];L_h^4) \times L^4([0,T];L_h^4)$ into $L^2([0,T];H^{-1,0})$ implies that it is enough to prove that

$$F^{\varepsilon,\eta} \stackrel{\text{def}}{=} v_{\text{app}}^{\varepsilon,\eta} \cdot \nabla v_{\text{app}}^{\varepsilon,\eta} - \overline{v}_{N_\eta} \cdot \nabla^h \overline{v}_{N_\eta}$$

goes to zero in $L^2([0,T];L^2(\Omega))$ as ε goes to zero, for any η. We recall that v_N is the solution of $(\text{SE}_{\nu,\beta})$ associated with $\overline{v}_{0,N}$ and $\overline{f}_N$ as defined in (7.1.26). By definition of $F^{\varepsilon,\eta}$, we have, omitting the index η in order to avoid excessive heaviness,

$$F^{\varepsilon,\eta} = \sum_{j=1}^{3} F_j^{\varepsilon,\eta} \quad \text{with}$$

$$F_1^{\varepsilon,\eta} \stackrel{\text{def}}{=} v_{\text{app}}^{\varepsilon,h} \cdot \nabla^h (v_{\text{app}}^\varepsilon - \overline{v}_N)$$
$$F_2^{\varepsilon,\eta} \stackrel{\text{def}}{=} (v_{\text{app}}^{\varepsilon,h} - \overline{v}_N) \cdot \nabla^h \overline{v}_N$$
$$F_3^{\varepsilon,\eta} \stackrel{\text{def}}{=} v_{\text{app}}^{\varepsilon,3} \partial_3 v_{\text{app}}^\varepsilon,$$

recalling that $v^3_{0,\mathrm{int}} = \overline{v}^3_N = 0$. But

$$\|F_1^{\varepsilon,\eta}\|_{L^2([0,T];L^2)} \leq \|v^{\varepsilon,h}_{\mathrm{app}}\|_{L^\infty([0,T];L^\infty)}\|\nabla^h(v^\varepsilon_{\mathrm{app}} - \overline{v}_N)\|_{L^2([0,T];L^2)}.$$

Using the estimate (7.1.25), we infer that

$$\limsup_{\varepsilon\to 0}\|F_1^{\varepsilon,\eta}\|_{L^2([0,T];L^2)} \leq C\limsup_{\varepsilon\to 0}\|\nabla^h(v^\varepsilon_{\mathrm{app}} - \overline{v}_N)\|_{L^2([0,T];L^2)},$$

and it is just a matter of using (7.1.28) to conclude. The result on $F_2^{\varepsilon,\eta}$ is proved along the same lines.

In order to estimate $F_3^{\varepsilon,\eta}$, let us write

$$\|F_3^{\varepsilon,\eta}\|_{L^2([0,T];L^2)} \leq \|v^{\varepsilon,3}_{\mathrm{app}}\|_{L^\infty([0,T];L^\infty)}\|\partial_3 v^\varepsilon_{\mathrm{app}}\|_{L^2([0,T];L^2)}.$$

By definition (7.1.11), we have

$$\|v^{\varepsilon,3}_{\mathrm{app}}\|_{L^\infty([0,T];L^\infty)} \leq C_\eta\varepsilon.$$

We infer that

$$\|F_3^{\varepsilon,\eta}\|_{L^2([0,T];L^2)} \leq C_\eta\varepsilon\|\partial_3 v^\varepsilon_{\mathrm{app}}\|_{L^2([0,T];L^2)}.$$

The energy estimate (7.1.25) implies that $\varepsilon^{\frac{1}{2}}\|\partial_3 v^\varepsilon_{\mathrm{app}}\|_{L^2([0,T];L^2)}$ is uniformly bounded, which implies the expected result on $F_3^{\varepsilon,\eta}$ and the lemma is proved. □

Now let us prove the following result.

Lemma 7.3 *Let $\overline{v}$ be a solution of $(\mathrm{SE}_{\nu,\beta})$ and let η be a positive real number. Denote by $(v^{\varepsilon,\eta}_{\mathrm{app}})_{\varepsilon>0}$ the families given by Lemma 7.1. Then for any vector field δ belonging to $L^\infty([0,T];\mathcal{H}) \cap L^2([0,T];\mathcal{V}_\sigma)$, we have*

$$\int_0^T \left(\delta(t)\cdot\nabla\delta(t)|v^{\varepsilon,\eta}_{\mathrm{app}}(t)\right)_{L^2} dt \leq \left(C_\eta\varepsilon^{\frac{1}{2}} + \frac{1}{4}\right)E^\varepsilon_T(\delta) + \frac{C}{\nu}\int_0^T \|\nabla^h\overline{v}(t)\|^2_{L^2}\|\delta(t)\|^2_{L^2}\,dt.$$

Proof The proof of the result relies on the explicit expression of $(v^{\varepsilon,\eta}_{\mathrm{app}})_{\varepsilon>0}$, jointly with the following lemma.

Lemma 7.4 *Let δ be a vector field in $\mathcal{V}_\sigma$ and let w be a bounded vector field. Then*

$$(\delta\cdot\nabla\delta|w)_{L^2} \leq C\|\nabla^h\delta\|_{L^2}\|\partial_3\delta\|_{L^2}\|d^{\frac{1}{2}}(\cdot)w\|_{L^2(]0,1[;L^\infty_h)},$$

where d denotes the distance to the boundary.

Proof Let us define

$$I_{j,k} = \int_\Omega \delta^k \partial_k \delta^j w^j \, dx,$$

for j and k in $\{1,2,3\}$. In the case $k \neq 3$, since δ vanishes at the boundary we can write that

$$\begin{aligned} |\delta^k(x_h, x_3)| &= \left| \int_0^{x_3} \partial_3 \delta^k(x_h, y_3) \, dy_3 \right| \\ &\leq x_3^{\frac{1}{2}} \|\partial_3 \delta\|_{L^2(]0,1[)}, \end{aligned}$$

and similarly for the upper boundary $x_3 = 1$. We therefore have

$$|\delta^k(x_h, x_3)| \leq d(x_3)^{\frac{1}{2}} \|\partial_3 \delta(x_h, \cdot)\|_{L^2(]0,1[)}.$$

We infer that for $k \neq 3$,

$$\begin{aligned} |I_{j,k}| &\leq \int_\Omega \|\partial_3 \delta(x_h, \cdot)\|_{L^2(]0,1[)} |\nabla^h \delta(x)| \, d(x_3)^{\frac{1}{2}} \, |w(x)| \, dx \\ &\leq \int_\Omega \|\partial_3 \delta(x_h, \cdot)\|_{L^2(]0,1[)} |\nabla^h \delta(x)| \, d(x_3)^{\frac{1}{2}} \, \|w(\cdot, x_3)\|_{L_h^\infty} \, dx. \end{aligned}$$

The Cauchy–Schwarz inequality implies finally that for $k \neq 3$,

$$|I_{j,k}| \leq \|\partial_3 \delta\|_{L^2} \|\nabla^h \delta\|_{L^2} \|d^{\frac{1}{2}}(\cdot) w\|_{L^2(]0,1[;L_h^\infty)}.$$

Now let us turn to the case $k = 3$. Using the fact that δ is divergence-free, we get

$$\begin{aligned} \delta^3(x_h, x_3) &= \int_0^{x_3} \partial_3 \delta^3(x_h, y_3) \, dy_3 \\ &= -\int_0^{x_3} \operatorname{div}_h \delta^h(x_h, y_3) \, dy_3. \end{aligned}$$

It follows as above that

$$|\delta^3(x_h, x_3)| \leq d(x_3)^{\frac{1}{2}} \|\nabla^h \delta(x_h, \cdot)\|_{L^2(]0,1[)},$$

which implies that

$$|I_{j,3}| \leq \int_\Omega \|\nabla^h \delta(x_h, \cdot)\|_{L^2(]0,1[)} |\partial_3 \delta(x)| \, d(x_3)^{\frac{1}{2}} \, |w(x)| \, dx.$$

The estimate is now the same as in the previous case $k \neq 3$. Thus the lemma is proved. □

Now let us go back to the proof of Lemma 7.3. We recall that by the definition of $v_{\text{app}}^{\varepsilon,\eta}$, we have

$$\|v_{\text{app}}^{\varepsilon,\eta} - \overline{v}_{N_\eta} - v_{0,\text{BL}}\|_{L^\infty([0,T];L^\infty(\Omega))} \leq C_\eta \varepsilon.$$

It follows that

$$\int_0^T \big(\delta(t)\cdot\nabla\delta(t)|(v_{\rm app}^{\varepsilon,\eta}-\overline{v}_{N_\eta}-v_{0,\rm BL})(t)\big)\,dt \le C_\eta\varepsilon\int_0^T \|\delta(t)\|_{L^2}\|\nabla\delta(t)\|_{L^2}\,dt$$

which, using the fact that $\varepsilon^{\frac12}\|\partial_3\delta\|_{L^2([0,T]\times\Omega)}\le E_T^\varepsilon(\delta)$, yields

$$\int_0^T \big(\delta(t)\cdot\nabla\delta(t)|(v_{\rm app}^{\varepsilon,\eta}-\overline{v}_{N_\eta}-v_{0,\rm BL})(t)\big)\,dt \le C_\eta\varepsilon^{\frac12}E_T^\varepsilon(\delta). \tag{7.2.1}$$

Moreover

$$|v_{0,\rm BL}(x)|\le C|\overline{v}_{N_\eta}(x_h)|\left(\exp\left(-\frac{x_3}{\varepsilon\sqrt{2\beta}}\right)+\exp\left(-\frac{1-x_3}{\varepsilon\sqrt{2\beta}}\right)\right).$$

Thus

$$\begin{aligned} &d(x_3)^{\frac12}\|v_{0,\rm BL}(\cdot,x_3)\|_{L_h^\infty}\\ &\le CN_\eta\|\overline{v}_{N_\eta}\|_{L^2}\left(x_3^{\frac12}\exp\left(-\frac{x_3}{\varepsilon\sqrt{2\beta}}\right)+(1-x_3)^{\frac12}\exp\left(-\frac{1-x_3}{\varepsilon\sqrt{2\beta}}\right)\right).\end{aligned}$$

A classical L^2 energy estimate on $(\mathrm{SE}_{\nu,\beta})$ implies that

$$\|\overline{v}_{N_\eta}\|_{L^\infty(\mathbf{R}^+;L^2)}^2\le\|\overline{v}_0\|_{L^2}^2+\frac{2}{\nu}\|\overline{f}\|_{L^2([0,T];H^{-1,0})}^2,$$

so we deduce that

$$\|d(x_3)^{\frac12}v_{0,\rm BL}\|_{L^2(]0,1[;L^\infty)}^2\le 2C_0N_\eta^2\varepsilon^2\int\frac{x_3}{\varepsilon}\exp\left(-2\frac{x_3}{\varepsilon\sqrt{2\beta}}\right)\frac{dx_3}{\varepsilon},$$

where C_0 denotes any $O\left(\|\overline{v}_0\|_{L^2}^2+\frac{2}{\nu}\|\overline{f}\|_{L^2([0,T];H^{-1,0})}^2\right)$. Finally

$$\|d(x_3)^{\frac12}v_{0,\rm BL}\|_{L^2(]0,1[;L^\infty)}^2\le C_0N_\eta^2\varepsilon^2. \tag{7.2.2}$$

Now let us write that

$$\int_0^T \big|(\delta(t)\cdot\nabla\delta(t)|v_{\rm app}^{\varepsilon,\eta}(t))_{L^2}\big|\,dt\le\sum_{j=1}^3\mathcal{V}_j(T)\quad\text{where}$$

$$\mathcal{V}_1(T)\stackrel{\rm def}{=}\int_0^T\Big|\big(\delta(t)\cdot\nabla\delta(t)|v_{\rm app}^{\varepsilon,\eta}-\overline{v}_{N_\eta}-v_{0,\rm BL}\big)_{L^2}\Big|\,dt,$$

$$\mathcal{V}_2(T)\stackrel{\rm def}{=}\int_0^T\big|(\delta(t)\cdot\nabla\delta(t)|v_{0,\rm BL}(t))_{L^2}\big|\,dt\quad\text{and}$$

$$\mathcal{V}_3(T)\stackrel{\rm def}{=}\int_0^T\Big|\big(\delta(t)\cdot\nabla\delta(t)|\overline{v}_{N_\eta}(t)\big)_{L^2}\Big|\,dt.$$

Estimate (7.2.1) claims exactly that

$$\mathcal{V}_1(T) \leq C_\eta \varepsilon^{\frac{1}{2}} E_T^\varepsilon(\delta).$$

Lemma 7.4, together with (7.2.2), imply that

$$\begin{aligned}\mathcal{V}_2(T) &\leq C_\eta \varepsilon \int_0^T \|\nabla^h \delta(t)\|_{L^2} \|\partial_3 \delta(t)\|_{L^2} \|d(x_3)^{\frac{1}{2}} v_{0,\mathrm{BL}}(t)\|_{L^2(]0,1[;L^\infty)} dt \\ &\leq C_\eta \|\nabla^h \delta\|_{L^2([0,T];L^2)} \varepsilon^{\frac{1}{2}} \|\partial_3 \delta\|_{L^2([0,T];L^2)}.\end{aligned}$$

By definition of E_t^ε we infer, using the Cauchy–Schwarz inequality, that

$$\mathcal{V}_2(T) \leq C_\eta \varepsilon^{\frac{1}{2}} E_T^\varepsilon(\delta).$$

Thus we get that

$$\int_0^T \left|(\delta(t) \cdot \nabla \delta(t) | v_{\mathrm{app}}^{\varepsilon,\eta}(t))_{L^2}\right| dt \leq \mathcal{V}_3(T) + C_\eta \varepsilon^{\frac{1}{2}} E_T^\varepsilon(\delta).$$

Now we are left with the estimate of

$$\mathcal{V}_3(T) = \int_0^T |(\delta(t) \cdot \nabla \delta(t) | \overline{v}_{N_\eta}(t))_{L^2}| dt.$$

As δ is divergence-free and vanishes at the boundary, we have

$$\begin{aligned}(\delta \cdot \nabla \delta | \overline{v}_{N_\eta})_{L^2} &= -(\delta^h \cdot \nabla^h \overline{v}_{N_\eta} | \delta)_{L^2} \\ &\leq \|\nabla^h \overline{v}_{N_\eta}\|_{L^2} \int_0^1 \|\delta(\cdot, x_3)\|_{L_h^4}^2 dx_3.\end{aligned}$$

The two-dimensional Gagliardo–Nirenberg and Cauchy–Schwarz inequalities imply that

$$\begin{aligned}\mathcal{V}_3(T) &\leq C \int_0^T \|\nabla^h \overline{v}_{N_\eta}(t)\|_{L^2} \|\delta(t)\|_{L^2} \|\nabla^h \delta(t)\|_{L^2} dt \\ &\leq \frac{\nu}{2} \int_0^T \|\nabla^h \delta(t)\|_{L^2}^2 dt + \frac{C}{\nu} \int_0^T \|\nabla^h \overline{v}_{N_\eta}(t)\|_{L^2}^2 \|\delta(t)\|_{L^2}^2 dt \qquad (7.2.3)\end{aligned}$$

which ends the proof of Lemma 7.3. □

7.3 The convergence theorem in the well-prepared case

In this section we intend to prove Theorem 7.1. The idea is to apply Lemma 7.1 to $\overline{v} \stackrel{\text{def}}{=} \overline{u}$. Indeed $\overline{u}$ solves a system of the type $(\mathrm{SE}_{\nu,\beta})$, stating $\overline{f} \stackrel{\text{def}}{=} Q(\overline{u}, \overline{u})$ with the notation of Definition 2.7, page 44. We recall that according to Lemma 2.3, $Q(\overline{u}, \overline{u})$ is an element of $L^2(\mathbf{R}^+; \mathcal{V}_\sigma')$ as long as $\overline{u}$ is in the energy space.

Let $T > 0$ be given and let η be an arbitrarily small positive real number. According to Lemma 7.1, the theorem will be proved if we prove that

$$\sup_{t \in [0,T]} \|(u^\varepsilon - u_{\mathrm{app}}^{\varepsilon,\eta})(t)\|_{L^2(\Omega)}^2 + \nu \int_0^T \|\nabla^h (u^\varepsilon - u_{\mathrm{app}}^{\varepsilon,\eta})(t)\|_{L^2(\Omega)}^2 dt = \rho^{\varepsilon,\eta},$$

where $u^{\varepsilon,\eta}_{\text{app}}$ is the family given by Lemma 7.1, associated with η.

$$\overline{v} = \overline{u} \quad \text{and} \quad \overline{f} = Q(\overline{u},\overline{u}) = -\overline{u}\cdot\nabla^h\overline{u}.$$

The main step of the proof is purely algebraic and is common to this case and to the ill-prepared case (in the $\mathbf{R}^2$ case in Section 7.6 as well as in the periodic case in Section 7.7). It is based on the following lemma.

Lemma 7.5 *Let (u^ε) be a family of Leray solutions of* (NSC_ε) *with initial data u_0, and let $(\Psi^\varepsilon)_{\varepsilon>0}$ be a family of $C^1(\mathbf{R}^+;\mathcal{V}_\sigma(\Omega))$ functions. Then the function $\delta^\varepsilon \stackrel{\text{def}}{=} u^\varepsilon - \Psi^\varepsilon$ satisfies*

$$E^\varepsilon_t(\delta^\varepsilon) = E^\varepsilon_t(u^\varepsilon) + E^\varepsilon_t(\Psi^\varepsilon) - 2(u(0)|\Psi^\varepsilon(0))_{L^2} \tag{7.3.1}$$
$$-2\int_0^t (\delta^\varepsilon(t')\cdot\nabla\delta^\varepsilon(t')|\Psi^\varepsilon(t'))_{L^2}\,dt' + 2\int_0^t (G^\varepsilon(\Psi^\varepsilon)(t')|u^\varepsilon(t'))_{L^2}dt',$$

where $G^\varepsilon(\Psi^\varepsilon) \stackrel{\text{def}}{=} L^\varepsilon\Psi^\varepsilon + \Psi^\varepsilon\cdot\nabla\Psi^\varepsilon$.

Before proving the lemma, let us show why it leads to Theorem 7.1. We apply Lemma 7.5 with $\Psi^\varepsilon = u^{\varepsilon,\eta}_{\text{app}}$ and $u_0 = \overline{u}_0$. Note that preliminary smoothing in time is required, so as to have $u^{\varepsilon,\eta}_{\text{app}} \in C^1(\mathbf{R}^+,\mathcal{V}_\sigma)$. That procedure should by now be familiar to the reader, and in order to avoid introducing additional notation we shall omit that time smoothing procedure in the following. As u^ε is a Leray solution of (NSC_ε), it satisfies

$$E^\varepsilon_t(u^\varepsilon) \leq \|\overline{u}_0\|^2_{L^2}.$$

Furthermore, estimate (7.1.25) yields

$$E^\varepsilon_t(u^{\varepsilon,\eta}_{\text{app}}) \leq \|\overline{u}_0\|^2_{L^2} + 2\int_0^t \langle\overline{u}(t')\cdot\nabla^h\overline{u}(t'),\overline{u}(t')\rangle dt' + \rho^{\varepsilon,\eta} = \|\overline{u}_0\|^2_{L^2} + \rho^{\varepsilon,\eta}$$

and since

$$\|u^{\varepsilon,\eta}_{\text{app}|t=0} - \overline{u}_0\|_{L^2} = \rho^{\varepsilon,\eta},$$

we get

$$E^\varepsilon_t(u^\varepsilon) + E^\varepsilon_t(u^{\varepsilon,\eta}_{\text{app}}) - 2(u^\varepsilon(0)|u^{\varepsilon,\eta}_{\text{app}|t=0})_{L^2} \leq \rho^{\varepsilon,\eta}. \tag{7.3.2}$$

Then Lemma 7.1 implies that

$$L^\varepsilon u^{\varepsilon,\eta}_{\text{app}} = -\overline{u}\cdot\nabla^h\overline{u} + R^{\varepsilon,\eta} + \nabla p^\varepsilon,$$

where $R^{\varepsilon,\eta}$ denotes from now on generically any vector field satisfying

$$\|R^{\varepsilon,\eta}\|_{L^2([0,T];H^{-1,0})} = \rho^{\varepsilon,\eta}.$$

It follows that one can write

$$G^{\varepsilon,\eta}(u^{\varepsilon,\eta}_{\text{app}}) = u^{\varepsilon,\eta}_{\text{app}}\cdot\nabla u^{\varepsilon,\eta}_{\text{app}} - \overline{u}_{N_\eta}\cdot\nabla^h\overline{u}_{N_\eta} + \overline{u}_{N_\eta}\cdot\nabla^h\overline{u}_{N_\eta}$$
$$-\overline{u}\cdot\nabla^h\overline{u} + R^{\varepsilon,\eta} + \nabla p^\varepsilon.$$

As Q is a continuous bilinear map from

$$L^4([0,T];L^4_h) \times L^4([0,T];L^4_h) \quad \text{into} \quad L^2([0,T];H^{-1}(\Omega_h))$$

we get

$$G^{\varepsilon,\eta}(u^{\varepsilon,\eta}_{\text{app}}) = u^{\varepsilon,\eta}_{\text{app}} \cdot \nabla u^{\varepsilon,\eta}_{\text{app}} - \overline{u}_{N_\eta} \cdot \nabla^h \overline{u}_{N_\eta} + R^{\varepsilon,\eta} + \nabla p^\varepsilon.$$

By Lemma 7.2 we find

$$G^{\varepsilon,\eta}(u^{\varepsilon,\eta}_{\text{app}}) = F^{\varepsilon,\eta} + R^{\varepsilon,\eta} + \nabla p^\varepsilon \tag{7.3.3}$$

with $\lim_{\varepsilon \to 0} \|F^{\varepsilon,\eta}\|_{L^2([0,T],L^2(\Omega))} = 0$. Since

$$\int_0^t (R^{\varepsilon,\eta}(t')|u^\varepsilon(t'))_{L^2}\,dt' \leq \|R^{\varepsilon,\eta}\|_{L^2([0,T],H^{-1,0})} \|\nabla^h u^\varepsilon\|_{L^2([0,T],L^2)} = \rho^{\varepsilon,\eta}, \tag{7.3.4}$$

and

$$\int_0^t (F^{\varepsilon,\eta}(t')|u^\varepsilon(t'))_{L^2}\,dt' \leq t^{\frac{1}{2}} \|\overline{u}_0\|_{L^2} \|F^{\varepsilon,\eta}\|_{L^2([0,T];L^2)} = \rho^{\varepsilon,\eta}, \tag{7.3.5}$$

plugging both inequalities into (7.3.3) yields

$$\int_0^T (G^{\varepsilon,\eta}(u^{\varepsilon,\eta}_{\text{app}})(t')|u^\varepsilon(t'))_{L^2}\,dt' \leq \rho^{\varepsilon,\eta}. \tag{7.3.6}$$

Putting (7.3.2) and (7.3.6) into (7.3.1) yields finally

$$E^\varepsilon_t(\delta^\varepsilon) \leq C \left| \int_0^t \left(\delta^\varepsilon(t') \cdot \nabla \delta^\varepsilon(t') | u^{\varepsilon,\eta}_{\text{app}}(t')\right)_{L^2}\,dt' \right| + \rho^{\varepsilon,\eta}.$$

It is now just a matter of using Lemma 7.3 to obtain, for ε small enough compared to N, that

$$E^\varepsilon_t(\delta^\varepsilon) \leq \rho^{\varepsilon,\eta} + \frac{C}{\nu} \int_0^t \|\nabla^h \overline{v}(t')\|^2_{L^2} \|\delta^\varepsilon(t')\|^2_{L^2}\,dt',$$

and the result follows from Gronwall's inequality: we find

$$E^\varepsilon_t(\delta^\varepsilon) \leq \rho^{\varepsilon,\eta} \exp \frac{C_0}{\nu^2} = \rho^{\varepsilon,\eta}$$

and Theorem 7.1 is proved. □

Proof of Lemma 7.5 This is a typical weak–strong type argument along the lines of those followed in Chapter 3 in the proof of the stability Theorem 3.3 (see pages 59–63). We have

$$E^\varepsilon_t(\delta^\varepsilon) = E^\varepsilon_t(u^\varepsilon) + E^\varepsilon_t(\Psi^\varepsilon) - 2B^\varepsilon_t(u^\varepsilon, \Psi^\varepsilon), \tag{7.3.7}$$

where

$$B_t^\varepsilon(a,b) \stackrel{\text{def}}{=} (a(t)|b(t))_{L^2} + 2\nu\int_0^t \left(\nabla^h a(t')|\nabla^h b(t')\right)_{L^2} dt' \\ + 2\varepsilon\beta\int_0^t (\partial_3 a(t')|\partial_3 b(t'))_{L^2}\, dt'.$$

Let us compute $B_t^\varepsilon(u^\varepsilon, \Psi^\varepsilon)$. Using Ψ^ε as a test function in Definition 2.5, page 42, we get

$$(u^\varepsilon(t)|\Psi^\varepsilon(t))_{L^2} = (u^\varepsilon(0)|\Psi^\varepsilon(0))_{L^2} - \nu\int_0^t (\nabla^h u^\varepsilon(t')|\nabla^h\Psi^\varepsilon(t'))_{L^2} dt' \\ - \varepsilon\beta\int_0^t (\partial_3 u^\varepsilon(t')|\partial_3\Psi^\varepsilon(t'))_{L^2} dt' - \int_0^t (u^\varepsilon(t')\cdot\nabla u^\varepsilon(t')|\Psi^\varepsilon(t'))_{L^2} dt' \\ + \int_0^t (u^\varepsilon(t')|\partial_t\Psi^\varepsilon(t'))_{L^2} dt' - \frac{1}{\varepsilon}\int_0^t (Ru^\varepsilon(t')|\Psi^\varepsilon(t'))_{L^2} dt'.$$

By definition of $L^\varepsilon\Psi^\varepsilon$, we have

$$\partial_t\Psi^\varepsilon = L^\varepsilon\Psi^\varepsilon + \left(\nu\Delta_h + \varepsilon\beta\partial_3^2 - \frac{1}{\varepsilon}R\right)\Psi^\varepsilon,$$

hence by an integration by parts allowed by the fact that u^ε and Ψ^ε vanish at the boundary, we infer that

$$(u^\varepsilon|\partial_t\Psi^\varepsilon)_{L^2} = -\nu(\nabla^h u^\varepsilon|\nabla^h\Psi^\varepsilon)_{L^2} - \varepsilon\beta(\partial_3 u^\varepsilon|\partial_3\Psi^\varepsilon)_{L^2} \\ - \frac{1}{\varepsilon}(u^\varepsilon|R\Psi^\varepsilon)_{L^2} + (u^\varepsilon|L^\varepsilon\Psi^\varepsilon)_{L^2}.$$

It follows that

$$(u^\varepsilon(t)|\Psi^\varepsilon(t))_{L^2} = (u^\varepsilon(0)|\Psi^\varepsilon(0))_{L^2} - 2\nu\int_0^t (\nabla^h u^\varepsilon(t')|\nabla^h\Psi^\varepsilon(t'))_{L^2} dt' \\ - 2\varepsilon\beta\int_0^t (\partial_3 u^\varepsilon(t')|\partial_3\Psi^\varepsilon(t'))_{L^2} dt' - \int_0^t (u^\varepsilon\cdot\nabla u^\varepsilon(t')|\Psi^\varepsilon(t'))_{L^2} dt' \\ - \frac{1}{\varepsilon}\int_0^t (Ru^\varepsilon(t')|\Psi^\varepsilon(t'))_{L^2} dt' - \frac{1}{\varepsilon}\int_0^t (R\Psi^\varepsilon(t')|u^\varepsilon(t'))_{L^2} dt' \\ + \int_0^t (u^\varepsilon(t')|L^\varepsilon\Psi^\varepsilon(t'))_{L^2} dt'$$

so

$$B_t^\varepsilon(u^\varepsilon,\Psi^\varepsilon) = (u^\varepsilon(0)|\Psi^\varepsilon(0))_{L^2} \tag{7.3.8}$$
$$- \int_0^t (u^\varepsilon\cdot\nabla u^\varepsilon(t')|\Psi^\varepsilon(t'))_{L^2}dt' + \int_0^t (u^\varepsilon(t')|L^\varepsilon\Psi^\varepsilon(t'))_{L^2}dt'.$$

But relation (3.2.2) page 57, implies that

$$\begin{aligned}-(u^\varepsilon\cdot\nabla u^\varepsilon|\Psi^\varepsilon)_{L^2} &= (Q(u^\varepsilon,u^\varepsilon)|\Psi^\varepsilon - u^\varepsilon)_{L^2}\\ &= (Q(\Psi^\varepsilon,\Psi^\varepsilon)|\delta^\varepsilon)_{L^2} + (Q(\delta^\varepsilon,u^\varepsilon)|\delta^\varepsilon)_{L^2}\\ &= -(\Psi^\varepsilon\cdot\nabla\Psi^\varepsilon|\delta^\varepsilon)_{L^2} + (\delta^\varepsilon\cdot\nabla\delta^\varepsilon|u^\varepsilon)_{L^2}.\end{aligned} \tag{7.3.9}$$

We then just have to put together (7.3.8), (7.3.9), and (7.3.7) to prove the lemma. □

7.4 The ill-prepared linear problem

The goal of this section is to construct approximate solutions to

$$(\mathrm{SC}^\varepsilon_\beta)\begin{cases}\partial_t v^\varepsilon - \nu\Delta_h v^\varepsilon - \beta\varepsilon\partial_3^2 v^\varepsilon + \dfrac{e^3\wedge v^\varepsilon}{\varepsilon} = -\nabla p^\varepsilon\\ \mathrm{div}\ v^\varepsilon = 0\\ v^\varepsilon_{|t=0} = v_0\\ v^\varepsilon_{|\partial\Omega} = 0\end{cases}$$

in the case when the initial data v_0 does depend on the vertical variable x_3. We suppose throughout this section that $\Omega_h = \mathbf{R}^2$ or $\mathbf{T}^2$ and in the periodic case we also suppose that the horizontal mean of v^ε vanishes. We introduce the following notation: $\mathcal{H}_0(\mathbf{T}^2\times[0,1])$ denotes the space of functions in $\mathcal{H}(\mathbf{T}^2\times[0,1])$ of vanishing horizontal mean. In the following, to simplify notation we will denote by ξ_h the horizontal Fourier variable (which in the periodic case belongs to $\mathbf{Z}^2\setminus\{0\}$).

Let us recall that

$$L^\varepsilon v \stackrel{\mathrm{def}}{=} \begin{pmatrix}\partial_t v^h - \nu\Delta_h v^h + \dfrac{Rv^h}{\varepsilon} - \beta\varepsilon\partial_3^2 v^h\\ \partial_t v^3 - \nu\Delta_h v^3 - \beta\varepsilon\partial_3^2 v^3\end{pmatrix}$$

and that R denotes rotation by angle $\pi/2$ in the horizontal plane.

The vector field v^ε, which of course also depends on the vertical variable x_3, does not belong to the kernel of $\mathbf{P}(e^3\wedge v^\varepsilon)$. Thus we have to deal with very fast time oscillations.

As in the well-prepared case, we will need to truncate functions in frequency space, hence to consider divergence vector fields which are a finite sum of trigonometric functions and the horizontal Fourier transform of which is supported in rings. Let us start by proving the following lemma.

Lemma 7.6 *Let us consider the space $\mathcal{B} = \bigcup_{N=1}^{\infty} \mathcal{B}_N$ of vector fields on Ω of the form*

$$v(x_h, x_3) = \begin{pmatrix} v^{0,h}(x_h) \\ 0 \end{pmatrix} + \sum_{k_3=1}^{N} \begin{pmatrix} v^{k_3,h}(x_h)\cos(k_3\pi x_3) \\ -\dfrac{1}{k_3\pi}\operatorname{div}_h v^{k_3,h}(x_h)\sin(k_3\pi x_3) \end{pmatrix}$$

where $(v^{k_3,h})_{1\leq k_3\leq N}$ are two-dimensional vector fields on $\mathbf{R}^2$ (resp. on $\mathbf{T}^2$), the Fourier transform of which has its support included in the ring $\mathcal{C}(1/N, N)$ of $\mathbf{R}^2$, and such that $\operatorname{div}_h v^{0,h} = 0$. The space $\mathcal{B}$ is dense in $\mathcal{H}(\mathbf{R}^2 \times [0,1])$ (resp. $\mathcal{H}_0(\mathbf{T}^2 \times [0,1])$).

Proof In order to prove this lemma, let us prove that the orthogonal space of $\mathcal{B}$ in $\mathcal{H}(\mathbf{R}^2 \times [0,1])$ (resp. $\mathcal{H}_0(\mathbf{T}^2 \times [0,1])$) is $\{0\}$. Let us consider a vector field w in the orthogonal space of $\mathcal{B}$. In particular, for any function φ in $L^2(\Omega_h)$ the Fourier transform of which is supported in a ring $\mathcal{C}$ of $\mathbf{R}^2$ (and of vanishing horizontal mean if $\Omega_h = \mathbf{T}^2$), we have, for any $k_3 \in \mathbf{N}\setminus\{0\}$,

$$\mathcal{P}_{k_3}(\varphi) \stackrel{\text{def}}{=} \left(w \Big| (\nabla^h\varphi\cos(k_3\pi x_3), -\frac{1}{k_3\pi}\Delta_h\varphi\sin(k_3\pi x_3))\right)_{L^2} = 0.$$

By integration by parts and using the fact that w^3 vanishes in the boundary, we get that

$$\begin{aligned}\mathcal{P}_{k_3}(\varphi) = &-\int_\Omega \operatorname{div}_h w^h(x_h, x_3)\varphi(x_h)\cos(k_3\pi x_3)dx_h dx_3 \\ &- \frac{1}{(k_3\pi)^2}\int_\Omega \partial_3 w^3(x_h, x_3)\Delta_h\varphi(x_h)\cos(k_3\pi x_3)dx_h dx_3.\end{aligned}$$

Using the fact that w is divergence-free gives that

$$\begin{aligned}\mathcal{P}_{k_3}(\varphi) = &\int_{\mathbf{R}^2}\left(\frac{1}{(k_3\pi)^2}\Delta_h\varphi - \varphi\right)(x_h) \\ &\times\left(\int_0^1 \operatorname{div}_h w^h(x_h, x_3)\cos(k_3\pi x_3)dx_3\right)dx_h.\end{aligned}$$

For any positive integer k_3, the operator $(k_3\pi)^{-2}\Delta_h - \mathrm{Id}$ is an isomorphism in the space of functions belonging to $L^2(\Omega_h)$ and the Fourier transform of which is included in the ring $\mathcal{C}_N \stackrel{\text{def}}{=} \{\eta \in \mathbf{R}^2 \,/\, |\eta| \in [N^{-1}, N]\}$. Thus if w belongs to the orthogonal space of $\mathcal{B}$, for any given positive integer k_3, we have for any function ψ such that $\operatorname{Supp}\widehat{\psi} \subset \mathcal{C}_N$ and for any positive integer k_3 such that $k_3 \leq N$,

$$\int_{\Omega_h}\psi(x_h)\left(\int_0^1 \operatorname{div}_h w^h(x_h, x_3)\cos(k_3\pi x_3)dx_3\right)dx_h = 0.$$

When $\Omega_h = \mathbf{R}^2$, we note that the space of functions of $L^2(\mathbf{R}^2)$, the Fourier transform of which is included in a ring, is a dense subspace of $L^2(\mathbf{R}^2)$. When Ω_h

is equal to $\mathbf{T}^2$, that still holds for $L^2(\mathbf{T}^2)$ restricted to functions the horizontal mean of which vanishes. So in both cases we get that, for any positive integer k_3,

$$\int_0^1 \operatorname{div}_h w^h(x_h, x_3)\cos(k_3\pi x_3)dx_3 = 0.$$

As $w^3_{|\partial\Omega} = 0$, we have, thanks to (7.0.1), that

$$\int_0^1 \operatorname{div}_h w^h(x_h, x_3)dx_3 = 0.$$

Thus for any non-negative integer k_3, we get

$$\int_0^1 \operatorname{div}_h w^h(x_h, x_3)\cos(k_3\pi x_3)dx_3 = 0$$

and thus $\operatorname{div} w^h(x_h, x_3) = 0$ on Ω. As the vector field w is divergence-free. we have $\partial_3 w^3 = 0$. As $w^3(x_h, 1) = w^3(x_h, 0) = 0$, we have $w^3 \equiv 0$ on Ω.

But w is orthogonal to $\mathcal{B}$; so for any function φ on $\mathbf{R}^2$, the Fourier transform of which is included in the ring $\mathcal{C}_N$ (and of vanishing horizontal mean if $\Omega_h = \mathbf{T}^2$), we have for any non-negative integer k_3

$$\big(w|(\nabla^{h,\perp}\varphi\cos(k_3\pi x_3), 0)\big)_{L^2} = 0.$$

This means exactly that, for any non-negative integer k_3.

$$\int_{\mathbf{R}^2}\varphi(x_h)\left(\int_0^1 \operatorname{curl}_h w^h(x_h, x_3)\cos(k_3\pi x_3)dx_3\right)dx_h.$$

Thus we have $\operatorname{curl}_h w^h = 0$ on Ω. The lemma is proved. □

Remarks • The choice of the basis $(\cos(k_3\pi x_3))_{k_3\in\mathbf{N}}$ for the horizontal component ensures that the boundary condition $v^3{}_{|\partial\Omega} = 0$ is satisfied because the vertical component is of the type $\sin(k_3\pi x_3)$ thanks to the divergence-free condition.

• Let us note that for any v in $\overline{\mathcal{B}}$, we have

$$\|v\|^2_{\mathcal{H}} = \sum_{k_3=0}^{\infty}\|v^{k_3,h}\|^2_{L^2(\Omega_h)} + \sum_{k_3=1}^{\infty}\frac{1}{(k_3\pi)^2}\|\operatorname{div}_h v^{k_3,h}\|^2_{L^2(\Omega_h)}. \tag{7.4.1}$$

This leads us naturally to the following definition. We define the sets

$$G = \left\{w\in L^2(\Omega_h;\mathbf{R}^2)\ \big/\ \operatorname{div}_h w\in L^2(\Omega_h)\right\}\quad\text{and}$$

$$G_0 = G\cap\left\{w\in L^2(\mathbf{T}^2;\mathbf{R}^2)\ \big/\ \int_{\mathbf{T}^2} w(x_h)\,dx_h = 0\right\}.$$

Then for any $w \in G$ (resp. in G_0) and any $v \in \mathcal{H}$ (resp. in $\mathcal{H}_0$), we define

$$\|w\|_{k_3,\Omega_h}^2 = \|w\|_{L^2(\Omega_h)}^2 + \frac{1}{(k_3\pi)^2}\|\operatorname{div}_h w\|_{L^2(\Omega_h)}^2 \quad \text{and}$$

$$v^{k_3,h}(x_h) = \int_0^1 v^h(x_h,x_3)\cos(k_3\pi x_3)\,dx_3.$$

Then Lemma 7.6 implies that

$$v(x) = \begin{pmatrix} \displaystyle\sum_{k_3\in\mathbf{N}} v^{k_3,h}(x_h)\cos(k_3\pi x_3) \\ \displaystyle -\sum_{k_3\geq 1}\frac{1}{k_3\pi}\operatorname{div}_h v^{k_3,h}(x_h)\sin(k_3\pi x_3) \end{pmatrix}$$

as well as the two easy but useful properties:

$$\|v\|_{L^2(\Omega)}^2 = \sum_{k_3\in\mathbf{N}} \|v^{k_3,h}\|_{k_3,\Omega_h}^2 \quad \text{and} \tag{7.4.2}$$

$$\|v\|_{L^2(\Omega)}^2 + \|\partial_3 v\|_{L^2(\Omega)}^2 = \sum_{k_3\in\mathbf{N}} (1+(k_3\pi)^2)\|v^{k_3,h}\|_{k_3,\Omega_h}^2. \tag{7.4.3}$$

Now let us start the construction of the approximate system. Let us consider a vector field v in the closure of $C^\infty([0,T];\mathcal{B})$ for the energy norm

$$\|v\|_{L^\infty([0,T];L^2)}^2 + 2\nu\int_0^T \|\nabla^h v(t')\|_{L^2}^2\,dt'.$$

We want to construct a family $v_{\text{app}}^\varepsilon$ (smooth, divergence-free and vanishing at the boundary), and an operator $\underline{L}$ such that

$$L^\varepsilon v_{\text{app}}^\varepsilon = \underline{L}v,$$

up to small remainder terms. As we shall see later on, the operator $\underline{L}$ is related to the linear part of the system $(\text{NSE}_{\nu,\varepsilon})$, page 158. More precisely, we want to construct, for any $\eta > 0$, a family $(v_{\text{app}}^{\varepsilon,\eta})_{\varepsilon>0}$ and a family $(N_\eta)_{\eta>0}$ of integers such that

$$L^\varepsilon v_{\text{app}}^{\varepsilon,\eta} = \underline{L}v + R^{\varepsilon,\eta}, \quad \text{with}$$

$$\lim_{\eta\to 0}\limsup_{\varepsilon\to 0}\|R^{\varepsilon,\eta}\|_{L^2([0,T];H^{-1,0})} = 0$$

while the family $(N_\eta)_{\eta>0}$ is given by the condition that

$$\|v - v_{N_\eta}\|^2_{L^\infty([0,T];L^2)} + 2\nu \int_0^T \|\nabla^h (v - v_{N_\eta})(t')\|^2_{L^2}\, dt' \leq \eta.$$

It seems reasonable to think that $\partial_t v^\varepsilon_{\text{app}}$ will be of size $1/\varepsilon$. This leads us to look for solutions of the form

$$\begin{aligned} v^\varepsilon_{\text{app}} &= v_{0,\text{int}} + v_{0,\text{BL}} + \varepsilon v_{1,\text{int}} + \varepsilon v_{1,\text{BL}} + \cdots, \\ p^\varepsilon &= \frac{1}{\varepsilon} p_{-1,\text{int}} + \frac{1}{\varepsilon} p_{-1,\text{BL}} + p_{0,\text{int}} + p_{0,\text{BL}} + \cdots, \end{aligned} \tag{7.4.4}$$

where $v_{j,\text{int}}$, $v_{j,\text{BL}}$, $p_{j,\text{int}}$, and $p_{j,\text{BL}}$ are respectively of the form

$$\sum_{k_3 \leq N_\eta} \left(v^{k_3,h}_{j,\text{int}}(\tau, t, x_h) \cos(k_3 \pi x_3), -\frac{1}{k_3 \pi} \operatorname{div}_h v^{k_3,h}_{j,\text{int}}(\tau, t, x_h) \sin(k_3 \pi x_3) \right),$$

$$f_j\left(\tau, t, x_h, \frac{x_3}{\varepsilon}\right) + g_j\left(\tau, t, x_h, \frac{1 - x_3}{\varepsilon}\right),$$

$$\sum_{k_3 \leq N_\eta} p^{k_3}_{j,\text{int}}(\tau, t, x_h) \cos(k_3 \pi x_3)$$

and

$$p_j\left(\tau, t, x_h, \frac{x_3}{\varepsilon}\right) + \widetilde{p}_j\left(\tau, t, x_h, \frac{1 - x_3}{\varepsilon}\right),$$

with $\tau = t/\varepsilon\cdot$

The main difference with the well-prepared case comes from the terms of order ε^{-1} in the interior. They vanish if, for any $k_3 \leq N_\eta$,

$$\begin{cases} \partial_\tau v^{k_3,h}_{0,\text{int}} + R v^{k_3,h}_{0,\text{int}} = -\nabla^h p^{k_3}_{-1,\text{int}} \\ \partial_\tau v^{k_3,h}_{0,\text{int}} = -k_3 \pi p^{k_3}_{-1,\text{int}}. \end{cases} \tag{7.4.5}$$

Let us introduce the following new unknown W defined by

$$W = \begin{pmatrix} W^h \\ W^3 \end{pmatrix} \stackrel{\text{def}}{=} \begin{pmatrix} \mathcal{F}_h \operatorname{div}_h v^h \\ \mathcal{F}_h \operatorname{curl}_h v^h \\ \mathcal{F}_h v^3 \end{pmatrix},$$

and we also denote $\widehat{p} \stackrel{\text{def}}{=} \mathcal{F}_h p$. The unknown v is recovered from W by

$$v = \mathcal{F}_h^{-1} \begin{pmatrix} A(\xi_h) W^h \\ W^3 \end{pmatrix}, \tag{7.4.6}$$

where $A(\xi_h)$ is defined by

$$A(\xi_h) \stackrel{\text{def}}{=} \begin{pmatrix} \xi_1 |\xi_h|^{-2} & -\xi_2 |\xi_h|^{-2} \\ \xi_2 |\xi_h|^{-2} & \xi_1 |\xi_h|^{-2} \end{pmatrix}. \tag{7.4.7}$$

We recall that ξ_h is in a fixed ring of $\mathbf{R}^2$. Let us notice that the divergence free condition on v becomes, expressed in terms of W,

$$W^1 + \partial_3 W^3 = 0. \tag{7.4.8}$$

Moreover the system $(\mathrm{SC}^\varepsilon_\beta)$ turns out to be

$$\widetilde{L}^\varepsilon = \begin{pmatrix} |\xi_h|^2 \widehat{p} \\ 0 \\ -\partial_3 \widehat{p} \end{pmatrix} \quad \text{with} \tag{7.4.9}$$

$$\widetilde{L}^\varepsilon \stackrel{\text{def}}{=} \begin{pmatrix} \partial_t W^{\varepsilon,h} + \nu|\xi_h|^2 W^{\varepsilon,h} - \beta\varepsilon\partial_3^2 W^{\varepsilon,h} + \dfrac{1}{\varepsilon} R W^{\varepsilon,h} \\ \partial_t W^{\varepsilon,3} + \nu|\xi_h|^2 W^{\varepsilon,3} - \beta\varepsilon\partial_3^2 W^{\varepsilon,3} \end{pmatrix}.$$

Remark All the computations from now on will be carried out on W rather than on v, and in particular in the case $k_3 = 0$. We recommend to the reader the exercise of rewriting the proof of Lemma 7.1 in that new formulation and recovering the formulas we will derive here (in the case $k_3 = 0$).

The Ansatz (7.4.4) becomes

$$\begin{aligned} W^\varepsilon &= W_{0,\text{int}} + W_{0,\text{BL}} + \varepsilon W_{1,\text{int}} + \varepsilon W_{1,\text{BL}} + \cdots, \\ \widehat{p^\varepsilon} &= \frac{1}{\varepsilon}\widehat{p}_{-1,\text{int}} + \frac{1}{\varepsilon}\widehat{p}_{-1,\text{BL}} + \widehat{p}_{0,\text{int}} + \widehat{p}_{0,\text{BL}} + \cdots, \end{aligned} \tag{7.4.10}$$

where $W_{j,\text{int}}$, $W_{j,\text{BL}}$, $\widehat{p}_{j,\text{int}}$, and $\widehat{p}_{j,\text{BL}}$ are respectively of the form

$$\sum_{k_3 \leq N_\eta} \left(W^{k_3,h}_{j,\text{int}}(\tau,t,\xi_h)\cos(k_3\pi x_3), -\frac{1}{k_3\pi} W^{k_3,1}_{j,\text{int}}(\tau,t,\xi_h)\sin(k_3\pi x_3) \right),$$

$$f_j\left(\tau,t,\xi_h,\frac{x_3}{\varepsilon}\right) + g_j\left(\tau,t,\xi_h,\frac{1-x_3}{\varepsilon}\right),$$

$$\sum_{k_3 \leq N_\eta} p^{k_3}_{j,\text{int}}(\tau,t,\xi_h)\cos(k_3\pi x_3)$$

and

$$\widehat{p}_j\left(\tau,t,\xi_h,\frac{x_3}{\varepsilon}\right) + \widetilde{p}_j\left(\tau,t,\xi_h,\frac{1-x_3}{\varepsilon}\right),$$

with $\tau = t/\varepsilon \cdot$ Now the relation (7.4.5) turns out to be equivalent to

$$\forall k_3 \leq N_\eta, \quad \begin{cases} \partial_\tau W^{k_3,h}_{0,\text{int}} + R W^{k_3,h}_{0,\text{int}} = \begin{pmatrix} |\xi_h|^2 \widehat{p}^{k_3}_{-1,\text{int}} \\ 0 \end{pmatrix} \\ \partial_\tau W^{k_3,3}_{0,\text{int}} = -k_3\pi \widehat{p}^{k_3}_{-1,\text{int}}. \end{cases} \tag{7.4.11}$$

The divergence-free condition as expressed in (7.4.8) gives

$$\forall k_3 \leq N_\eta\,, \quad W^{k_3,1}_{0,\text{int}} + k_3\pi W^{k_3,3}_{0,\text{int}} = 0. \tag{7.4.12}$$

It determines as usual the pressure which is given here by

$$\widehat{p}^{k_3}_{-1,\text{int}} = -\frac{1}{|\xi_h|^2 + (k_3\pi)^2} W^{k_3,2}_{0,\text{int}}.$$

By (7.4.12), $W^{k_3,3}_{0,\text{int}}$ is totally determined by $W^{k_3,1}_{0,\text{int}}$ and the relation (7.4.11) becomes

$$\partial_\tau W^{k_3,h}_{0,\text{int}} = R_{k_3} W^{k_3,h}_{0,\text{int}} \tag{7.4.13}$$

with

$$R_{k_3} \stackrel{\text{def}}{=} \begin{pmatrix} 0 & -\lambda^2_{k_3} \\ 1 & 0 \end{pmatrix} \quad \text{and} \quad \lambda_{k_3} \stackrel{\text{def}}{=} \left(\frac{(k_3\pi)^2}{|\xi_h|^2 + (k_3\pi)^2} \right)^{\frac{1}{2}} \cdot \tag{7.4.14}$$

Remark Although it does not appear in the notation, to avoid unnecessary heaviness, one should keep in mind that λ_{k_3} depends on the horizontal frequency and not on k_3 alone.

Thus, if $\mathcal{L}_{k_3}$ is the matrix defined by

$$\mathcal{L}_{k_3}(\tau) \stackrel{\text{def}}{=} \begin{pmatrix} \cos\tau_{k_3} & \lambda_{k_3}\sin\tau_{k_3} \\ -\dfrac{1}{\lambda_{k_3}}\sin\tau_{k_3} & \cos\tau_{k_3} \end{pmatrix} \quad \text{with} \quad \tau_{k_3} \stackrel{\text{def}}{=} \lambda_{k_3}\tau, \tag{7.4.15}$$

we get that $W^{k_3,h}_{0,\text{int}}$ can be put into the form

$$W^{k_3,h}_{0,\text{int}} = \mathcal{L}_{k_3}(\tau)\widetilde{W}^{k_3}(t,\xi_h). \tag{7.4.16}$$

where

$$\widetilde{W}^{k_3}(t,\xi_h) = \mathcal{F}_h \begin{pmatrix} \text{div}_h\, v^{k_3,h}_{N_\eta} \\ \text{curl}_h v^{k_3,h}_{N_\eta} \end{pmatrix} (t,\xi_h). \tag{7.4.17}$$

Let us notice that obviously $\mathcal{L}_{k_3}(\tau)$ satisfies

$$\dot{\mathcal{L}}_{k_3} + R_{k_3}\mathcal{L}_{k_3} = 0 \quad \text{with} \quad \mathcal{L}_{k_3}(0) = \text{Id}\,.$$

To sum up, we have

$$W^h_{0,\text{int}} = \sum_{k_3 \leq N_\eta} \mathcal{L}_{k_3}(\tau)\widetilde{W}^{k_3}(t,\xi_h)\cos(k_3\pi x_3) \tag{7.4.18}$$

and, because of the divergence-free condition as expressed in (7.4.12),

$$W^3_{0,\text{int}} = -\sum_{k_3 \leq N_\eta} \frac{1}{k_3\pi}\Pi^1\mathcal{L}_{k_3}(\tau)\widetilde{W}^{k_3}(t,\xi_h)\sin(k_3\pi x_3) \tag{7.4.19}$$

where Π^1 denotes the projection on the first coordinates in the horizontal plane.

Remark The terms $(\widetilde{W}^{k_3})_{1\le k_3\le N_\eta}$ can be understood as the filtered part of $W_{0,\mathrm{int}}$ in the sense of Chapter 6.

From now on, we shall follow exactly the same lines as in the well-prepared case. Let us recall them:

- We first ensure the boundary condition by introducing $W^{k_3}_{0,\mathrm{BL}}$.
- Then, as $W^{k_3}_{0,\mathrm{BL}}$ violates the divergence-free condition, we introduce the boundary layer of size ε $W^{k_3,3}_{1,\mathrm{BL}}$ in order to ensure that divergence-free condition.
- This introduction of $W^{k_3,1}_{1,\mathrm{BL}}$ destroys the boundary condition which we restore by introducing $W^{k_3}_{1,\mathrm{int}}$.

The existence of time oscillations will make the operations a little bit more delicate, especially in the third step where really new phenomena will appear because of fast time oscillations.

Step 1: The boundary layer of size ε^0
Let us study the term of size ε^0 for the horizontal component of the boundary layer. As the third component of the interior solution of size ε^0 is 0 on the boundary, then the third component of the boundary layer of size ε^0 is identically 0. Thus $\partial_3\widehat{p}_{-1,\mathrm{BL}}=0$. We meet again the well-known fact that the pressure does not vary in the boundary layers.

As $\cos(k_3\pi)=(-1)^{k_3}$ we look for the boundary layer in the form

$$W^{k_3,h}_{0,\mathrm{BL}}\stackrel{\mathrm{def}}{=}M_{k_3}\left(\frac{x_3}{\varepsilon}\right)\mathcal{L}_{k_3}\left(\frac{t}{\varepsilon}\right)\widetilde{W}^{k_3}+(-1)^{k_3}M_{k_3}\left(\frac{1-x_3}{\varepsilon}\right)\mathcal{L}_{k_3}\left(\frac{t}{\varepsilon}\right)\widetilde{W}^{k_3}$$

and

$$W^{k_3,3}_{0,\mathrm{BL}}=0.$$

The term of size ε^0:

$$\widetilde{L}^{\varepsilon}W^{k_3,h}_{0,\mathrm{BL}}=\partial_t W^{k_3,h}_{0,\mathrm{BL}}-\beta\varepsilon\partial_3^2W^{k_3,h}_{0,\mathrm{BL}}+\frac{1}{\varepsilon}RW^{k_3,h}_{0,\mathrm{BL}}$$

must be zero, so we infer that

$$M_{k_3}\partial_\tau\mathcal{L}_{k_3}-\beta M''_{k_3}\mathcal{L}_{k_3}+RM_{k_3}\mathcal{L}_{k_3}=0.$$

Let us note that in the case when $k_3=0$, we have $\mathcal{L}_{k_3}=\mathrm{Id}$. This case corresponds to the well-prepared case and the above relation corresponds to (7.1.4). Thus we have

$$W^{0,h}_{0,\mathrm{BL}}=\left(M_0\left(\frac{x_3}{\sqrt{E}}\right)+M_0\left(\frac{1-x_3}{\sqrt{E}}\right)\right)W^{0,h}_{0,\mathrm{int}} \tag{7.4.20}$$

where M_0 is given by (7.1.5).

Now let us assume that $k_3 \geq 1$. As $\partial_\tau \mathcal{L}_{k_3} = -R_{k_3}\mathcal{L}_{k_3}$, it turns out that the equation on the boundary layer is

$$\begin{cases} -\beta M''_{k_3} = M_{k_3}R_{k_3} - RM_{k_3} \\ M_{k_3}(0) = -\operatorname{Id} \\ M_{k_3}(+\infty) = 0. \end{cases}$$

This is a linear differential equation of order 2 with an initial and a final condition. The solution is given by

$$M_{k_3}(\zeta) = -\sum_{\pm} \frac{1}{2}\mu^\pm_{k_3} \exp(-\zeta^\pm_{k_3}) M^\pm_{k_3}(\zeta^\pm_{k_3}) \quad \text{with}$$

$$M^\pm_{k_3}(\theta) \stackrel{\text{def}}{=} \begin{pmatrix} \cos\theta & \mp\lambda_{k_3}\sin\theta \\ -\sin\theta & \mp\lambda_{k_3}\cos\theta \end{pmatrix}$$

and

$$\zeta^\pm_{k_3} \stackrel{\text{def}}{=} \frac{\zeta}{\sqrt{2\beta^\pm_{k_3}}}, \quad \beta^\pm_{k_3} \stackrel{\text{def}}{=} \frac{\beta}{1 \pm \lambda_{k_3}} \quad \text{and} \quad \mu^\pm_{k_3} \stackrel{\text{def}}{=} 1 \mp \frac{1}{\lambda_{k_3}} \cdot \tag{7.4.21}$$

So stating $E^\pm_{k_3} \stackrel{\text{def}}{=} 2\varepsilon^2\beta^\pm_{k_3}$, we infer by the definition of $\mathcal{L}_{k_3}$ that

$$\begin{aligned} W^{k_3,h}_{0,\mathrm{BL}} = &-\frac{1}{2}\sum_{\pm} \mu^\pm_{k_3} \exp\left(-\frac{x_3}{\sqrt{E^\pm_{k_3}}}\right) M^\pm_{k_3}\left(\frac{x_3}{\sqrt{E^\pm_{k_3}}} \mp \frac{\lambda_{k_3}t}{\varepsilon}\right) \widetilde{W}^{k_3} \\ &- \frac{(-1)^{k_3}}{2}\sum_{\pm} \mu^\pm_{k_3} \exp\left(-\frac{1-x_3}{\sqrt{E^\pm_{k_3}}}\right) M^\pm_{k_3}\left(\frac{1-x_3}{\sqrt{E^\pm_{k_3}}} \mp \frac{\lambda_{k_3}t}{\varepsilon}\right) \widetilde{W}^{k_3}, \end{aligned}$$

recalling the definition of $\widetilde{W}^{k_3}$ in (7.4.17). To sum up this step, if we have

$$W^h_{0,\mathrm{BL}} = \sum_{k_3=0}^{N_\eta} W^{k_3,h}_{0,\mathrm{BL}} \quad \text{and} \quad W^3_{0,\mathrm{BL}} = 0, \tag{7.4.22}$$

where $W^{0,h}_{0,\mathrm{BL}}$ is given by (7.4.20) and $W^{k_3,h}_{0,\mathrm{BL}}$ for positive k_3 by the above equation, then

$$\|\widetilde{L}^\varepsilon W_{0,\mathrm{BL}}\|_{L^2([0,T];H^{-1,0})} \leq C_\eta \varepsilon^{\frac{1}{2}}. \tag{7.4.23}$$

Remarks

- The eigenvectors of $M_{k_3}^{\pm}$ and those of $\mathcal{L}_{k_3}$ are different and time oscillations introduce a phase shift in the boundary layer.
- The fact that we work with a vector field, the horizontal Fourier transform of which has its support in a ring, prevents λ_{k_3} from being too close to 1.
- Let us note that the boundary layer operators M_{k_3} depend strongly on (ξ_h, k_3).

Step 2: The divergence-free condition for boundary layers
As in the well-prepared case, the fact that the boundary layer must be divergence-free implies that we have to introduce a vertical component of the boundary layer of size ε. When $k_3 = 0$, which corresponds to the well-prepared case, thanks to (7.4.20), we find the analog of (7.1.11) in terms of the unknown W, i.e.

$$W_{1,\mathrm{BL}}^{0,3} = \sqrt{2\beta}\left(f\left(\frac{x_3}{\sqrt{E}}\right) - f\left(\frac{1-x_3}{\sqrt{E}}\right)\right) W_{0,\mathrm{int}}^2, \tag{7.4.24}$$

still with $f(\zeta) = -\frac{1}{2}e^{-\zeta}(\sin\zeta + \cos\zeta)$.

Now let us assume that k_3 is positive. The divergence-free condition as expressed in (7.4.12) gives $\varepsilon\partial_3 W_{1,\mathrm{BL}}^{k_3,3} = -\Pi^1 W_{0,\mathrm{BL}}^{k_3,h}$. So we get

$$\begin{aligned}
\varepsilon\partial_3 W_{1,\mathrm{BL}}^{k_3,3} &= \frac{1}{2}\sum_{\pm}\mu_{k_3}^{\pm}\exp\left(-\frac{x_3}{\sqrt{E_{k_3}^{\pm}}}\right)\\
&\times\left(\cos\left(\frac{x_3}{\sqrt{E_{k_3}^{\pm}}} \mp \frac{\lambda_{k_3}t}{\varepsilon}\right)\widetilde{W}^{k_3,1} \mp \lambda_{k_3}\sin\left(\frac{x_3}{\sqrt{E_{k_3}^{\pm}}} \mp \frac{\lambda_{k_3}t}{\varepsilon}\right)\widetilde{W}^{k_3,2}\right)\\
&+\frac{(-1)^{k_3}}{2}\sum_{\pm}\mu_{k_3}^{\pm}\exp\left(-\frac{1-x_3}{\sqrt{E_{k_3}^{\pm}}}\right)\\
&\times\left(\cos\left(\frac{1-x_3}{\sqrt{E_{k_3}^{\pm}}} \mp \frac{\lambda_{k_3}t}{\varepsilon}\right)\widetilde{W}^{k_3,1} \mp \lambda_{k_3}\sin\left(\frac{1-x_3}{\sqrt{E_{k_3}^{\pm}}} \mp \frac{\lambda_{k_3}t}{\varepsilon}\right)\widetilde{W}^{k_3,2}\right).
\end{aligned}$$

We get, by integration and with the notation

$$\mathrm{cs}^{\pm} \stackrel{\mathrm{def}}{=} \cos \pm \sin \quad \text{and} \quad \gamma_{k_3}^{\pm} \stackrel{\mathrm{def}}{=} \mu_{k_3}^{\pm}\sqrt{2\beta_{k_3}^{\pm}}, \tag{7.4.25}$$

$$
\begin{aligned}
W_{1,\mathrm{BL}}^{k_3,3} = & -\frac{1}{4}\sum_{\pm}\gamma_{k_3}^{\pm}\exp\left(-\frac{x_3}{\sqrt{E_{k_3}^{\pm}}}\right)\\
& \times\left(\mathrm{cs}^{-}\left(\frac{x_3}{\sqrt{E_{k_3}^{\pm}}}\mp\frac{\lambda_{k_3}t}{\varepsilon}\right)\widetilde{W}^{k_3,1}\mp\lambda_{k_3}\mathrm{cs}^{+}\left(\frac{x_3}{\sqrt{E_{k_3}^{\pm}}}\mp\frac{\lambda_{k_3}t}{\varepsilon}\right)\widetilde{W}^{k_3,2}\right)\\
& +\frac{(-1)^{k_3}}{4}\sum_{\pm}\gamma_{k_3}^{\pm}\exp\left(-\frac{1-x_3}{\sqrt{E_{k_3}^{\pm}}}\right)\\
& \times\left(\mathrm{cs}^{-}\left(\frac{1-x_3}{\sqrt{E_{k_3}^{\pm}}}\mp\frac{\lambda_{k_3}t}{\varepsilon}\right)\widetilde{W}^{k_3,1}\mp\lambda_{k_3}\mathrm{cs}^{+}\left(\frac{1-x_3}{\sqrt{E_{k_3}^{\pm}}}\mp\frac{\lambda_{k_3}t}{\varepsilon}\right)\widetilde{W}^{k_3,2}\right).
\end{aligned}
$$

It is clear that this boundary layer is a sum of a rapidly decreasing function of x_3/ε and of a rapidly decreasing function of $(1-x_3)/\varepsilon$. But it is obvious that these two functions do not vanish, respectively, at $x_3 = 0$ and $x_3 = 1$. To sum up, we have

$$
W_{1,\mathrm{BL}}^{3} = \sum_{k_3=0}^{N_\eta} W_{1,\mathrm{BL}}^{k_3,3} \tag{7.4.26}
$$

where $W_{1,\mathrm{BL}}^{0,3}$ is defined by (7.4.24) and the $W_{1,\mathrm{BL}}^{k_3,3}$ for positive k_3 are defined by the above formula. The horizontal boundary layer is not given at this stage: it will be introduced only to ensure that the boundary condition of size ε is satisfied.

Step 3: The boundary condition of size ε

In the case when $k_3 = 0$, we have, as in (7.1.12) and (7.1.13), up to exponentially small terms,

$$
W_{1,\mathrm{BL}|x_3=0}^{0,3} = -W_{1,\mathrm{BL}|x_3=1}^{0,3} = -D_0\left(\frac{t}{\varepsilon}\right)\widetilde{W}^{0}
$$

where $D_0 : \mathbf{R}^+ \to \mathcal{L}(\mathbf{R}^2;\mathbf{R})$ is defined by

$$
D_0(\tau)\begin{pmatrix} A^1 \\ A^2 \end{pmatrix} \stackrel{\mathrm{def}}{=} \frac{1}{2}\sqrt{2\beta}\,A^2. \tag{7.4.27}
$$

Obviously D_0 does not in fact depend on τ, but we use that notation to be consistent with the $k_3 \neq 0$ case below.

In the case when k_3 is positive, up to exponentially small terms we have, stating again $\tau_{k_3} = \lambda_{k_3}\tau$,

$$
W_{1,\mathrm{BL}|x_3=0}^{k_3,3} = -(-1)^{k_3}W_{1,\mathrm{BL}|x_3=1}^{k_3,3} = -D_{k_3}\left(\frac{t}{\varepsilon}\right)\widetilde{W}^{k_3}
$$

where D_{k_3} is the linear form on $\mathbf{R}^2$ defined by

$$D_{k_3}(\tau)\begin{pmatrix}A^1\\A^2\end{pmatrix}\stackrel{\text{def}}{=}\frac{1}{4}\sum_{\pm}\gamma_{k_3}^{\pm}\left(\mathrm{cs}^{\pm}(\tau_{k_3})A^1\mp\lambda_{k_3}\mathrm{cs}^{\mp}(\tau_{k_3})A^2\right). \tag{7.4.28}$$

As in the well-prepared case (see page 162), let us lift the boundary value of $W_{1.\mathrm{BL}}^{k_3,3}$ by introducing the divergence-free vector field $\underline{W}_{1,\mathrm{int}}$ defined by

$$\underline{W}_{1,\mathrm{int}}(\tau)\stackrel{\text{def}}{=}\sum_{\ell=0}^{N_\eta}\left(D_\ell(\tau)\widetilde{W}^\ell\right)\begin{pmatrix}\delta_\ell\\0\\r_\ell(x_3)\end{pmatrix} \tag{7.4.29}$$

with $\delta_\ell=1+(-1)^\ell$ and $r_\ell(x_3)=1$ when ℓ is odd, $r_\ell(x_3)=1-2x_3$ when ℓ is even. Then let us look for $W_{1.\mathrm{int}}$ in the form

$$W_{1,\mathrm{int}}=\underline{W}_{1,\mathrm{int}}+\sum_{k_3=0}^{M_\eta}\begin{pmatrix}W_{1,\mathrm{int}}^{k_3,h}\cos(k_3\pi x_3)\\-\dfrac{1}{k_3\pi}W_{1,\mathrm{int}}^{k_3,1}\sin(k_3\pi x_3)\end{pmatrix}$$

where M_η is an integer (greater than N_η) which will be chosen later on. Let us note that the boundary condition on the vertical component together with the divergence-free condition are satisfied, namely

$$(W_{1,\mathrm{BL}}^3+W_{1,\mathrm{int}}^3)_{|\partial\Omega}=0\quad\text{and}\quad W_{1,\mathrm{int}}^1+\partial_3W_{1,\mathrm{int}}^3=0.$$

Let us compute up to a gradient the terms of size ε^0 of

$$\widetilde{L}^\varepsilon(W_{0,\mathrm{int}}+\varepsilon W_{1,\mathrm{int}}).$$

By definition of $W_{0,\mathrm{int}}$, we have

$$\mathbf{P}\widetilde{L}^\varepsilon W_{0,\mathrm{int}}=\sum_{k_3=0}^{N_\eta}\begin{pmatrix}\mathcal{L}_{k_3}\left(\dfrac{t}{\varepsilon}\right)(\partial_t+\nu|\xi_h|^2)\widetilde{W}^{k_3}\cos(k_3\pi x_3)\\-\dfrac{1}{k_3\pi}\Pi^1\left(\mathcal{L}_{k_3}\left(\dfrac{t}{\varepsilon}\right)(\partial_t+\nu|\xi_h|^2)\widetilde{W}^{k_3}\right)\sin(k_3\pi x_3)\end{pmatrix}.$$

By definition of $W_{1,\mathrm{int}}$, we have

$$\varepsilon\mathbf{P}\widetilde{L}^\varepsilon W_{1,\mathrm{int}}=\sum_{k_3=0}^{M_\eta}\begin{pmatrix}(\partial_\tau+R)W_{1,\mathrm{int}}^{k_3,h}\cos(k_3\pi x_3)\\-\dfrac{1}{k_3\pi}\partial_\tau W_{1,\mathrm{int}}^{k_3,1}\sin(k_3\pi x_3)\end{pmatrix}+\begin{pmatrix}(\partial_\tau+R)\underline{W}_{1,\mathrm{int}}^h\\\partial_\tau\underline{W}_{1,\mathrm{int}}^3\end{pmatrix}.$$

Let us approximate $\underline{W}_{1,\mathrm{int}}^3$ (and thus $\partial_\tau\underline{W}_{1,\mathrm{int}}^3$) by a sum of $\sin(k_3\pi x_3)$. This is allowed by the following proposition.

Proposition 7.1 *For any η, an integer M_η greater than N_η exists such that, if $\underline{W}_{1,\mathrm{int},M_\eta}$ is defined by*

$$\underline{W}_{1,\mathrm{int},M_\eta}=\sum_{\ell=0}^{N_\eta}\left(D_\ell(\tau)\widetilde{W}^\ell\right)\begin{pmatrix}\delta_\ell\\0\\\displaystyle\sum_{k_3=1}^{M_\eta}r_{\ell,k_3}\sin(k_3\pi x_3)\end{pmatrix}, \tag{7.4.30}$$

with $r_{\ell,k_3} \stackrel{\text{def}}{=} \frac{1}{k_3\pi}(1+(-1)^{\ell+k_3})$. *then we have*

$$\forall T>0\,,\ \lim_{\eta\to 0}\left\|\mathcal{F}^{-1}\partial_\tau\left(\underline{W}^3_{1,\text{int}}-\underline{W}^3_{1,\text{int},M_\eta}\right)\right\|_{L^2([0,T];H^{-1,0})}=0.$$

Proof Let us first observe that, by definition of the r_{ℓ,k_3}, we have

$$r_\ell(x_3)=\sum_{k_3\geq 0} r_{\ell,k_3}\sin(k_3\pi x_3).$$

Thus, by definition of $\underline{W}_{1,\text{int},M_\eta}$, we have

$$\partial_\tau(\underline{W}^3_{1,\text{int}}-\underline{W}^3_{1,\text{int},M_\eta})=\sum_{\ell=0}^{N_\eta}\sum_{k_3\geq M_\eta+1} r_{\ell,k_3}\partial_\tau D_\ell(\tau)\widetilde{W}^\ell\sin(k_3\pi x_3).$$

By definition of the $H^{-1,0}$ norm, we have

$$\begin{aligned}\left\|\partial_\tau(\underline{W}^3_{1,\text{int}}-\underline{W}^3_{1,int,M_\eta})\right\|^2_{H^{-1,0}} &= \left\|\sum_{\ell=0}^{N_\eta}\partial_\tau D_\ell(\tau)\widetilde{W}^\ell\right\|^2_{H^{-1}(\Omega_h)}\\ &\quad\times\left\|\sum_{k_3\geq M_\eta} r_{\ell,k_3}\sin(k_3\pi x_3)\right\|^2_{L^2(]0,1[)}\\ &\leq \frac{C}{M_\eta}\left\|\sum_{\ell=0}^{N_\eta}\partial_\tau D_\ell(\tau)\widetilde{W}^\ell\right\|^2_{H^{-1}(\Omega_h)}\\ &\leq \frac{C_\eta}{M_\eta}\sum_{\ell=0}^{N_\eta}\|\widetilde{W}^\ell\|^2_{H^{-1}(\Omega_h)}.\end{aligned}$$

Thanks to (7.4.19), the family is assumed to be bounded in $H^{-1,0}$ (this is the L^2 energy estimate on the original vector field v). Thus we have

$$\left\|\partial_\tau(\underline{W}^3_{1,\text{int}}-\underline{W}^3_{1,\text{int},M_\eta})\right\|^2_{H^{-1,0}}\leq\frac{C_\eta}{M_\eta}.$$

Proposition 7.1 is now proved. □

Remarks

- It is obvious that, for any M_η, $\underline{W}^h_{1,\text{int}}=\underline{W}^h_{1,\text{int},M_\eta}$.
- Until now, the time oscillations produced more complicated formulas compared to the well-prepared case, but no real different phenomena. The real difference will appear now. Indeed, in the previous computations, time oscillations of frequency λ_ℓ were coupled only with the vertical mode of frequency ℓ. This is not the case anymore because $\underline{W}^3_{1,\text{int}}$ contains time oscillations of frequency λ_ℓ for any $\ell\in\{1,\ldots,N_\eta\}$.

- The precision of the whole approximation will be limited by the above proposition.
- The number of vertical modes M_η used for the approximation may be much greater than N_η.

Let us go back to algebraic computations. Thanks to the above Proposition 7.1, we have

$$\varepsilon \widetilde{L}^\varepsilon W_{1,\mathrm{int}} = \sum_{k_3=0}^{M_\eta} \begin{pmatrix} (\partial_\tau + R) W_{1,\mathrm{int}}^{k_3,h} \cos(k_3 \pi x_3) \\ \partial_\tau \left(-\dfrac{W_{1,\mathrm{int}}^{k_3,1}}{k_3\pi} + \displaystyle\sum_{\ell=0}^{N_\eta} r_{\ell,k_3} D_\ell \widetilde{W}^\ell \right) \sin(k_3 \pi x_3) \end{pmatrix}$$
$$+ \begin{pmatrix} (\partial_\tau + R) \underline{W}_{1,\mathrm{int}}^h \\ 0 \end{pmatrix} + \begin{pmatrix} -|\xi_h|^2 \widehat{p}_{0,\mathrm{int}} \\ 0 \\ \partial_3 \widehat{p}_{0,\mathrm{int}} \end{pmatrix} + R^{\varepsilon,\eta}.$$

As usual, the divergence-free condition determines the pressure. Here, it seems natural to look for $\widehat{p}_{0,\mathrm{int}}$ in the form

$$\widehat{p}_{0,\mathrm{int}} = \sum_{k_3=0}^{M_\eta} \widehat{p}_{0,\mathrm{int}}^{k_3} \cos(k_3 \pi x_3).$$

We find

$$\widehat{p}_{0,\mathrm{int}}^0 = \frac{1}{|\xi_h|^2} \left(\partial_\tau (W_{1,\mathrm{int}}^{0,1} + \underline{W}_{1,\mathrm{int}}^1) - W_{1,\mathrm{int}}^{0,2} \right) \quad \text{and}$$

$$\widehat{p}_{0,\mathrm{int}}^{k_3} = \frac{1}{|\xi_h|^2 + (k_3\pi)^2} \left(\sum_{\ell \le N_\eta} k_3 \pi r_{\ell,k_3} \frac{dD_\ell}{d\tau} \widetilde{W}^\ell - W_{1,\mathrm{int}}^{k_3,2} \right) \quad \text{for } k_3 \neq 0.$$

This gives

$$\varepsilon \mathbf{P} \widetilde{L}^\varepsilon W_{1,\mathrm{int}} = \sum_{k_3=0}^{M_\eta} \begin{pmatrix} L_{k_3}^{\varepsilon,h} \cos(k_3 \pi x_3) \\ -\dfrac{L_{k_3}^{\varepsilon,1}}{k_3\pi} \sin(k_3 \pi x_3) \end{pmatrix} + R^{\varepsilon,\eta} \quad \text{with}$$

$$L_0^{\varepsilon,h} = \begin{pmatrix} 0 \\ \displaystyle\sum_{\ell=0}^{N_\eta} \delta_\ell D_\ell(\tau) \widetilde{W}^\ell \end{pmatrix} + \begin{pmatrix} 0 \\ \partial_\tau W_{1,\mathrm{int}}^{0,2} + W_{1,\mathrm{int}}^{0,1} \end{pmatrix}$$

and, for $k_3 \neq 0$,

$$L_{k_3}^{\varepsilon,h} = (\partial_\tau + R_{k_3})W^{k_3,h} - F_{k_3} \quad \text{with}$$

$$F_{k_3}(\tau) \stackrel{\text{def}}{=} \frac{1-\lambda_{k_3}^2}{4} \sum_{\ell \leq N_\eta} k_3 \pi r_{\ell,k_3} \lambda_\ell \sum_{\pm} \gamma_\ell^{\pm} M_\ell^{\pm}(\tau) \widetilde{W}^\ell \quad \text{and} \tag{7.4.31}$$

$$M_\ell^{\pm}(\tau) \stackrel{\text{def}}{=} \begin{pmatrix} -\sin\tau_\ell \pm \cos\tau_\ell & \pm\lambda_\ell \sin\tau_\ell + \lambda_\ell \cos\tau_\ell \\ 0 & 0 \end{pmatrix}. \tag{7.4.32}$$

We infer that

$$\mathbf{P}\widetilde{L}^\varepsilon(W_{0,\text{int}} + \varepsilon W_{1,\text{int}}) = \sum_{k_3=0}^{M_\eta} \begin{pmatrix} \mathcal{W}^{k_3,h} \cos(k_3\pi x_3) \\ -\dfrac{\mathcal{W}^{k_3,1}}{k_3\pi} \sin(k_3 \pi x_3) \end{pmatrix}$$

with, for $k_3 = 0$,

$$\mathcal{W}^{0,h} = (\partial_t + \nu|\xi_h|^2) W_{0,\text{int}}^{0,h} + \begin{pmatrix} 0 \\ \displaystyle\sum_{\ell=0}^{N_\eta} \delta_\ell D_\ell(\tau) \widetilde{W}^\ell \end{pmatrix} + \begin{pmatrix} 0 \\ \partial_\tau W_{1,\text{int}}^{0,2} + W_{1,\text{int}}^{0,1} \end{pmatrix},$$

for $k_3 \in \{1, \dots, N_\eta\}$,

$$\mathcal{W}^{k_3,h} = \mathcal{L}_{k_3}\left(\frac{t}{\varepsilon}\right)(\partial_t + \nu|\xi_h|^2)\widetilde{W}^{k_3} + (\partial_\tau + R_{k_3})W^{k_3,h} - F_{k_3}$$

and, for $k_3 \in \{N_\eta + 1, \dots, M_\eta\}$,

$$\mathcal{W}^{k_3,h} = (\partial_\tau + R_{k_3})W^{k_3,h} - F_{k_3}.$$

The following lemma sorts all the time oscillations which are contained in the term F_{k_3}.

Lemma 7.7 *We have the following identity:*

$$F_{k_3}(\tau) = \mathcal{L}_{k_3}(\tau)\left(-B_{k_3}\widetilde{W}^{k_3} + \sum_{\ell=1}^{N_\eta} B_{k_3,\ell}(\tau)\widetilde{W}^\ell\right) \quad \text{if} \quad k_3 \in \{1, \dots, N_\eta\}$$

$$F_{k_3}(\tau) = \mathcal{L}_{k_3}(\tau)\sum_{\ell=1}^{N_\eta} B_{k_3,\ell}(\tau)\widetilde{W}^\ell \quad \text{if} \quad k_3 \in \{N_\eta + 1, \dots, M_\eta\}$$

where $B_{k_3,\ell}(\tau)$ are matrices, the coefficients of which are cosine or sine functions of $(\lambda_{k_3} \pm \lambda_\ell)\tau$ for $\ell \neq k_3$ and of $2\lambda_{k_3}\tau$ when $\ell = k_3$, and where, for $k_3 \leq N_\eta$,

$$B_{k_3} \stackrel{\text{def}}{=} \frac{(1-\lambda_{k_3}^2)\lambda_{k_3}}{4} \begin{pmatrix} \gamma_{k_3}^- - \gamma_{k_3}^+ & -\lambda_{k_3}(\gamma_{k_3}^+ + \gamma_{k_3}^-) \\ \dfrac{\gamma_{k_3}^+ + \gamma_{k_3}^-}{\lambda_{k_3}} & \gamma_{k_3}^- - \gamma_{k_3}^+ \end{pmatrix}.$$

Proof It is elementary. Let us observe that, by the definition of $\mathcal{L}_{k_3}$ and $M_\ell^\pm$, we have

$$\mathcal{L}_{k_3}^{-1}(\tau)M_\ell^\pm = \begin{pmatrix} \cos\tau_{k_3}(-\sin\tau_\ell \pm \cos\tau_\ell) & \lambda_\ell\cos\tau_{k_3}(\cos\tau_\ell \pm \sin\tau_\ell) \\ \dfrac{1}{\lambda_{k_3}}\sin\tau_{k_3}(-\sin\tau_\ell \pm \cos\tau_\ell) & \dfrac{\lambda_\ell}{\lambda_{k_3}}\sin\tau_{k_3}(\cos\tau_\ell \pm \sin\tau_\ell) \end{pmatrix}$$

Using the formulas which transform a product of two sines or cosines into a sum, we infer that, when $\ell \neq k_3$,

$$\mathcal{L}_{k_3}^{-1}(\tau)M_\ell^\pm(\tau) = B_{k_3,\ell}(\tau).$$

When $\ell = k_3$, we have

$$\mathcal{L}_{k_3}^{-1}(\tau)M_{k_3}^\pm = \frac{1}{2}\begin{pmatrix} \pm 1 & \lambda_{k_3} \\ -\dfrac{1}{\lambda_{k_3}} & \pm 1 \end{pmatrix} + B_{k_3,k_3}.$$

This proves the lemma. □

Now we can write the terms $\mathcal{W}^{k_3,h}$ in a simpler way. We have, for $k_3 = 0$,

$$\mathcal{W}^{0,h} = (\partial_t + \nu|\xi_h|^2)W_{0,\text{int}}^{0,h} + \begin{pmatrix} 0 \\ W_{1,\text{int}}^{0,1} \end{pmatrix} + \begin{pmatrix} 0 \\ \partial_\tau W_{1,\text{int}}^{0,2} + \sum_{\ell=0}^{N_\eta} \delta_\ell D_\ell(\tau)\widetilde{W}^\ell \end{pmatrix},$$

for $k_3 \in \{1, \ldots, N_\eta\}$,

$$\begin{aligned}\mathcal{W}^{k_3,h} &= \mathcal{L}_{k_3}(\tau)(\partial_t + \nu|\xi_h|^2) + B_{k_3}\widetilde{W}^{k_3} \\ &\quad + (\partial_\tau + R_{k_3})W_{1,\text{int}}^{k_3,h} - \mathcal{L}_{k_3}(\tau)\sum_{\ell=0}^{N_\eta} B_{k_3,\ell}(\tau)\widetilde{W}^\ell,\end{aligned}$$

and for $k_3 \in \{N_\eta + 1, \ldots, M_\eta\}$,

$$\mathcal{W}^{k_3,h} = (\partial_\tau + R_{k_3})W_{1,\text{int}}^{k_3,h} - \mathcal{L}_{k_3}(\tau)\sum_{\ell=0}^{N_\eta} B_{k_3,\ell}(\tau)\widetilde{W}^\ell.$$

Let us define

$$W_{1.\text{int}}^{0,h} = \begin{pmatrix} 0 \\ \sum_{\ell=1}^{N_\eta} \delta_\ell \sum_\pm \dfrac{\gamma_\ell^\pm}{4\lambda_\ell}\left(\mp \text{cs}^\mp(\tau_\ell)\widetilde{W}^{\ell,1} + \lambda_\ell \text{cs}^\pm(\tau_\ell)\right)\widetilde{W}^{\ell,2} \end{pmatrix} \tag{7.4.33}$$

and, for $k_3 \in \{1, \ldots, M_\eta\}$,

$$W_{1,\text{int}}^{k_3,h} = \mathcal{L}_{k_3}\left(\frac{t}{\varepsilon}\right)\sum_{\ell=1}^{N_\eta} C_{k_3,\ell}\left(\frac{t}{\varepsilon}\right)\widetilde{W}^\ell(t) \tag{7.4.34}$$

where the $C_{k_3,\ell}$ are (2×2 matrix valued) smooth bounded functions of τ, the derivatives of which are the oscillating functions $B_{k_3,\ell}$.

This gives

$$
\begin{aligned}
&\mathbf{P}\widetilde{L}^\varepsilon(W_{0,\text{int}}+\varepsilon W_{1,\text{int}})\\
&\quad=\begin{pmatrix} 0\\ (\partial_t+\nu|\xi_h|^2+\sqrt{2\beta})\,W^0_{0,\text{int}}\\ 0\end{pmatrix} \qquad (7.4.35)\\
&\quad+\begin{pmatrix} \mathcal{L}_{k_3}\left(\dfrac{t}{\varepsilon}\right)(\partial_t+\nu|\xi_h|^2+B_{k_3})\,\widetilde{W}^{k_3}\cos(k_3\pi x_3)\\ -\dfrac{1}{k_3\pi}\Pi^1\mathcal{L}_{k_3}\left(\dfrac{t}{\varepsilon}\right)(\partial_t+\nu|\xi_h|^2+B_{k_3})\,\widetilde{W}^{k_3}\sin(k_3\pi x_3)\end{pmatrix}+R^{\varepsilon,\eta}.
\end{aligned}
$$

Note that the horizontal components do not satisfy the Dirichlet boundary condition. As usual that is taken care of by introducing a boundary layer $W^{k_3,h}_{1,\text{BL}}$ which will cancel out the value of $W^{k_3,h}_{1,\text{int}}$ at the boundary. We therefore define $W^{k_3,2}_{1,\text{BL}}$ exponentially decreasing away from the boundary, such that

$$W^{k_3,h}_{1,\text{BL}|\partial\Omega}=-W^{k_3,h}_{1,\text{int}|\partial\Omega}.$$

We will not give the explicit value of $W^{k_3,h}_{1,\text{BL}}$ here as it is of no use in the following. As in the well-prepared case, the introduction of $W^{k_3,h}_{1,\text{BL}}$ requires introducing an additional boundary layer of order 2 called $W^{k_3,h}_{2,\text{BL}}$ in the decomposition, to take care of the divergence-free condition. Again explicit values are unnecessary and are omitted.

Remark It can be useful to notice that solutions of

$$\left(\frac{d}{dt}+\nu|\xi_h|^2+B_{k_3}\right)\widetilde{W}^{k_3}=0$$

are written

$$\widetilde{W}^{k_3}(t)=e^{-\nu|\xi_h|^2t-P_{k_3}t}\begin{pmatrix}\cos\delta_{k_3}t & \lambda_{k_3}\sin\delta_{k_3}t\\ -\dfrac{1}{\lambda_{k_3}}\sin\delta_{k_3}t & \cos\delta_{k_3}t\end{pmatrix}\widetilde{W}^{k_3}(0) \qquad (7.4.36)$$

with

$$\delta_{k_3}\stackrel{\text{def}}{=}\frac{(1-\lambda_{k_3}^2)\lambda_{k_3}}{4}(\gamma^+_{k_3}+\gamma^-_{k_3})\quad\text{and}\quad P_{k_3}\stackrel{\text{def}}{=}\frac{(1-\lambda_{k_3}^2)\lambda_{k_3}}{4}(\gamma^-_{k_3}-\gamma^+_{k_3}).$$

By formulas (7.4.21) and (7.4.25), we have

$$\gamma^-_{k_3}-\gamma^+_{k_3}=\frac{\sqrt{2\beta}}{\lambda_{k_3}}\left(\frac{1+\lambda_{k_3}}{\sqrt{1-\lambda_{k_3}}}+\frac{1-\lambda_{k_3}}{\sqrt{1+\lambda_{k_3}}}\right)>0. \qquad (7.4.37)$$

Let us now sum up the above construction in order to state (and prove) the analog of Lemma 7.1, page 164. We are going to define the approximation

$$v^\varepsilon_{\text{app}} = v_{0,\text{int}} + v_{0,\text{BL}} + \varepsilon v_{1,\text{int}} + \varepsilon v_{1,\text{BL}} + \varepsilon^2 v_{2,\text{BL}}$$

such that there is a smooth function v_{exp}, exponentially decreasing near $\partial\Omega$, satisfying

$$\operatorname{div}(v^\varepsilon_{\text{app}} - v_{\text{exp}}) = 0 \quad \text{and} \quad (v^\varepsilon_{\text{app}} - v_{\text{exp}})_{|\partial\Omega} = 0.$$

Let us give the explicit expression of each term of the decomposition (up to the boundary layers of order 1 and more, and the exponentially decreasing remainder). In order to do so, we need to define a one-parameter group of unitary operators on L^2 we shall denote by $(\mathcal{L}(\tau))_{\tau\in\mathbf{R}}$. By the density Lemma 7.6, defining $(\mathcal{L}(\tau))_{\tau\in\mathbf{R}}$ on $\mathcal{B}$ will give a definition on $\mathcal{H}$. So let us state, for any $v \in \mathcal{B}$,

$$(\mathcal{L}(\tau)v)(x_h, x_3) \stackrel{\text{def}}{=} \begin{pmatrix} v^{0,h}(x_h) \\ 0 \end{pmatrix}$$

$$+ \mathcal{F}_h^{-1} \sum_{k_3=1}^{\infty} \begin{pmatrix} A(\xi_h)\mathcal{L}_{k_3}(\tau)A^{-1}(\xi_h)\widehat{v}^{k_3,h}(\xi_h)\cos(k_3\pi x_3) \\ \dfrac{i}{k_3\pi}\xi_h \cdot A(\xi_h)\mathcal{L}_{k_3}(\tau)A^{-1}(\xi_h)\widehat{v}^{k_3,h}(\xi_h)\sin(k_3\pi x_3) \end{pmatrix} \quad (7.4.38)$$

where $A(\xi_h)$ is defined by (7.4.7) and $\mathcal{L}_{k_3}(\tau)$ by (7.4.15).

Let us remark that when $k_3 = 0$, (this corresponds to the well-prepared case studied in the previous section), we get $\mathcal{L}_{k_3} = \text{Id}$ because in that case $\lambda_{k_3} = 0$. Let us note that $(\mathcal{L}(\tau))_{\tau\in\mathbf{R}}$ is a group of unitary operators on $H^{s,0}$ for any real number s. Moreover, the restriction of $\mathcal{L}(\tau)$ to functions which do not depend on the third variable (which corresponds to the well-prepared case) is the identity. We will call the group $(\mathcal{L}(\tau))_{\tau\in\mathbf{R}}$ the Poincaré group.

Similarly we shall need the definition of the following "Ekman operator", which once again we define on $\mathcal{B}$:

$$(\mathcal{E}v)(x_h, x_3) \stackrel{\text{def}}{=} \sqrt{2\beta} \begin{pmatrix} v^{0,h}(x_h) \\ 0 \end{pmatrix}$$

$$+ \mathcal{F}_h^{-1} \sum_{k_3=1}^{\infty} \begin{pmatrix} A(\xi_h)B_{k_3}A^{-1}(\xi_h)\widehat{v}^{k_3,h}(\xi_h)\cos(k_3\pi x_3) \\ \dfrac{i}{k_3\pi}\xi_h \cdot A(\xi_h)B_{k_3}A^{-1}(\xi_h)\widehat{v}^{k_3,h}(\xi_h)\sin(k_3\pi x_3) \end{pmatrix} \quad (7.4.39)$$

with

$$B_{k_3} = \alpha_{k_3}\text{Id} + \frac{1-\lambda_{k_3}^2}{4} R_{k_3} \quad (7.4.40)$$

where

$$\alpha_{k_3} = \frac{(1-\lambda_{k_3}^2)\lambda_{k_3}}{4}(\gamma_{k_3}^- - \gamma_{k_3}^+) > 0 \quad \text{and} \quad \gamma_{k_3}^\pm \stackrel{\text{def}}{=} \left(1 \mp \frac{1}{\lambda_{k_3}}\right)\sqrt{\frac{2\beta}{1 \pm \lambda_{k_3}}}\cdot$$

Let us note that the restriction of $\mathcal{E}$ on functions which do not depend on the third variable (which again corresponds to the well-prepared case) is $\sqrt{2\beta}\,\text{Id}$.

Proposition 7.2 *The operator $\mathcal{E}$ defined by the above formula (7.4.39) is a non-negative, bounded operator on $\overline{\mathcal{B}}$ for the L^2 norm.*

Proof Let us start by proving that $\mathcal{E}$ is non-negative. Using (7.4.2), we have for any $u \in \mathcal{H}$ (resp. in $\mathcal{H}_0$)

$$(\mathcal{E}u \mid u)_{L^2(\Omega)} = \sum_{k_3 \in \mathbf{N}} (\mathcal{E}_{k_3} u^{k_3,h} \mid u^{k_3,h})_{k_3,\Omega_h},$$

where

$$\mathcal{E}_{k_3} w = \mathcal{F}_h^{-1}(A(\xi_h) B_{k_3} A^{-1}(\xi_h) \widehat{w}(\xi_h)).$$

So by (7.4.40) we have

$$\mathcal{E}_{k_3} = \mathcal{F}_h^{-1}\left(\alpha_{k_3}\mathrm{Id} + \frac{1-\lambda_{k_3}^2}{4} A(\xi_h) R_{k_3} A^{-1}(\xi_h)\right).$$

We know that the evolution group generated by R_{k_3} is an isometry of G (resp. of G_0) because $\mathcal{L}$ is an isometry of $\mathcal{H}$. It follows that

$$(\mathcal{E}_{k_3} u^{k_3,h} \mid u^{k_3,h})_{k_3,\Omega_h} \geq 0.$$

Now let us prove the L^2 boundedness. According to (7.4.1), we have

$$\|\mathcal{E}v\|_{\mathcal{H}}^2 = 2\beta \|v^{0,h}\|_{L^2(\Omega_h)}^2 + \sum_{k_3=1}^{\infty} 4\pi^2 \|A(\xi_h) B_{k_3} A^{-1}(\xi_h) v^{k_3,h}\|_{L^2(\Omega_h, d\xi_h)}^2$$
$$+ \sum_{k_3=1}^{\infty} \frac{4\pi}{k_3^2} \|\xi_h \cdot A(\xi_h) B_{k_3} A^{-1}(\xi_h) \widehat{v}^{k_3,h}(\xi_h)\|_{L^2(\Omega_h, d\xi_h)}^2.$$

We leave to the reader the fact that

$$\sup_{\xi_h, k_3} \|B_{k_3}\|_{\mathcal{L}(\mathbf{R}^2;\mathbf{R}^2)} < +\infty.$$

As $A(\xi_h)$ is homogeneous of degree -1, this implies that

$$\sup_{\xi_h, k_3} \|A(\xi_h) B_{k_3} A^{-1}(\xi_h)\|_{\mathcal{L}(\mathbf{R}^2;\mathbf{R}^2)} = C < +\infty.$$

Thus we have that

$$\sum_{k_3=1}^{\infty} 4\pi^2 \|A(\xi_h) B_{k_3} A^{-1}(\xi_h) v^{k_3,h}\|_{L^2}^2 \leq C \sum_{k_3=1}^{\infty} \|\widehat{v}^{k_3,h}\|_{L_h^2}^2.$$

By definition of $A(\xi_h)$, we have $\xi_h \cdot A(\xi_h) W = W^1$. Thus we infer

$$\frac{i}{k_3\pi} \xi_h \cdot A(\xi_h) B_{k_3} A^{-1}(\xi_h) \widehat{v}^{k_3,h}(\xi_h) = \frac{1}{k_3\pi} \Pi^1 (B_{k_3} A^{-1}(\xi_h) \widehat{v}^{k_3,h}(\xi_h)).$$

Let us observe that, by definition of B_{k_3} and $A^{-1}(\xi_h)$, we have

$$
\begin{aligned}
V^{k_3}(\xi_h) &\stackrel{\text{def}}{=} \frac{i}{k_3\pi}\Pi^1 B_{k_3} A^{-1}(\xi_h)\widehat{v}^{k_3,h}(\xi_h)\\
&= i\frac{(1-\lambda_{k_3}^2)}{4k_3\pi}\lambda_{k_3}(\gamma_{k_3}^- - \gamma_{k_3}^+)(A^{-1}(\xi_h)\widehat{v}^{k_3,h})^1\\
&\quad - i\frac{(1-\lambda_{k_3}^2)}{4k_3\pi}\lambda_{k_3}^2(\gamma_{k_3}^- + \gamma_{k_3}^+)(A^{-1}(\xi_h)\widehat{v}^{k_3,h})^2\\
&= i\frac{(1-\lambda_{k_3}^2)}{4}\lambda_{k_3}(\gamma_{k_3}^- - \gamma_{k_3}^+)\frac{1}{k_3\pi}(\mathcal{F}_h \operatorname{div}_h v^{k_3,h})(\xi_h)\\
&\quad - \frac{(1-\lambda_{k_3}^2)}{4k_3\pi}\lambda_{k_3}^2(\gamma_{k_3}^- + \gamma_{k_3}^+)\mathcal{F}_h(\operatorname{curl}_h v^{k_3,h}).
\end{aligned}
$$

Straightforward computations left to the reader imply that a constant C exists such that

$$
\left|\frac{1-\lambda_{k_3^2}}{4}\lambda_{k_3}(\gamma_{k_3}^- - \gamma_{k_3}^+)\right| \leq C \quad\text{and}\quad \frac{(1-\lambda_{k_3}^2)}{4k_3\pi}\lambda_{k_3}^2(\gamma_{k_3}^- + \gamma_{k_3}^+) \leq \frac{C}{|\xi_h|}.
$$

Thus we have

$$
\frac{1}{k_3\pi}\left\|\Pi^1 B_{k_3} A^{-1}(\xi_h)\widehat{v}^{k_3,h}(\xi_h)\right\|_{L_h^2}^2 \leq \frac{C}{(k_3\pi)^2}\|\operatorname{div}_h v^{k_3,h}\|_{L_h^2}^2 + C\|v^{k_3,h}\|_{L_h^2}^2. \tag{7.4.41}
$$

Then we deduce that

$$
\begin{aligned}
\|\mathcal{E}v\|_{\mathcal{H}}^2 &\leq C\sum_{k_3=0}^{\infty}\|v^{k_3,h}\|_{k_3,\Omega_h}^2\\
&\leq C\|v\|_{\mathcal{H}}^2.
\end{aligned}
$$

Thus Proposition 7.2 is proved. □

In order to solve the limit system $(\mathrm{NSE}_{\nu,\varepsilon})$ that will appear in the case of periodic horizontal boundary conditions, we will need the following result.

Proposition 7.3 *The operator $\mathcal{E}$ defined by the above formula (7.4.39) is a non-negative, bounded operator on $\mathcal{H}\cap H^{0,1}$.*

Proof Using (7.4.3) it is a straightforward adaptation of the proof of Proposition 7.2: indeed the orthonormal basis used for the description of $\mathcal{E}$ is a basis of diagonalization of ∂_3^2. □

Now let us consider a time-dependent family $(v_{N_\eta})_{\eta>0}$ in $C^1([0,T];\mathcal{B}_{N_\eta})$. We define first

$$
v_{0,\mathrm{int}} = \mathcal{L}\left(\frac{t}{\varepsilon}\right)v_{N_\eta}. \tag{7.4.42}
$$

Then we state

$$v_{0,\mathrm{BL}} = \mathcal{F}_h^{-1}\begin{pmatrix} A(\xi_h)W_{0,\mathrm{BL}}^h \\ 0 \end{pmatrix}, \tag{7.4.43}$$

where $W_{0,\mathrm{BL}}^h$ is defined by (7.4.22). Next we define

$$v_{1,\mathrm{int}} = \mathcal{F}_h^{-1}\begin{pmatrix} A(\xi_h)W_{1,\mathrm{int}}^h \\ W_{1,\mathrm{int}}^3 \end{pmatrix}, \tag{7.4.44}$$

with $W_{1,\mathrm{int}}^h$ defined by (7.4.33) and (7.4.34). Now we define

$$\widetilde{v}_{\mathrm{app}}^{\varepsilon,\eta} = v_{0,\mathrm{int}} + v_{0,\mathrm{BL}} + \varepsilon v_{1,\mathrm{int}} \tag{7.4.45}$$

and as noted before, there are vector fields $v_{1,\mathrm{BL}}$, $v_{2,\mathrm{BL}}$ and v_{exp}, smooth and exponentially decreasing near $\partial\Omega$ such that

$$v_{\mathrm{app}}^{\varepsilon,\eta} \stackrel{\mathrm{def}}{=} \widetilde{v}_{\mathrm{app}}^{\varepsilon,\eta} + \varepsilon v_{1,\mathrm{BL}}^h + \varepsilon^2 v_{2,\mathrm{BL}} + v_{\mathrm{exp}} \tag{7.4.46}$$

is divergence-free and vanishes at the boundary. Thanks to (7.4.23) we know that $L^\varepsilon v_{0,\mathrm{BL}}$ is a remainder term, and of course the same goes for $\varepsilon L^\varepsilon v_{1,\mathrm{BL}}$, for $\varepsilon^2 L^\varepsilon v_{2,\mathrm{BL}}$ and for $L^\varepsilon v_{\mathrm{exp}}$. By definition of $\mathcal{L}$ and $\mathcal{E}$, the formula (7.4.35) becomes therefore

$$\mathbf{P}L^\varepsilon v_{\mathrm{app}}^{\varepsilon,\eta} = (\partial_t - \nu\Delta_h + \mathcal{E})v_{N_\eta} + R^{\varepsilon,\eta}. \tag{7.4.47}$$

Now we are ready to state the key lemma. We denote the limit system by

$$(\mathrm{SE}_{\nu,\mathcal{E}}) \begin{cases} \partial_t v - \nu\Delta_h v + \mathcal{E}\, v = f - \nabla p \\ \operatorname{div} v = 0 \\ v_{|t=0} = v_0\,. \end{cases}$$

The proof of the following elementary proposition is left to the reader.

Proposition 7.4 *Let T be in $\overline{\mathbf{R}^+}$, and let $v_0 \in \mathcal{H}(\Omega)$ be given. Consider f in the space $L^2([0,T];H^{-1,0})$. Then there is a unique vector field v belonging to $C([0,T];\mathcal{H}(\Omega)) \cap L^2([0,T];H^{1,0})$ solution of $(\mathrm{SE}_{\nu,\mathcal{E}})$. It satisfies, moreover, for all $t \in [0,T]$,*

$$\frac{1}{2}\|v(t)\|_{L^2}^2 + \nu\int_0^t \|\nabla^h v(t')\|_{L^2}^2 dt' + \int_0^t (\mathcal{E}v(t')|v(t'))\,dt' = \frac{1}{2}\|v_0\|_{L^2}^2 + \int_0^t \langle f(t'), v(t')\rangle\,dt'.$$

Lemma 7.8 *Let T be in $\overline{\mathbf{R}^+}$, and let $v_0 \in \mathcal{H}(\Omega)$ be given. Consider f in the space $L^2([0,T];H^{-1,0})$, and denote by v the solution of $(\mathrm{SE}_{\nu,\mathcal{E}})$. Then for any* *positive η, there exists a family $(v^{\varepsilon,\eta}_{\text{app}})_{\varepsilon>0}$ of smooth divergence-free vector fields on Ω, vanishing on the boundary, such that*

$$\left\| \mathbf{P} L^\varepsilon v^{\varepsilon,\eta}_{\text{app}} - \mathcal{L}\left(\frac{t}{\varepsilon}\right) f \right\|_{L^\infty([0,T];H^{-1,0})} = \rho^{\varepsilon,\eta}, \quad \textit{and} \tag{7.4.48}$$

$$E_t\left(v^{\varepsilon,\eta}_{\text{app}} - \mathcal{L}\left(\frac{t}{\varepsilon}\right) v\right) = \rho^{\varepsilon,\eta}. \tag{7.4.49}$$

Moreover the energy of $v^{\varepsilon,\eta}_{\text{app}}$ is controlled, in the sense that

$$E^\varepsilon_t(v^{\varepsilon,\eta}_{\text{app}}) \leq \|v_0\|^2_{L^2} + \int_0^t \langle f(t'), v(t')\rangle dt' + \rho^{\varepsilon,\eta}. \tag{7.4.50}$$

Proof Calling $\mathbf{P}_N$ the orthogonal projection from $\mathcal{H}$ onto $\mathcal{B}_N$, let us define

$$v_{0,N} \stackrel{\text{def}}{=} \mathbf{P}_N v_0 \quad \text{and} \quad f_N \stackrel{\text{def}}{=} \mathbf{P}_N f.$$

Then since

$$\lim_{N\to\infty} v_{0,N} = v_0 \quad \text{in} \quad \mathcal{H},$$

and

$$\lim_{N\to\infty} f_N = f \quad \text{in} \quad L^2([0,T];H^{-1,0}),$$

we can choose $N_\eta \in \mathbf{N}$ such that

$$\|v_{0,N_\eta} - v_0\|_{\mathcal{H}} + \|f_N - f\|_{L^2([0,T];H^{-1,0})} \leq \eta.$$

We will call v_{N_η} the associate solution of $(\text{SE}_{\nu,\mathcal{E}})$, which since $\mathbf{P}_N\mathcal{E} = \mathcal{E}\mathbf{P}_N$ satisfies

$$\|v_{N_\eta} - v\|^2_{L^\infty([0,T];\mathcal{H}(\Omega))} + 2\nu \int_0^T \|\nabla^h(v_{N_\eta} - v)(t)\|^2_{L^2(\Omega)} dt \leq C\eta^2.$$

Inequality (7.4.49) will therefore be proved if we show that

$$\limsup_{\varepsilon\to 0} \sup_{t\in[0,T]} E_t\left(v^{\varepsilon,\eta}_{\text{app}} - \mathcal{L}\left(\frac{t}{\varepsilon}\right) v_{N_\eta}\right) = 0. \tag{7.4.51}$$

Let us write

$$v_{N_\eta}(x_h,x_3) = (v^{0,h}(x_h),0) + \sum_{1\leq k_3\leq N_\eta} \begin{pmatrix} v^{k_3,h}_{N_\eta}(x_h)\cos k_3\pi x_3 \\ -\dfrac{1}{k_3\pi}\operatorname{div}_h v^{k_3,h}_{N_\eta}(x_h)\sin k_3\pi x_3 \end{pmatrix}. \tag{7.4.52}$$

Then, as in (7.4.45) and (7.4.46), we can define an approximate solution $v^{\varepsilon,\eta}_{\text{app}}$, divergence-free and vanishing at the boundary, such that by (7.4.47)

$$\mathbf{P} L^\varepsilon v^{\varepsilon,\eta}_{\text{app}} - (\partial_t - \nu\Delta_h + \mathcal{E}) v_{N_\eta} = R^{\varepsilon,\eta},$$

which directly yields (7.4.48).

Let us prove the estimate (7.4.51). According to the definition (7.4.42) of $v_{0,\rm int}$ and to the relations (7.4.45) and (7.4.46), we have

$$v^{\varepsilon,\eta}_{\rm app} = \mathcal{L}\left(\frac{t}{\varepsilon}\right) v_{N_\eta} + v_{0,\rm BL} + \varepsilon v_{1,\rm int} + \varepsilon v_{1,\rm BL} + \varepsilon^2 v_{2,\rm BL} + v_{\rm exp}$$

so to prove (7.4.51), we need to check that

$$E_t(v_{0,\rm BL} + \varepsilon v_{1,\rm int} + \varepsilon v_{1,\rm BL} + \varepsilon^2 v_{2,\rm BL} + v_{\rm exp}) = \rho^{\varepsilon,\eta}.$$

This is achieved, just like in the well-prepared case, by noticing that

$$\sum_{j=0}^{2} \varepsilon^j \|v^\varepsilon_{j,\rm BL}(t)\|_{L^2} + \varepsilon\|v_{1,\rm int}(t)\|_{L^2} + \|v_{\rm exp}(t)\|_{L^2} \leq C_\eta \varepsilon^{\frac{1}{2}}.$$

In order to prove estimate (7.4.50), let us start from (7.4.48) which claims that

$$\mathbf{P}L^\varepsilon v^{\varepsilon,\eta}_{\rm app} = \mathcal{L}\left(\frac{t}{\varepsilon}\right) f + R^{\varepsilon,\eta}.$$

An energy estimate implies that

$$\begin{aligned} E^\varepsilon_t(v^{\varepsilon,\eta}) &= \|v^{\varepsilon,\eta}_{\rm app}(t)\|^2_{L^2} + 2\nu\int_0^t \|\nabla_h v^{\varepsilon,\eta}(t')\|^2_{L^2}dt' + 2\beta\int_0 t\|\partial_3 v^{\varepsilon,\eta}\|^2_{L^2}dt' \\ &= 2\int_0^t \left\langle \mathcal{L}\left(\frac{t'}{\varepsilon}\right) f(t'), v^{\varepsilon,\eta}(t')\right\rangle dt' + 2\int_0^t \langle R^{\varepsilon,\eta}(t'), v^{\varepsilon,\eta}(t')\rangle dt'. \end{aligned}$$

Thanks to (7.4.51), we have

$$\left\| v^{\varepsilon,\eta} - \mathcal{L}\left(\frac{t}{\varepsilon}\right) v \right\|_{L^2([0,T];H^{1,0})} = \rho^{\varepsilon,\eta}.$$

This implies that

$$2\int_0^t \left\langle \mathcal{L}\left(\frac{t'}{\varepsilon}\right) f(t'), v^{\varepsilon,\eta}(t')\right\rangle dt' \leq 2\int_0^t \left\langle \mathcal{L}\left(\frac{t'}{\varepsilon}\right) f(t'), \mathcal{L}\left(\frac{t'}{\varepsilon}\right) v(t')\right\rangle dt' + \rho^{\varepsilon,\eta}.$$

As $\mathcal{L}$ is a group of isometries of L^2, we have ${}^t\mathcal{L}\mathcal{L} = \mathrm{Id}$. This gives

$$2\int_0^t \left\langle \mathcal{L}\left(\frac{t'}{\varepsilon}\right) f(t'), v^{\varepsilon,\eta}(t')\right\rangle dt' \leq 2\int_0^t \langle f(t'), v(t')\rangle dt' + \rho^{\varepsilon,\eta}.$$

Thanks to (7.4.49), we have

$$2\int_0^t \langle R^{\varepsilon,\eta}(t'), v^{\varepsilon,\eta}(t')\rangle\, dt' \leq 2\|R^{\varepsilon,\eta}\|_{L^2([0,T];H^{-1,0})}\|v^{\varepsilon,\eta}\|_{L^2([0,T];H^{1,0})} = \rho^{\varepsilon,\eta}.$$

The whole of Lemma 7.8 is proved. □

7.5 Non-linear estimates in the ill-prepared case

As in the well-prepared case discussed in Section 7.2, we shall see in this section that Ekman boundary layers do not affect non-linear terms. The two estimates proved here (Lemmas 7.9 and 7.10) will be essential in the proof of Theorems 7.2 and 7.3 in the following sections.

Let us start by proving the following lemma, which is the generalization of Lemma 7.2 to the case of ill-prepared data. It claims that Ekman boundary layers disappear in non-linear terms.

Lemma 7.9 *Let us consider v and f two vector fields satisfying the hypotheses of Lemma 7.8. Let us consider any family of approximations $(v^{\varepsilon,\eta}_{\rm app})_{\varepsilon>0}$ given by Lemma 7.8; we have*

$$\lim_{\varepsilon\to 0}\left\| v^{\varepsilon,\eta}_{\rm app}\cdot\nabla v^{\varepsilon,\eta}_{\rm app} - \mathcal{L}\left(\frac{t}{\varepsilon}\right)v_{N_\eta}\cdot\nabla\mathcal{L}\left(\frac{t}{\varepsilon}\right)v_{N_\eta}\right\|_{L^2([0,T];L^2)} = 0.$$

Proof Recalling that $\mathcal{L}(t/\varepsilon)\,v_{N_\eta} = v^\varepsilon_{0,{\rm int}}$, let us write (omitting to mention the dependance on η to avoid excessive heaviness)

$$v^\varepsilon_{\rm app}\cdot\nabla v^\varepsilon_{\rm app} - v^\varepsilon_{0,{\rm int}}\cdot\nabla v^\varepsilon_{0,{\rm int}} = \sum_{j=1}^4 B^\varepsilon_j \quad\text{with}$$

$$B^\varepsilon_1 \stackrel{\rm def}{=} (v^\varepsilon_{\rm app} - v^\varepsilon_{0,{\rm int}})^h\cdot\nabla^h v^\varepsilon_{\rm app},$$

$$B^\varepsilon_2 \stackrel{\rm def}{=} v^{\varepsilon,h}_{0,{\rm int}}\cdot\nabla^h(v^\varepsilon_{\rm app} - v^\varepsilon_{0,{\rm int}}),$$

$$B^\varepsilon_3 \stackrel{\rm def}{=} (v^\varepsilon_{\rm app} - v^\varepsilon_{0,{\rm int}})^3\partial_3 v^\varepsilon_{\rm app},$$

$$B^\varepsilon_4 \stackrel{\rm def}{=} v^{\varepsilon,3}_{0,{\rm int}}\partial_3(v^\varepsilon_{\rm app} - v^\varepsilon_{0,{\rm int}})$$

We have

$$\|B^\varepsilon_1\|_{L^2([0,T];L^2)} \leq \|v^\varepsilon_{\rm app} - v^\varepsilon_{0,{\rm int}}\|_{L^\infty([0,T];L^2)}\|\nabla^h v^\varepsilon_{\rm app}\|_{L^2([0,T];L^\infty)}.$$

By definition of $v^\varepsilon_{\rm app}$, the difference $v^\varepsilon_{\rm app} - v^\varepsilon_{0,{\rm int}}$ consists of terms of order ε and of a boundary layer of order 0. Obviously, we get that

$$\|v^\varepsilon_{\rm app} - v^\varepsilon_{0,{\rm int}}\|_{L^\infty([0,T];L^2)} \leq C_\eta\varepsilon^{\frac12}. \tag{7.5.1}$$

Using the fact that the horizontal Fourier transform of $v^\varepsilon_{\rm app}$ is supported in $B(0,N)$, we get, using the energy estimate (7.4.50), that

$$\|\nabla^h v^\varepsilon_{\rm app}\|_{L^2([0,T];L^\infty)} \leq C_\eta \quad\text{and thus}\quad \lim_{\varepsilon\to 0}\|B^\varepsilon_1\|_{L^2([0,T];L^2)} = 0.$$

The estimate on B^ε_2 is analogous. The terms B^ε_3 and B^ε_4 in which vertical derivatives are involved require different arguments. We have that

$$\|B^\varepsilon_3\|_{L^2([0,T];L^2)} \leq \|v^{\varepsilon,3}_{\rm app} - v^{\varepsilon,3}_{\rm int}\|_{L^\infty([0,T];L^\infty)}\|\partial_3 v^\varepsilon_{\rm app}\|_{L^2([0,T];L^2)}.$$

As $v_{\rm int}^{\varepsilon,3}$ vanishes on the boundary, there is no boundary layer term of order zero on the vertical component of $v_{\rm app}^{\varepsilon}$. Thus the term $v_{\rm app}^{\varepsilon,3} - v_{0,\rm int}^{\varepsilon,3}$ consists of a finite sum of bounded terms (uniformly with respect to ε) of order ε. Thus we have

$$\|v_{\rm app}^{\varepsilon,3} - v_{\rm int}^{\varepsilon,3}\|_{L^\infty([0,T];L^\infty)} \leq \varepsilon.$$

This implies that

$$\|B_3^\varepsilon\|_{L^2([0,T];L^2)} \leq C_\eta \varepsilon \|\partial_3 v_{\rm app}^\varepsilon\|_{L^2([0,T];L^2)}.$$

The energy estimate (7.4.50) claims in particular that

$$\varepsilon^{\frac{1}{2}} \|\partial_3 v_{\rm app}^\varepsilon\|_{L^2([0,T];L^2)} = C_0^{\frac{1}{2}}.$$

This gives that

$$\lim_{\varepsilon \to 0} \|B_3^\varepsilon\|_{L^2([0,T];L^2)} = 0.$$

In order to estimate B_4^ε, let us study $v_{0,\rm int}^{\varepsilon,3}$. By definition of $v_{0,\rm int}^{\varepsilon,3}$, it consists of a finite sum of smooth functions which vanish on the boundary. It follows that

$$\|v_{0,\rm int}^{\varepsilon,3}(t,\cdot,x_3)\|_{L_T^\infty(L_h^\infty)} \leq C_\eta d(x_3)$$

where $d(x_3)$ denotes the distance to the boundary of $]0,1[$. As

$$\|B_4^\varepsilon\|_{L^2([0,T]\times\Omega)}^2 \leq \int_{[0,T]\times[0,1]} \|v_{0,\rm int}^{\varepsilon,3}(t,\cdot,x_3)\|_{L_h^\infty}^2 \|\partial_3(v_{\rm app}^{\varepsilon,3} - v_{0,\rm int}^{\varepsilon,3})(t,\cdot,x_3)\|_{L_h^2}^2 dx_3 dt,$$

we infer that

$$\|B_4^\varepsilon\|_{L^2([0,T]\times\Omega)}^2 \leq C_\eta \int_{[0,T]\times[0,1]} d^2(x_3) \|\partial_3(v_{\rm app}^\varepsilon - v_{0,\rm int}^\varepsilon)(t,\cdot,x_3)\|_{L_h^2}^2 dx_3 dt.$$

The worse term of $\partial_3(v_{\rm app}^\varepsilon - v_{0,\rm int}^\varepsilon)$ is $\partial_3 v_{0,\rm BL}^{\varepsilon,h}$. More precisely, using (7.4.45) and (7.4.46), we get

$$\int_0^1 d^2(x_3)\|\partial_3(v_{\rm app}^\varepsilon - v_{0,\rm int}^\varepsilon)(t,\cdot,x_3)\|_{L_h^2}^2 dx_3$$
$$\leq \int_0^1 d^2(x_3)\|\partial_3 v_{0,\rm BL}^\varepsilon(t,\cdot,x_3)\|_{L_h^2}^2 dx_3 + \rho^{\varepsilon,\eta}.$$

By definition of $v_{0,\rm BL}^\varepsilon$, we have

$$\|\partial_3 v_{0,\rm BL}^\varepsilon(t,\cdot,x_3)\|_{L^2(\Omega_h)} \leq \frac{C_\eta}{\varepsilon} \sum_{k_3=0}^N \|\widetilde{v}_N^{k_3,h}(t)\|_{L_h^2}$$
$$\times \sum_\pm \sup_{\xi_h \in \mathcal{C}_N} \left(\exp\left(-\frac{x_3}{\varepsilon\sqrt{2\beta_{k_3}^\pm}}\right) + \exp\left(-\frac{1-x_3}{\varepsilon\sqrt{2\beta_{k_3}^\pm}}\right)\right),$$

where recall that $\mathcal{C}_N = \mathcal{C}(1/N, N)$. So we infer that

$$\int_0^1 d^2(x_3)\|\partial_3 v_{0,\mathrm{BL}}^\varepsilon(t,\cdot,x_3)\|_{L_h^2}^2\,dx_3 \leq C_\eta \varepsilon \sum_{k_3=0}^{N} \|v_N^{k_3,h}(t,\cdot)\|_{L_h^2}^2$$
$$\times \int_0^1 \frac{d^2(x_3)}{\varepsilon^2} \exp\left(-\frac{C_\eta d(x_3)}{\varepsilon}\right) \frac{dx_3}{\varepsilon}.$$

Computing the integral gives

$$\int_0^1 d^2(x_3)\|\partial_3 v_{0,\mathrm{BL}}^\varepsilon(t,\cdot,x_3)\|_{L_h^2}^2\,dx_3 \leq C_\eta \varepsilon \sum_{k_3=0}^{N} \|v_N^{k_3,h}(t,\cdot)\|_{L_h^2}^2.$$

By definition of v_N, using the result that the family $(\sin(k_3\pi x_3))$ is orthogonal in L^2, we infer that

$$\int_0^1 d^2(x_3)\|\partial_3 v_{0,\mathrm{BL}}^\varepsilon(t,\cdot,x_3)\|_{L_h^2}^2\,dx_3 \leq C_\eta \varepsilon \|v_N\|_{L^2}^2. \tag{7.5.2}$$

The lemma is proved. □

The following result is the analog of Lemma 7.3, in the ill-prepared case. We have defined, for any vector field v,

$$v = \overline{v} + \widetilde{v}, \quad \text{where } \overline{v} = \int_0^1 v(x_h, x_3)\,dx_3.$$

Lemma 7.10 *Let v be a solution of $(\mathrm{SE}_{\nu,\varepsilon})$ and let η be a positive real number. Denote by $(v_{\mathrm{app}}^{\varepsilon,\eta})_{\varepsilon>0}$ the family given by Lemma 7.8. Then for any vector field δ belonging to $L^\infty([0,T];\mathcal{H}) \cap L^2([0,T];\mathcal{V}_\sigma)$, we have*

$$\int_0^T \left(\delta(t)\cdot\nabla\delta(t)|v_{\mathrm{app}}^\varepsilon(t)\right)_{L^2} dt \leq \left(C_\eta \varepsilon^{\frac{1}{2}} + \frac{1}{4}\right) E_T^\varepsilon(\delta)$$
$$+ \int_0^T \left(\delta(t)\cdot\nabla\delta(t)|\mathcal{L}\left(\frac{t}{\varepsilon}\right)\widetilde{v}_{N_\eta}(t)\right)_{L^2} dt + \frac{C}{\nu}\int_0^T \|\nabla^h \overline{v}(t)\|_{L^2}^2 \|\delta(t)\|_{L^2}^2 dt.$$

Proof The proof follows the lines of the proof of Lemma 7.3 in the well-prepared case, so we will skip some details. We can decompose

$$v_{\mathrm{app}}^\varepsilon = (v_{\mathrm{app}}^\varepsilon - v_{N_\eta} - v_{0,\mathrm{BL}}) + \overline{v}_{N_\eta} + \mathcal{L}\left(\frac{t}{\varepsilon}\right)\widetilde{v}_{N_\eta} + v_{0,\mathrm{BL}},$$

and as for (7.2.1) we have $\|v_{\mathrm{app}}^\varepsilon - v_{N_\eta} - v_{0,\mathrm{BL}}\|_{L^\infty([0,T]\times\Omega)} \leq C_\eta \varepsilon$. So we infer as in the case of (7.2.1) that

$$\int_0^T \left(\delta(t)\cdot\nabla\delta(t)|(v_{\mathrm{app}}^\varepsilon - v_{N_\eta} - v_{0,\mathrm{BL}})(t)\right)_{L^2} dt \leq C_\eta \varepsilon^{\frac{1}{2}} E_t^\varepsilon(\delta).$$

Similarly we have as in (7.2.3)

$$\int_0^T \left(\delta(t)\cdot\nabla\delta(t)|\overline{v}_{N_\eta}(t)\right)_{L^2} dt \leq \frac{\nu}{2}\int_0^T \|\nabla^h\delta(t)\|_{L^2}^2 dt + \frac{C}{\nu}\int_0^T \|\nabla^h\overline{v}_{N_\eta}(t)\|_{L^2}^2 \|\delta(t)\|_{L^2}^2 dt,$$

so we are left with the term containing the boundary layers, which as in the well-prepared case we want to prove it can be neglected. We recall that

$$v_{0,\mathrm{BL}} = \mathcal{F}_h^{-1}\left(A(\xi_h)W_{0,\mathrm{BL}}^h, W_{0,\mathrm{BL}}^3\right) \quad \text{where } W_{0,\mathrm{BL}}^h = \sum_{k_3=0}^{N_\eta} W_{0,\mathrm{BL}}^{k_3,h}$$

and where $W_{0,\mathrm{BL}}^{0,h}$ is given by the boundary layer of the well-prepared case (hence its contribution can be neglected just like in the well-prepared case), and where $W_{0,\mathrm{BL}}^{k_3,h}$ is given by (7.4.22). Writing that

$$|v_{0,\mathrm{BL}}(x)| \leq C_{N_\eta} \sup_{\xi_h\in\mathcal{C}_{N_\eta}} \sum_{k_3=0}^{N_\eta}\left(\exp\left(-\frac{x_3}{\varepsilon\sqrt{2\beta_{k_3}^{\pm}}}\right) + \exp\left(-\frac{1-x_3}{\varepsilon\sqrt{2\beta_{k_3}^{\pm}}}\right)\right),$$

the lemma follows exactly as in the well-prepared case. □

7.6 The convergence theorem in the whole space

The goal of this section is to prove Theorem 7.2. Let us start by considering $\overline{u}$, the solution of

$$(\overline{\mathrm{NSE}_{\nu,\beta}})\begin{cases}\partial_t\overline{u} + \mathrm{div}_h(\overline{u}\otimes\overline{u}) - \nu\Delta_h\overline{u} + \sqrt{2\beta}\,\overline{u}^h = -(\nabla^h\overline{p},0)\\ \mathrm{div}_h\,\overline{u} = 0\\ \overline{u}_{|t=0} = \overline{u}_0^h \stackrel{\mathrm{def}}{=} \displaystyle\int_0^1 u_0^h(x_h,x_3)dx_3.\end{cases}$$

Then let us consider u, the solution of the forced equation

$$(\mathrm{FSE}_{\nu,\mathcal{E}})\begin{cases}\partial_t u - \nu\Delta_h u + \mathcal{E}u = (Q(\overline{u},\overline{u}),0)\\ \mathrm{div}\,u = 0\\ u_{|t=0} = u_0.\end{cases}$$

Let us note that $\overline{u}\displaystyle\int_0^1 u(x_h,x_3)dx_3$ is the solution of $(\mathrm{FSE}_{\nu,\mathcal{E}})$. The main step of the proof of Theorem 7.2 consists in proving the following lemma.

Lemma 7.11 *Let $\eta>0$ be given, and consider $(u_{\mathrm{app}}^{\varepsilon,\eta})_{\varepsilon>0}$, the family of approximations given by Lemma 7.8 associated with η and u. Then*

$$\sup_{t\in[0,T]} E_t^\varepsilon(u^\varepsilon - u_{\mathrm{app}}^{\varepsilon,\eta}) = \rho^{\varepsilon,\eta}. \tag{7.6.1}$$

Remark Lemma 7.11 does not imply Theorem 7.2 directly as it remains to prove that $u^{\varepsilon,\eta}_{\text{app}}$ converges towards $\overline{u}$ in some sense. That is the purpose of Lemma 7.13 below.

Proof of Lemma 7.11 Let us start by recalling that

$$u^{\varepsilon,\eta}_{0,\text{int}} = \overline{u}_{N_\eta} + \mathcal{L}\left(\frac{t}{\varepsilon}\right)\widetilde{u}_{N_\eta} \quad \text{with}$$

$$\overline{u}_{N_\eta} \stackrel{\text{def}}{=} \mathcal{F}^{-1}\left(\mathbf{1}_{\mathcal{C}(\frac{1}{N_\eta},N_\eta)}\mathcal{F}\overline{u}\right) \quad \text{and}$$

$$\widetilde{u}_{N_\eta} \stackrel{\text{def}}{=} \sum_{k_3=1}^{N_\eta}\begin{pmatrix} \widetilde{u}^{k_3,h}(x_h)\cos(k_3\pi x_3) \\ -\dfrac{1}{k_3\pi}\operatorname{div}_h\widetilde{u}^{k_3,h}(x_h)\sin(k_3\pi x_3)\end{pmatrix},$$

the support of the horizontal Fourier transform of $\widetilde{u}^{k_3,h}$ being included in the ring $\mathcal{C}(1/N_\eta, N_\eta)$.

In order to prove Lemma 7.11, we will follow the same method as in the well-prepared case, namely a weak–strong uniqueness argument. Let us indeed apply Lemma 7.5 to $u^{\varepsilon,\eta}_{\text{app}}$ (omitting again the smoothing in time). We find

$$\begin{aligned} E^\varepsilon_t(\delta^\varepsilon) = E^\varepsilon_t(u^\varepsilon) &+ E^\varepsilon_t(u^{\varepsilon,\eta}_{\text{app}}) - 2(u_0|u^{\varepsilon,\eta}_{\text{app}|t=0})_{L^2} \\ &- 2\int_0^t \left(\delta^\varepsilon(t')\cdot\nabla\delta^\varepsilon(t')|u^{\varepsilon,\eta}_{\text{app}}(t')\right)_{L^2}dt' \\ &+ 2\int_0^t (G^\varepsilon(u^{\varepsilon,\eta}_{\text{app}})(t')|u^\varepsilon(t'))_{L^2}dt', \end{aligned} \tag{7.6.2}$$

where according to Lemma 7.8,

$$\begin{aligned} G^\varepsilon(u^{\varepsilon,\eta}_{\text{app}}) &= u^{\varepsilon,\eta}_{\text{app}}\cdot\nabla u^{\varepsilon,\eta}_{\text{app}} + L^\varepsilon u^{\varepsilon,\eta}_{\text{app}} \\ &= u^{\varepsilon,\eta}_{\text{app}}\cdot\nabla u^{\varepsilon,\eta}_{\text{app}} - \overline{u}^h\cdot\nabla^h\overline{u} + R^{\varepsilon,\eta} + \nabla p^\varepsilon. \end{aligned} \tag{7.6.3}$$

We have as usual

$$E^\varepsilon_t(u^\varepsilon) \leq \|u_0\|^2_{L^2}. \tag{7.6.4}$$

Moreover by Lemma 7.8 we have

$$E^\varepsilon_t(u^{\varepsilon,\eta}_{\text{app}}) \leq \|u_0\|^2_{L^2} + 2\int_0^t \langle\overline{u}^h\cdot\nabla^h\overline{u}, u\rangle(t')dt' + \rho^{\varepsilon,\eta}.$$

Let us define $\widetilde{u} = u - \overline{u}$. Then by definition

$$\partial_t\widetilde{u} - \nu\Delta_h\widetilde{u} + \mathcal{E}\widetilde{u} = 0, \quad \text{with} \quad \widetilde{u}_{|t=0} = u_0 - (\overline{u}^h_0, 0),$$

which in particular implies that

$$\forall t \geq 0, \quad \int_0^1 \widetilde{u}^h(t, x_h, x_3)dx_3 = 0.$$

Writing $\langle \overline{u} \cdot \nabla^h \overline{u}, u \rangle = \langle \overline{u} \cdot \nabla^h \overline{u}, \overline{u} \rangle + \langle \overline{u} \cdot \nabla^h \overline{u}, \widetilde{u}^h \rangle$, we find that

$$\int_0^t \langle \overline{u} \cdot \nabla^h \overline{u}, u \rangle (t')dt' = 0$$

hence

$$E_t^\varepsilon(u_{\text{app}}^{\varepsilon,\eta}) \leq \|u_0\|_{L^2}^2 + \rho^{\varepsilon,\eta}. \tag{7.6.5}$$

Finally clearly $\|u^\varepsilon_{\text{app}|t=0} - u_0\|_{L^2} = \rho^{\varepsilon,\eta}$. So with (7.6.4) and (7.6.5) we get

$$E_t^\varepsilon(u^\varepsilon) + E_t^\varepsilon(u_{\text{app}}^{\varepsilon,\eta}) - 2(u_0|u^{\varepsilon,\eta}_{\text{app}|t=0})_{L^2} = \rho^{\varepsilon,\eta}. \tag{7.6.6}$$

Then Lemma 7.10 implies that

$$\int_0^t \left(\delta^\varepsilon(t') \cdot \nabla \delta^\varepsilon(t') | u_{\text{app}}^{\varepsilon,\eta}(t')\right)_{L^2} dt \leq \left(\rho^{\varepsilon,\eta} + \frac{1}{4}\right) E_t^\varepsilon(\delta^\varepsilon) + \sum_{j=1}^2 \mathcal{U}_j(t) \tag{7.6.7}$$

with

$$\mathcal{U}_1(t) \stackrel{\text{def}}{=} \int_0^t \left(\delta^\varepsilon(t') \cdot \nabla \delta^\varepsilon(t') | \mathcal{L}\left(\frac{t'}{\varepsilon}\right) \widetilde{u}_N(t')\right)_{L^2} dt' \quad \text{and}$$

$$\mathcal{U}_2(t) \stackrel{\text{def}}{=} \frac{C}{\nu} \int_0^t \|\nabla^h \overline{u}(t')\|_{L^2}^2 \|\delta^\varepsilon(t')\|_{L^2}^2 dt'.$$

Finally since Q is continuous from $L_T^4(L_h^4) \times L_T^4(L_h^4)$ into $L_T^2(H^{-1}(\Omega_h))$ (denoting L_h^p for $L^p(\Omega_h)$), we can write

$$G^\varepsilon(u_{\text{app}}^{\varepsilon,\eta}) = F_1^{\varepsilon,\eta} + F_2^{\varepsilon,\eta} + R^{\varepsilon,\eta} + \nabla p^\varepsilon \quad \text{with}$$

$$F_1^{\varepsilon,\eta} \stackrel{\text{def}}{=} u_{\text{app}}^{\varepsilon,\eta} \cdot \nabla u_{\text{app}}^{\varepsilon,\eta} - u_{0,\text{int}}^{\varepsilon,\eta} \cdot \nabla u_{0,\text{int}}^{\varepsilon,\eta} \quad \text{and}$$

$$F_2^{\varepsilon,\eta} \stackrel{\text{def}}{=} u_{0,\text{int}}^{\varepsilon,\eta} \cdot \nabla u_{0,\text{int}}^{\varepsilon,\eta} - \overline{u}_{N_\eta} \cdot \nabla \overline{u}_{N_\eta}.$$

Notice that by Lemma 7.9 we have $\lim_{\varepsilon \to 0} \|F_1^{\varepsilon,\eta}\|_{L^2([0,T];L^2)} = 0$. As in the case of (7.3.4) and (7.3.5), we have

$$\int_0^t \left(R^{\varepsilon,\eta}(t') + F_1^{\varepsilon,\eta}(t') | u^\varepsilon(t')\right)_{L^2} dt' = \rho^{\varepsilon,\eta},$$

so finally

$$\int_0^t (G^\varepsilon(u_{\text{app}}^{\varepsilon,\eta})(t')|u^\varepsilon(t'))_{L^2}dt' \leq \int_0^t (F_2^{\varepsilon,\eta}(t')|u^\varepsilon(t'))_{L^2}dt' + \rho^{\varepsilon,\eta}. \tag{7.6.8}$$

Plugging (7.6.6), (7.6.7) and (7.6.8) into (7.6.2) implies that for ε small enough,

$$E_t^\varepsilon(\delta^\varepsilon) \leq \rho^{\varepsilon,\eta} + \frac{C}{\nu}\int_0^t \|\nabla^h \overline{u}(t')\|_{L^2}^2 \|\delta^\varepsilon(t')\|_{L^2}^2 dt' + \sum_{j=1}^2 \mathcal{E}_j(t) \tag{7.6.9}$$

with

$$\mathcal{E}_1(t) \stackrel{\text{def}}{=} C\left|\int_0^t (F_2^{\varepsilon,\eta}(t')|u^\varepsilon(t'))_{L^2}dt'\right|$$

$$\mathcal{E}_2(t) \stackrel{\text{def}}{=} C\left|\int_0^t \left(\delta^\varepsilon(t')\cdot\nabla\delta^\varepsilon(t')|\mathcal{L}\left(\frac{t'}{\varepsilon}\right)\widetilde{u}_N\right)dt'\right|.$$

Let us note that for the moment no dispersive effects have been used, and the computations hold in the periodic setting as well as in $\mathbf{R}^2$. However to continue the proof we will argue differently depending on the setting.

As x_h is in $\mathbf{R}^2$ in this section, we will be able to use Strichartz estimates to get rid of $F_2^{\varepsilon,\eta}$ and of $\mathcal{L}(t'/\varepsilon)\widetilde{u}_N$ in estimate (7.6.9). Let us write

$$\begin{aligned} F_2^{\varepsilon,\eta} &= u_{0,\text{int}}^\varepsilon \cdot \nabla(u_{0,\text{int}}^\varepsilon - \overline{u}_N) + (u_{0,\text{int}}^\varepsilon - \overline{u}_N)\cdot\nabla\overline{u}_N \\ &= u_{0,\text{int}}^\varepsilon \cdot \nabla\mathcal{L}\left(\frac{t}{\varepsilon}\right)\widetilde{u}_N + \mathcal{L}\left(\frac{t}{\varepsilon}\right)\widetilde{u}_N\cdot\nabla\overline{u}_N. \end{aligned}$$

Using the fact that $\widetilde{u}_N$ and thus $\mathcal{L}(t/\varepsilon)\widetilde{u}_N$ is a finite sum of products of functions of x_h, the horizontal Fourier transform of which is supported in $B(0,N)$ by $\cos k_3\pi x_3$ or $\sin k_3\pi x_3$, we have

$$\left\|u_{0,\text{int}}^\varepsilon\cdot\nabla\mathcal{L}\left(\frac{t}{\varepsilon}\right)\widetilde{u}_N\right\|_{L^2([0,T];L^2)} \leq C_\eta\|u_{0,\text{int}}^\varepsilon\|_{L^\infty([0,T];L^2)}\left\|\mathcal{L}\left(\frac{t}{\varepsilon}\right)\widetilde{u}_N\right\|_{L^2([0,T];L^\infty)}.$$

Let us postpone the proof of the following lemma which is a consequence of dispersive effects.

Lemma 7.12 *For every $N\in\mathbf{N}$ and every $T>0$, the following estimate holds:*

$$\lim_{\varepsilon\to 0}\left\|\mathcal{L}\left(\frac{t}{\varepsilon}\right)\widetilde{u}_N\right\|_{L^2([0,T];L^\infty)} = \lim_{\varepsilon\to 0}\left\|\nabla\mathcal{L}\left(\frac{t}{\varepsilon}\right)\widetilde{u}_N\right\|_{L^2([0,T];L^\infty)} = 0.$$

This lemma implies immediately that

$$\lim_{\varepsilon\to 0}\left\|u_{0,\text{int}}^\varepsilon\cdot\nabla\mathcal{L}\left(\frac{t}{\varepsilon}\right)\widetilde{u}_N\right\|_{L^2([0,T];L^2)} = 0.$$

As above, we can write that

$$\left\|\mathcal{L}\left(\frac{t}{\varepsilon}\right)\widetilde{u}_N\cdot\nabla\overline{u}_N\right\|_{L^2([0,T];L^2)}\leq C_\eta\left\|\mathcal{L}\left(\frac{t}{\varepsilon}\right)\widetilde{u}_N\right\|_{L^2([0,T];L^\infty)}\|\overline{u}_N\|_{L^\infty([0,T];L^2)}.$$

Using Lemma 7.12, we deduce that $\lim_{\varepsilon\to 0}\|F_2^{\varepsilon,\eta}\|_{L^2([0,T];L^2)}=0$, so that

$$\int_0^t (F_2^{\varepsilon,\eta}(t')|u^\varepsilon(t'))_{L^2}\,dt'\leq\rho^{\varepsilon,\eta}. \tag{7.6.10}$$

As δ^ε vanishes at the boundary, we have, by integration by parts,

$$\int_0^t\left(\delta^\varepsilon(t')\cdot\nabla\delta^\varepsilon(t')\Big|\mathcal{L}\left(\frac{t'}{\varepsilon}\right)\widetilde{u}_N(t')\right)_{L^2}dt'$$
$$=\int_0^t\left(\delta^\varepsilon(t')\otimes\delta^\varepsilon(t')\Big|\nabla\mathcal{L}\left(\frac{t'}{\varepsilon}\right)\widetilde{u}_N(t')\right)_{L^2}dt'.$$

We immediately infer that

$$\int_0^t\left(\delta^\varepsilon(t')\cdot\nabla\delta^\varepsilon(t')\Big|\mathcal{L}\left(\frac{t'}{\varepsilon}\right)\widetilde{u}_N(t')\right)_{L^2}dt'$$
$$\leq C\int_0^t\|\delta^\varepsilon(t')\|_{L^2}^2\left\|\nabla\mathcal{L}\left(\frac{t'}{\varepsilon}\right)\widetilde{u}_N(t')\right\|_{L^\infty}dt'.$$

By the Cauchy–Schwarz inequality, we get

$$\int_0^t\left(\delta^\varepsilon(t')\cdot\nabla\delta^\varepsilon(t')\Big|\mathcal{L}\left(\frac{t'}{\varepsilon}\right)\widetilde{u}_N(t')\right)_{L^2}dt'$$
$$\leq C_\eta t^{\frac{1}{2}}E_t^\varepsilon(\delta^\varepsilon)\left\|\nabla\mathcal{L}\left(\frac{t}{\varepsilon}\right)\widetilde{u}_N\right\|_{L^2([0,T];L^\infty)}.$$

Lemma 7.12 therefore implies that

$$\left|\int_0^t\left(\delta^\varepsilon(t')\cdot\nabla\delta^\varepsilon(t')|\mathcal{L}\left(\frac{t'}{\varepsilon}\right)\widetilde{u}_N(t')\right)_{L^2}dt'\right|\leq\rho^{\varepsilon,\eta}. \tag{7.6.11}$$

Plugging (7.6.10) and (7.6.11) into (7.6.9) and using a Gronwall inequality finally yields

$$E_t^\varepsilon(\delta^\varepsilon)\leq\rho^{\varepsilon,\eta}\exp\left(\frac{C}{\nu}\int_0^t\|\nabla^h\overline{u}(t')\|_{L^2}^2dt'\right)$$
$$\leq\rho^{\varepsilon,\eta}\exp\left(\frac{C}{\nu^2}\|\overline{u}_0\|_{L^2}^2\right)=\rho^{\varepsilon,\eta}.$$

That proves Lemma 7.11, provided we prove Lemma 7.12. □

Proof of Lemma 7.12 We shall prove a slightly better result, namely that

$$\left\|\mathcal{L}\left(\frac{t}{\varepsilon}\right)\widetilde{u}_N\right\|_{L^1_T(L^\infty)}\leq C_\eta\varepsilon^{\frac{1}{2}}\|u_0\|_{L^2}. \tag{7.6.12}$$

Since $\mathcal{L}(t/\varepsilon)\widetilde{u}_N$ is uniformly bounded in the space $L^\infty([0,T];L^2)$, it is controlled by C_η in $L^\infty([0,T];L^\infty)$ by Bernstein's lemma. Then Lemma 7.12 follows from (7.6.12).

So let us prove (7.6.12). This is typically a Strichartz-type estimate, of the type of those derived in Chapter 5, Theorem 5.3, page 95. However the setting here is slightly different (the vertical variable is in $[0,1]$ instead of $\mathbf{R}$) so we will give the details of the proof.

By definition, $\widetilde{U}_N^\varepsilon \stackrel{\text{def}}{=} \mathcal{L}(t/\varepsilon)\widetilde{u}_N$ satisfies

$$\partial_t \widetilde{U}_N^\varepsilon - \nu\Delta_h \widetilde{U}_N^\varepsilon + \frac{1}{\varepsilon}R\widetilde{U}_N^\varepsilon = -\nabla p_N^\varepsilon + f_N^\varepsilon \quad \text{with}$$

$$\widetilde{U}_{N|t=0}^\varepsilon = u_{0,N} - \overline{u}_{0,N} \quad \text{and}$$

$$f_N^\varepsilon \stackrel{\text{def}}{=} -\mathcal{L}\left(\frac{t}{\varepsilon}\right)\mathcal{E}u_N \in L^1_{loc}(\mathbf{R}^+,\mathcal{H}).$$

Using Duhamel's formula (we omit the argument, see for instance Section 5.2), (7.6.12) follows from the following result. If the vector field v^ε solves the anisotropic Stokes–Coriolis system

$$\partial_t v^\varepsilon - \nu\Delta_h v^\varepsilon + \frac{1}{\varepsilon}Rv^\varepsilon = -\nabla p^\varepsilon, \quad v_{|t=0}^\varepsilon = v_0,$$

in $C(\mathbf{R}^+;\mathcal{B}_N)\cap L^2(\mathbf{R}^+;H^{1,0})$, then

$$\|v^\varepsilon\|_{L^1(\mathbf{R}^+;L^\infty)} \leq C_N\varepsilon^{\frac{1}{2}}\|v_0\|_{L^2}.$$

So let us prove that result. As $\mathcal{L}$ and Δ_h commute, we have

$$v^\varepsilon(t) = \mathcal{L}\left(\frac{t}{\varepsilon}\right)e^{\nu t\Delta_h}v_0 = e^{\nu t\Delta_h}\mathcal{L}\left(\frac{t}{\varepsilon}\right)v_0.$$

Now let us write

$$v_0 = \sum_{k_3=1}^N \begin{pmatrix} v_0^{k_3,h}(x_h)\cos(k_3\pi x_3) \\ -\dfrac{1}{k_3\pi}\operatorname{div}_h v_0^{k_3,h}(x_h)\sin(k_3\pi x_3) \end{pmatrix}$$

where $v_0^{k_3,h}$ has support included in the ring $\mathcal{C}(1/N,N)$.

We have

$$\mathcal{F}_h\left(\mathcal{L}\left(\frac{t}{\varepsilon}\right)v_0\right)(\xi_h) = \sum_{k_3=1}^N \Bigg(A(\xi_h)\mathcal{L}_{k_3}\left(\frac{t}{\varepsilon}\right)A^{-1}(\xi_h)\widehat{v}_0^{k_3,h}(\xi_h)\cos(k_3\pi x_3),$$

$$\frac{i}{k_3\pi}\xi_h\cdot A(\xi_h)\mathcal{L}_{k_3}\left(\frac{t}{\varepsilon}\right)A^{-1}(\xi_h)\widehat{v}_0^{k_3,h}(\xi_h)\sin(k_3\pi x_3)\Bigg).$$

Let us compute this expression more precisely. We have

$$A^{-1}(\xi_h)\widehat{v}_0^{k_3,h}(\xi_h) = \begin{pmatrix} \xi_h \cdot \widehat{v}_0^{k_3,h}(\xi_h) \\ \xi_h \wedge \widehat{v}_0^{k_3,h}(\xi_h) \end{pmatrix}.$$

Since

$$\mathcal{L}_{k_3}(\tau)\begin{pmatrix} \xi_h \cdot \widehat{v}_0^{k_3,h} \\ \xi_h \wedge \widehat{v}_0^{k_3,h} \end{pmatrix} = \begin{pmatrix} \cos(\tau_{k_3})\xi_h \cdot \widehat{v}_0^{k_3,h} + \lambda_{k_3}\sin(\tau_{k_3})\xi_h \wedge \widehat{v}_0^{k_3,h} \\ -\dfrac{1}{\lambda_{k_3}}\sin(\tau_{k_3})\xi_h \cdot \widehat{v}_0^{k_3,h} + \cos(\tau_{k_3})\xi_h \wedge \widehat{v}_0^{k_3,h} \end{pmatrix},$$

with the notation introduced in (7.4.15), it follows that

$$A(\xi_h)\mathcal{L}_{k_3}\left(\frac{t}{\varepsilon}\right)A^{-1}(\xi_h, k_3)e^{-\nu t|\xi_h|^2}\widehat{v}_0^{k_3,h}(\xi_h)$$

is a combination of terms of the following type (complex notation will be helpful for the computations below)

$$e^{\pm i\lambda_{k_3}\frac{t}{\varepsilon}}\widetilde{A}(\xi_h, k_3)e^{-\nu t|\xi_h|^2}\widehat{v}_0^{k_3,h}(\xi_h), \tag{7.6.13}$$

where $\widetilde{A}(\xi_h, k_3)$ is a 2×2 matrix, the coefficients of which are bounded by C_η.

Let us consider a radial function $\Psi \in \mathcal{D}(\mathbf{R}^2\setminus\{0\})$, the value of which is one near the ring $\mathcal{C}(1/N, N)$, and let us define

$$\mathcal{I}_{k_3}(t,\tau,x_h) \stackrel{\text{def}}{=} \int_{\mathbf{R}^2} e^{i(x_h|\xi_h)+i\lambda_{k_3}\tau-\nu t|\xi_h|^2}\Psi(\xi_h)d\xi_h.$$

Since $v^\varepsilon(t)$ is a finite combination of terms of the type

$$\mathcal{I}_{k_3}\left(t,\frac{t}{\varepsilon},\cdot\right) * \mathcal{F}_h^{-1}\widetilde{A}(\xi_h, k_3)\widehat{v}_0^{k_3,j},$$

the result will be proved if, for any $k_3 \le N$,

$$\left\|\mathcal{I}_{k_3}\left(t,\frac{t}{\varepsilon},\cdot\right) * \gamma\right\|_{L^1([0,T];L_h^\infty)} \le C_N\varepsilon^{\frac{1}{2}}\|\gamma\|_{L^2}. \tag{7.6.14}$$

By Lemma 5.2, page 94 we know that there exist constants C_η and c_η such that

$$\|\mathcal{I}_{k_3}(t,\tau,\cdot)\|_{L_h^\infty} \le \frac{C_\eta}{\tau^{\frac{1}{2}}}e^{-c_\eta t}. \tag{7.6.15}$$

Now we shall use a duality argument, exactly as in the proof of Theorem 5.3, page 95. We observe that

$$\|a\|_{L^1(\mathbf{R}^+;L_h^\infty)} = \sup_{\varphi\in\mathcal{G}}\int_{\mathbf{R}^+\times\mathbf{R}^2} a(t,x_h)\varphi(t,x_h)dx_h dt$$

with $\mathcal{G} \stackrel{\text{def}}{=} \left\{\varphi \in \mathcal{D}(\mathbf{R}^+ \times \mathbf{R}^2), \|\varphi\|_{L^\infty(\mathbf{R}^+;L^1_h)} \leq 1\right\}$. So we can write

$$\begin{aligned}\Gamma^\varepsilon_{k_3} &\stackrel{\text{def}}{=} \left\|\mathcal{I}_{k_3}\left(t, \frac{t}{\varepsilon}, \cdot\right) * \gamma\right\|_{L^1(\mathbf{R}^+;L^\infty_h)}\\ &= \sup_{\varphi\in\mathcal{G}} \int_{\mathbf{R}^+\times\mathbf{R}^4} \mathcal{I}_{k_3}\left(t, \frac{t}{\varepsilon}, x_h - y_h\right)\gamma(y_h)\varphi(t,x_h)dx_h dy_h dt\\ &= \sup_{\varphi\in\mathcal{G}} \int_{\mathbf{R}^+\times\mathbf{R}^2} \gamma(y_h)\left(\int_{\mathbf{R}^2} \mathcal{I}_{k_3}\left(t, \frac{t}{\varepsilon}, x_h - y_h\right)\varphi(t,x_h)dx_h\right)dy_h dt.\end{aligned}$$

The Cauchy–Schwarz inequality then yields

$$\Gamma^\varepsilon_{k_3} \leq \|\gamma\|_{L^2_h} \sup_{\varphi\in\mathcal{G}} \left\|\int_{\mathbf{R}^+} \check{\mathcal{I}}_{k_3}\left(t, \frac{t}{\varepsilon}, \cdot\right) * \varphi(t,\cdot)dt\right\|_{L^2_h}. \tag{7.6.16}$$

By the Fourier–Plancherel theorem, we have

$$\begin{aligned}\widetilde{\Gamma}_{k_3} &\stackrel{\text{def}}{=} \left\|\int_{\mathbf{R}^+} \check{\mathcal{I}}_{k_3}\left(t, \frac{t}{\varepsilon}, \cdot\right) * \varphi(t,\cdot)dt\right\|^2_{L^2_h}\\ &= (2\pi)^{-2}\left\|\int_{\mathbf{R}^+} \mathcal{F}_h\widehat{\mathcal{I}}_{k_3}\left(t, \frac{t}{\varepsilon}, \cdot\right)\widehat{\varphi}(t,\cdot)dt\right\|^2_{L^2_h}\end{aligned}$$

and we have that

$$\left\|\int_{\mathbf{R}^+} \mathcal{F}_h\widehat{\mathcal{I}}_{k_3}\left(t, \frac{t}{\varepsilon}, \cdot\right)\widehat{\varphi}(t,\cdot)dt\right\|^2_{L^2_h}$$

is less than or equal to

$$\int_{(\mathbf{R}^+)^2\times\mathbf{R}^2} \widehat{\mathcal{I}}_{k_3}\left(t, \frac{t}{\varepsilon}, -\xi_h\right)\overline{\widehat{\varphi}}(t,\xi_h)\overline{\widehat{\mathcal{I}}_{k_3}}\left(s, \frac{s}{\varepsilon}, -\xi_h\right)\widehat{\varphi}(s,\xi_h)d\xi_h dt ds.$$

But by definition of $\mathcal{I}_{k_3}$, we have

$$\widehat{\mathcal{I}}_{k_3}\left(t, \frac{t}{\varepsilon}, -\xi_h\right)\overline{\widehat{\mathcal{I}}_{k_3}}\left(s, \frac{s}{\varepsilon}, -\xi_h\right) = \widehat{\mathcal{I}}_{k_3}\left(t+s, \frac{t-s}{\varepsilon}, -\xi_h\right)\Psi(-\xi_h).$$

It follows that

$$\widetilde{\Gamma}_{k_3} \leq C\int_{(\mathbf{R}^+)^2\times\mathbf{R}^2} \mathcal{F}\left(\check{\mathcal{I}}_{k_3}\left(t+s, \frac{t-s}{\varepsilon}, \cdot\right) * \varphi(t,\cdot)\right)\overline{\widehat{\varphi}}(s,\xi_h)d\xi_h dt ds.$$

Now we use the Fourier–Plancherel theorem again to get

$$\begin{aligned}\widetilde{\Gamma}_{k_3} &\leq C\int_{(\mathbf{R}^+)^2\times\mathbf{R}^2} \left(\check{\mathcal{I}}_{k_3}\left(t+s, \frac{t-s}{\varepsilon}, \cdot\right) * \varphi(t,\cdot)\right)(x_h)\varphi(s,-x_h)dx_h dt ds\\ &\leq C\int_{(\mathbf{R}^+)^2} \left\|\check{\mathcal{I}}_{k_3}\left(t+s, \frac{t-s}{\varepsilon}, \cdot\right) * \varphi(t,\cdot)\right\|_{L^\infty_h} \|\varphi(s,\cdot)\|_{L^1_h} dt ds.\end{aligned}$$

Using the dispersion estimate (7.6.15), we get

$$\widetilde{\Gamma}_{k_3} \leq C_\eta \int_{(\mathbf{R}^+)^2} \frac{\varepsilon^{1/2}}{(t-s)^{1/2}} e^{c_\eta(t+s)} \|\varphi(t)\|_{L^1_h} \|\varphi(s)\|_{L^1_h} ds dt.$$

and the result follows by integration since $\|\varphi\|_{L^1_h}$ is bounded in time. The lemma is proved. □

Now to end the proof of Theorem 7.2 we still need to prove that $u^{\varepsilon,\eta}_{\text{app}}$ converges towards $\overline{u}$ in an appropriate way. That is the object of the following lemma.

Lemma 7.13 *Under the assumptions of Lemma 7.11, we have for any compact subset K of Ω*

$$\int_{[0,T]\times K} |u^{\varepsilon,\eta}_{\text{app}}(t,x) - (\overline{u}(t,x_h),0)|^2 dx_h dx_3 = \rho^{\varepsilon,\eta}.$$

Remark Putting together Lemmas 7.11 and 7.13 clearly completes the proof of Theorem 7.2. □

Proof of Lemma 7.13 If is enough to prove that for all $N > 0$,

$$\lim_{\varepsilon\to 0} \int_{[0,T]\times K} |u^{\varepsilon,\eta}_{\text{app}}(t,x) - \overline{u}_{N_\eta}(t,x_h)|^2 dx_h dx_3 = 0.$$

But by Lemma 7.8 we have

$$\lim_{\varepsilon\to 0} \left\| u^{\varepsilon,\eta}_{\text{app}} - \mathcal{L}\left(\frac{t}{\varepsilon}\right) u_{N_\eta} \right\|_{L^2([0,T]\times\Omega)} = 0$$

and by definition

$$\mathcal{L}\left(\frac{t}{\varepsilon}\right) u_N = \overline{u}_N + \mathcal{L}\left(\frac{t}{\varepsilon}\right) \widetilde{u}_N,$$

so the result simply follows from Lemma 7.12. □

7.7 The convergence theorem in the periodic case

7.7.1 *Proof of the theorem*

In this section we shall prove Theorem 7.3, dealing with the case when the horizontal variable x_h is no longer in the whole space $\mathbf{R}^2$ as in the previous section, but in $\mathbf{T}^2$ like in Chapter 6. As noted in the introduction of this chapter, the difference with the periodic case is of course the boundary condition at $x_3 = 0$ and $x_3 = 1$. We will, as in the previous section, work with the space $\mathcal{B}$. which is dense in $\mathcal{H}_0$, which means in particular that we restrict our attention to vector fields satisfying the symmetry condition

$$u(x_h, x_3) = (u^h(x_h, -x_3), -u^3(x_h, -x_3)).$$

The general approach to prove Theorem 7.3 is the same as in the $\mathbf{R}^2$ case, so we shall continuously be referring to the results and computations of the previous sections of this chapter. However the fact that the horizontal variables are no longer taken in $\mathbf{R}^2$ prevents one from using the dispersive effects pointed out in Chapter 5, and used in the proof of Theorem 7.2 above. The analysis becomes therefore somewhat more complicated, as the interaction of fast oscillating waves (which no longer disperse and disappear) has to be taken into account; for that reason, the methods developed in Chapter 6 will be used in this section again (namely in the construction of an approximate solution, where both the ideas of the previous section and of Chapter 6 will be used).

Let us start by considering u, the solution of

$$(\mathrm{NSE}_{\nu,\mathcal{E}}) \begin{cases} \partial_t u - \nu\Delta_h u + \mathcal{E}u = \mathcal{Q}(u,u) \\ \operatorname{div} u = 0 \\ u_{|t=0} = u_0. \end{cases}$$

We recall that $\mathcal{E}$ is defined in (7.4.39), and that the quadratic form $\mathcal{Q}$ was defined in Chapter 6, Proposition 6.1. In the following we will denote by $\mathbf{P}$ the projector onto divergence-free vector fields of the form (7.0.5). We have clearly $\mathbf{P}\mathcal{E} = \mathcal{E}$.

The existence and uniqueness of solutions to $(\mathrm{NSE}_{\nu,\mathcal{E}})$ was proved in Section 6.6, Chapter 6. From now on we will suppose that $\mathbf{T}^2$ satisfies condition $(\mathcal{A})$ and that

$$\int_{\mathbf{T}^2} u_0(x_h, x_3)\, dx_h = 0.$$

A positive real number η and a time $T > 0$ being given, let us denote by $(u^\varepsilon_{\mathrm{app}})_{\varepsilon>0}$ the family of approximations given by Lemma 7.8 with $v = u$ and $f = \mathcal{Q}(u,u)$. Notice that by Proposition 6.5, the horizontal mean of f is zero for all times.

Let us observe that

$$\begin{aligned} u^\varepsilon_{0,\mathrm{int}} &= \overline{u}_N + \mathcal{L}\left(\frac{t}{\varepsilon}\right)\widetilde{u}_N \quad \text{with} \\ \overline{u}_N &\stackrel{\mathrm{def}}{=} \mathcal{F}^{-1}(\mathbf{1}_{\mathcal{C}(1,N)}\mathcal{F}\overline{u}) \quad \text{and} \\ \widetilde{u}_N &\stackrel{\mathrm{def}}{=} \sum_{k_3=1}^{N} \left(\widetilde{u}^{k_3,h}(x_h)\cos(k_3\pi x_3), -\frac{1}{k_3\pi}\operatorname{div}_h \widetilde{u}^{k_3,h}(x_h)\sin(k_3\pi x_3)\right), \end{aligned}$$

the support of $\widetilde{u}^{k_3,h}$ being included in the ring $\mathcal{C}(1,N)$ (and N is large, depending on η).

As in the $\mathbf{R}^2$ case studied in the previous section, our aim is to apply Lemma 7.5. Unfortunately if we apply that lemma directly with $\Psi^\varepsilon = u^\varepsilon_{\mathrm{app}}$

as defined above, we will not be able to conclude, as the term $G^\varepsilon(u^\varepsilon_{\rm app})$ is not small in the periodic case (due to the absence of dispersion). So as in Chapter 6, we will get around that difficulty by introducing an additional corrector. Let us postpone the proof of the following lemma.

Lemma 7.14 *There is a smooth family of divergence-free vector fields $u^\varepsilon_{\widetilde{\rm app}}$ belonging to $C^1(\mathbf{R}^+;\mathcal{V}_\sigma)$ satisfying the following properties:*

$$u^\varepsilon_{\widetilde{\rm app}} = u^\varepsilon_{\rm app} + \rho^{\varepsilon,\eta} \quad in \quad L^\infty_{\rm loc}(\mathbf{R}^+;H^1(\Omega)) \tag{7.7.1}$$

and

$$L^\varepsilon u^\varepsilon_{\widetilde{\rm app}} + u^\varepsilon_{\widetilde{\rm app}}\cdot\nabla u^\varepsilon_{\widetilde{\rm app}} = H^{\varepsilon,\eta} + \nabla p^\varepsilon, \tag{7.7.2}$$

where

$$\int_0^t (H^{\varepsilon,\eta}(t')|u^\varepsilon(t'))\,dt' \le \rho^{\varepsilon,\eta}.$$

Let us continue with the proof of the theorem, applying Lemma 7.5 to the family $u^\varepsilon_{\widetilde{\rm app}}$. We find that $\delta^{\varepsilon,\eta} \stackrel{\rm def}{=} u^\varepsilon - u^\varepsilon_{\widetilde{\rm app}}$ satisfies

$$E^\varepsilon_t(\delta^{\varepsilon,\eta}) = E^\varepsilon_t(u^\varepsilon) + E^\varepsilon_t(u^\varepsilon_{\widetilde{\rm app}}) - 2(u(0)|u^\varepsilon_{\widetilde{\rm app}}(0))_{L^2} + \sum_{j=1}^2 R_j(t) \tag{7.7.3}$$

with

$$R_1(t) \stackrel{\rm def}{=} -2\int_0^t \left(\delta^{\varepsilon,\eta}(t')\cdot\nabla\delta^{\varepsilon,\eta}(t')|u^\varepsilon_{\widetilde{\rm app}}(t')\right)_{L^2} dt',$$

$$R_2(t) \stackrel{\rm def}{=} 2\int_0^t (G^\varepsilon(u^\varepsilon_{\widetilde{\rm app}})(t')|u^\varepsilon(t'))_{L^2} dt' \quad \text{and}$$

$$G^\varepsilon(u^\varepsilon_{\widetilde{\rm app}}) \stackrel{\rm def}{=} L^\varepsilon u^\varepsilon_{\widetilde{\rm app}} + u^\varepsilon_{\widetilde{\rm app}}\cdot\nabla u^\varepsilon_{\widetilde{\rm app}}.$$

As usual we have

$$E^\varepsilon_t(u^\varepsilon) \le \|u_0\|^2_{L^2}, \tag{7.7.4}$$

so let us compute the energy of $u^\varepsilon_{\widetilde{\rm app}}$. By (7.7.1) clearly

$$E^\varepsilon_t(u^\varepsilon_{\widetilde{\rm app}}) \le E^\varepsilon_t(u^\varepsilon_{\rm app}) + \rho^{\varepsilon,\eta},$$

and by Lemma 7.8 one has

$$E^\varepsilon_t(u^\varepsilon_{\rm app}) \le \|u_0\|^2_{L^2} + 2\int_0^t \langle\mathcal{Q}(u(t'),u(t')),u(t')\rangle dt' + \rho^{\varepsilon,\eta}.$$

The symmetry properties of $\mathcal{Q}$ imply that $E_t^\varepsilon(u_{\text{app}}^\varepsilon) \leq \|u_0\|_{L^2}^2 + \rho^{\varepsilon,\eta}$, hence

$$E_t^\varepsilon(u_{\widetilde{\text{app}}}^\varepsilon) \leq \|u_0\|_{L^2}^2 + \rho^{\varepsilon,\eta}. \tag{7.7.5}$$

Similarly by (7.7.1), we can write that

$$\|u_{\widetilde{\text{app}}|t=0}^\varepsilon - u_0\|_{L^2} = \rho^{\varepsilon,\eta}. \tag{7.7.6}$$

Putting (7.7.4), (7.7.5) and (7.7.6) together yields

$$E_t^\varepsilon(u^\varepsilon) + E_t^\varepsilon(u_{\widetilde{\text{app}}}^\varepsilon) - 2(u(0)|u_{\widetilde{\text{app}}}^\varepsilon(0))_{L^2} \leq \rho^{\varepsilon,\eta}. \tag{7.7.7}$$

Then according to (7.7.2), we have

$$R_2(t) = \rho^{\varepsilon,\eta}, \tag{7.7.8}$$

so plugging (7.7.7) and (7.7.8) into (7.7.4) yields

$$\begin{aligned}
E_t^\varepsilon(\delta^{\varepsilon,\eta}) &\leq \rho^{\varepsilon,\eta} + C\left|\int_0^t \left(\delta^{\varepsilon,\eta}(t')\cdot\nabla\delta^{\varepsilon,\eta}(t')|u_{\widetilde{\text{app}}}^\varepsilon(t')\right)_{L^2} dt'\right| \\
&\leq \rho^{\varepsilon,\eta} + C\left|\int_0^t \left(\delta^{\varepsilon,\eta}(t')\cdot\nabla\delta^{\varepsilon,\eta}(t')|u_{\text{app}}^\varepsilon(t')\right)_{L^2} dt'\right|
\end{aligned}$$

where we have used (7.7.1). To estimate the last term we use Lemma 7.10, which implies that

$$\left|\int_0^t (\delta^{\varepsilon,\eta}\cdot\nabla\delta^{\varepsilon,\eta}|u_{\text{app}}^\varepsilon)_{L^2}(t')dt'\right| \leq \left(C_\eta\varepsilon^{\frac{1}{2}} + \frac{1}{4}\right) E_t^\varepsilon(\delta)$$
$$+ \int_0^t \left(\delta^{\varepsilon,\eta}(t')\cdot\nabla\delta^{\varepsilon,\eta}(t')|\mathcal{L}\left(\frac{t'}{\varepsilon}\right)\widetilde{u}_N(t')\right)_{L^2} dt' + \frac{C}{\nu}\int_0^t \|\nabla^h\overline{u}(t')\|_{L^2}^2\|\delta(t')\|_{L^2}^2 dt'.$$

Applying Lemma 6.5 page 146 to $b = \mathcal{L}(t/\varepsilon)\widetilde{u}_N$ implies (since $\mathcal{L}$ is unitary) that

$$\begin{aligned}
E_t^\varepsilon(\delta^{\varepsilon,\eta}) &\leq \rho^{\varepsilon,\eta} + \frac{\nu}{2}\int_0^t \|\nabla^h\delta^{\varepsilon,\eta}(t')\|_{L^2}dt' \\
&\quad + \frac{C}{\nu}\int_0^t \|\delta^{\varepsilon,\eta}(t')\|_{L^2}^2\|\nabla^h\overline{u}(t')\|_{L^2}^2 dt' \\
&\quad + \frac{C}{\nu}\int_0^t \|\delta^{\varepsilon,\eta}(t')\|_{L^2}^2 \left(\|\nabla^h\widetilde{u}(t')\|_{L^2}^2 + \|\partial_3\widetilde{u}(t)\|_{L^2}^2\|\partial_3\nabla^h\widetilde{u}(t)\|_{L^2}^2\right) dt',
\end{aligned}$$

and Gronwall's lemma yields finally that $E_t^\varepsilon(\delta^{\varepsilon,\eta}) \leq \rho^{\varepsilon,\eta}$, using Proposition 6.5. The theorem follows, up to the proof of Lemmas 7.14. □

Proof of Lemma 7.14 Our aim is to find a smooth, divergence-free corrector to $u_{\text{app}}^\varepsilon$, vanishing at the boundary, such that equation (7.7.2) is satisfied.

By Lemma 7.8, we know that

$$L^\varepsilon u_{\text{app}}^\varepsilon + u_{\text{app}}^\varepsilon \cdot \nabla u_{\text{app}}^\varepsilon = u_{\text{app}}^\varepsilon \cdot \nabla u_{\text{app}}^\varepsilon + \mathcal{L}\left(\frac{t}{\varepsilon}\right) \mathcal{Q}(u,u) + R^{\varepsilon,\eta} + \nabla p^\varepsilon,$$

and by the continuity properties of $\mathcal{Q}$ stated in Proposition 6.6 we infer that

$$L^\varepsilon u_{\text{app}}^\varepsilon + u_{\text{app}}^\varepsilon \cdot \nabla u_{\text{app}}^\varepsilon = u_{\text{app}}^\varepsilon \cdot \nabla u_{\text{app}}^\varepsilon + \mathcal{L}\left(\frac{t}{\varepsilon}\right) \mathcal{Q}(u_N,u_N) + R^{\varepsilon,\eta} + \nabla p^\varepsilon.$$

Let us define

$$\begin{aligned} F_1^{\varepsilon,\eta} &= u_{\text{app}}^\varepsilon \cdot \nabla u_{\text{app}}^\varepsilon - \mathcal{L}\left(\frac{t}{\varepsilon}\right) u_N \cdot \nabla \mathcal{L}\left(\frac{t}{\varepsilon}\right) u_N \\ &= u_{\text{app}}^\varepsilon \cdot \nabla u_{\text{app}}^\varepsilon + \mathcal{L}\left(\frac{t}{\varepsilon}\right) \mathcal{Q}^\varepsilon(u_N,u_N). \end{aligned}$$

By Lemma 7.9 we have $\lim\limits_{\varepsilon\to 0} \|F_1^{\varepsilon,\eta}\|_{L^2([0,T];L^2)} = 0$, so as for (7.3.5) we have

$$\int_0^t (F_1^{\varepsilon,\eta}(t')|u^\varepsilon(t'))\,dt' \leq \rho^{\varepsilon,\eta}.$$

Note that by (7.3.4) one also has

$$\int_0^t (R^{\varepsilon,\eta}(t')|u^\varepsilon(t'))\,dt' \leq \rho^{\varepsilon,\eta}.$$

so we find that one can write

$$L^\varepsilon u_{\text{app}}^\varepsilon + u_{\text{app}}^\varepsilon \cdot \nabla u_{\text{app}}^\varepsilon = H^{\varepsilon,\eta} + \mathbf{P}_N \mathcal{L}\left(\frac{t}{\varepsilon}\right) (\mathcal{Q} - \mathcal{Q}^\varepsilon)(u_N,u_N) + \nabla p^\varepsilon,$$

where $H^{\varepsilon,\eta}$ satisfies

$$\int_0^t (H^{\varepsilon,\eta}(t')|u^\varepsilon(t'))dt' = \rho^{\varepsilon,\eta}.$$

Now let us prove the following lemma.

Lemma 7.15 *Let $\eta > 0$ be given. There is a family of divergence-free vector fields $\mathcal{F}^{\varepsilon,\eta}$, bounded in $L^\infty_{\text{loc}}(\mathbf{R}^+; H^s(\Omega))$ for all $s \geq 0$, such that*

$$\mathbf{P}_N \mathcal{L}\left(\frac{t}{\varepsilon}\right) (\mathcal{Q}^\varepsilon - \mathcal{Q})(u_N,u_N) = \varepsilon \mathbf{P} L^\varepsilon \mathcal{F}^{\varepsilon,\eta} + R^{\varepsilon,\eta}. \tag{7.7.9}$$

Proof We define the following function, as in the purely periodic case, page 143,

$$\mathcal{R}_N(\tau,t) \stackrel{\text{def}}{=} \mathcal{F}^{-1}\mathbf{1}_{|n|\leq N} \sum_{\sigma\in\{+,-\}^3} \sum_{k\notin\mathcal{K}_n^\sigma} \frac{e^{-i\tau\omega_n^\sigma}}{i\omega_n^\sigma} \Big(u_N^{\sigma_1}(t,k)\cdot(n-k) \\ \times u_N^{\sigma_2}(t,n-k), e^{\sigma_3}(n)\Big) e^{\sigma_3}(n).$$

Then we have

$$\partial_\tau \mathcal{R}_N = \mathbf{P}_N(\mathcal{Q}^\varepsilon - \mathcal{Q})(u_N, u_N). \tag{7.7.10}$$

Moreover $\mathcal{F}^{\varepsilon,\eta} \stackrel{\text{def}}{=} \mathcal{L}(t/\varepsilon)\mathcal{R}_N$ is clearly smooth and divergence-free. Now let us check (7.7.9). We have

$$\begin{aligned} \varepsilon \mathbf{P} L^\varepsilon \mathcal{F}^{\varepsilon,\eta} &= \left(\varepsilon\partial_t + \mathbf{P}R - \varepsilon\Delta_h - \varepsilon^2\beta\partial_3^2\right)\mathcal{L}\left(\frac{t}{\varepsilon}\right)\mathcal{R}_N \\ &= \mathcal{G}^{\varepsilon,\eta} + (\partial_\tau\mathcal{L})\left(\frac{t}{\varepsilon}\right)\mathcal{R} + \mathbf{P}R\mathcal{L}\left(\frac{t}{\varepsilon}\right)\mathcal{R} + \mathcal{L}\left(\frac{t}{\varepsilon}\right)\partial_\tau\mathcal{R}, \end{aligned}$$

where

$$\mathcal{G}^{\varepsilon,\eta} \stackrel{\text{def}}{=} -\varepsilon\Delta_h\mathcal{F}^{\varepsilon,\eta} - \varepsilon^2\beta\partial_3^2\mathcal{F}^{\varepsilon,\eta} + \varepsilon\mathcal{L}\left(\frac{t}{\varepsilon}\right)\partial_t\mathcal{R}.$$

The first two terms of $\mathcal{G}^{\varepsilon,\eta}$ are clearly remainder terms of the generic type $R^{\varepsilon,\eta}$, as long as N is chosen large enough in terms of η, so is the last term which is exactly the term $\varepsilon\mathcal{L}(t/\varepsilon)R_\eta^{\varepsilon,t}$ where $R_\eta^{\varepsilon,t}$ was defined on page 143. So $\mathcal{G}^{\varepsilon,\eta}$ is a remainder of the type $R^{\varepsilon,\eta}$. Since $\partial_\tau\mathcal{L} + \mathbf{P}R\mathcal{L} = 0$, we infer that

$$\varepsilon\mathbf{P}L^\varepsilon\mathcal{F}^{\varepsilon,\eta} = \mathcal{L}\left(\frac{t}{\varepsilon}\right)\mathbf{P}_N(\mathcal{Q}^\varepsilon - \mathcal{Q})(u_N, u_N) + R^{\varepsilon,\eta}.$$

and Lemma 7.15 is proved. □

It follows from Lemma 7.15 that

$$L^\varepsilon u^\varepsilon_{\text{app}} + u^\varepsilon_{\text{app}}\cdot\nabla u^\varepsilon_{\text{app}} = H^{\varepsilon,\eta} - \varepsilon L^\varepsilon\mathcal{F}^{\varepsilon,\eta} + \nabla p^\varepsilon.$$

Lemma 7.14 is almost proved, except for the fact that $\mathcal{F}^{\varepsilon,h}$ does not vanish on the boundary. So we shall finally correct $\mathcal{F}^{\varepsilon,\eta}$ by a boundary layer: we write $\mathcal{F}^{\varepsilon,\eta}$ in the form

$$\mathcal{F}^{\varepsilon,\eta} = \sum_{k_3=0}^{N} \left(\mathcal{F}^{\varepsilon,h,k_3}\cos(k_3\pi x_3), \mathcal{F}^{\varepsilon,3,k_3}\sin(k_3\pi x_3)\right),$$

and we define a smooth function b on $\mathbf{R}^+$ such that $b(0) = 1$ and $b(\zeta) = 0$ when $\zeta \geq 1/2$ and such that b is mean free. Define

$$B(x) = \int_0^x b(y)dy.$$

Finally consider

$$\widetilde{\mathcal{F}}^{\varepsilon,\eta} \stackrel{\text{def}}{=} \mathcal{F}^{\varepsilon,\eta} - \mathcal{F}^{\varepsilon,\eta}_{\text{BL}},$$

where $\mathcal{F}^{\varepsilon,\eta}_{\text{BL}} = \sum_{k_3=0}^{N} \left(\mathcal{F}^{\varepsilon,k_3,h}_{\text{BL}}, \mathcal{F}^{\varepsilon,k_3,3}_{\text{BL}}\right)$ with

$$\mathcal{F}^{\varepsilon,k_3,h}_{\text{BL}} = \mathcal{F}^{\varepsilon,h,k_3}_{|x_3=0}\, b\left(\frac{x_3}{\varepsilon}\right) + (-1)^{k_3} \mathcal{F}^{\varepsilon,h,k_3}_{|x_3=1}\, b\left(\frac{1-x_3}{\varepsilon}\right)$$

and

$$\mathcal{F}^{\varepsilon,k_3,3}_{\text{BL}} = -\varepsilon \operatorname{div}_h \mathcal{F}^{\varepsilon,h,k_3}_{|x_3=0} B\left(\frac{x_3}{\varepsilon}\right) + \varepsilon(-1)^{k_3} \operatorname{div}_h \mathcal{F}^{\varepsilon,h,k_3}_{|x_3=1} B\left(\frac{1-x_3}{\varepsilon}\right).$$

Clearly $\widetilde{\mathcal{F}}^{\varepsilon,\eta}$ is smooth, divergence-free, and vanishes on the boundary (because b is mean free). Finally let us define

$$u^{\varepsilon}_{\widetilde{\text{app}}} = u^{\varepsilon}_{\text{app}} + \varepsilon \widetilde{\mathcal{F}}^{\varepsilon,\eta}.$$

The vector field $u^{\varepsilon}_{\widetilde{\text{app}}}$ is smooth, divergence-free, and vanishes on the boundary. Clearly $u^{\varepsilon}_{\widetilde{\text{app}}}$ satisfies (7.7.1), and we leave the smoothing of $u^{\varepsilon}_{\widetilde{\text{app}}}$ in time to the reader. The only point to check is (7.7.2), but that is simply due to Lemma 7.15, along with the fact that the boundary layers contribute to the equation by negligible terms of the type $R^{\varepsilon,\eta}$, as they are of order 1 in ε. Lemma 7.14 is proved. □

8

References and remarks on rotating fluids

The problem investigated in this part can be seen as a particular case of the study of the asymptotic behavior (when ε tends to 0) of solutions of systems of the type

$$\partial_t u^\varepsilon - \Delta_\varepsilon u^\varepsilon + Q(u^\varepsilon, u^\varepsilon) + \frac{1}{\varepsilon} A u = 0$$

where Δ_ε is a non-negative operator of order 2 possibly depending on ε, and A is a skew-symmetric operator. This framework contains of course a lot of problems including hyperbolic cases when $\Delta_\varepsilon = 0$. Let us notice that, formally, any element of the weak closure of the family $(u^\varepsilon)_{\varepsilon>0}$ belongs to the kernel of A.

We can distinguish from the beginning two types of problems depending on the nature of the initial data. The first case, known as the well-prepared case, is the case when the initial data belong to the kernel of A. The second case, known as the ill-prepared case, is the general case.

In the well-prepared case, let us mention the pioneer paper [82] by S. Klainerman and A. Majda about the incompressible limit for inviscid fluids. A lot of work has been done in this case. In the more specific case of rotating fluids, let us mention the work by T. Beale and A. Bourgeois (see [10]) and T. Colin and P. Fabrie (see [37]).

In the case of ill-prepared data, the nature of the domain plays a crucial role. The first result in this case was established in 1994 in the pioneering work [112] by S. Schochet for periodic boundary conditions. In the context of general hyperbolic problems, he introduced the key concept of limiting system (see the definition on page 125). In the more specific case of viscous rotating fluids, E. Grenier proved in 1997 in [69] Theorem 6.3, page 125, of this book. At this point, it is impossible not to mention the role of the inspiration played by the papers by J.-L. Joly, G. Métivier and J. Rauch (see for instance [77] and [78]).

In spite of the fact that the corresponding theorems have been proved afterwards, the case of the whole space, the purpose of Chapter 5 of this book, appears to be simpler because of the dispersion phenomena. These phenomena were pointed out in 1986 by S. Ukai in [119] and by K. Asano in [1] to prove the convergence of weakly compressible fluids to incompressible fluids in the whole space.

These dispersion phenomena are related to Strichartz estimates. These types of estimates appeared in the context of the wave equation in the work [116] by

R. Strichartz, [13] and [14] by P. Brenner and [102] by H. Pecher. The reader who wants to become more familiar with Strichartz estimates can refer to [66] by J. Ginibre and G. Velo and [80] by M. Keel and T. Tao.

A huge literature exists concerning applications to non-linear problems. For the Schrödinger equations, the literature is numerous. The book [26] provides a nice introduction to the subject. For a recent example of such applications, we refer for instance to [36].

Concerning the non-linear wave equation, the reader can refer to [104] for semilinear equations, and [8] and [83] for quasilinear equations. Let us mention the approach of commuting vector fields developed in [81] which does not require us to write a parametrix. This type of inequality has been used in the context of the incompressible limit for viscous fluids by B. Desjardins and E. Grenier (see [47]) to prove the analog of Theorem 5.6, page 104. In [41], R. Danchin proved the analog of Theorem 5.7 for the incompressible limit.

In the context of rotating fluids, the use of these techniques comes from [32] where a weaker version of Theorems 5.6, page 104, and 5.7, page 108, are proved.

In the case of periodic boundary conditions, the first result of the type of Theorem 6.2, page 119, was proved in 1996 by A. Babin, A. Mahalov and B. Nicolaenko in [5] under a non-resonance condition (namely condition $(\mathcal{R})$ introduced in Definition 6.2, page 144). Then the same authors dropped this condition in 1999 (see [6]) and proved Theorem 7.2, page 157. Moreover, asymptotic expansions in ε have been proved by I. Gallagher in 1998 (see [59] and [60]). Let us note that in the context of the incompressible limit, there is no such non-resonance condition. In spite of that, N. Masmoudi proved in [96] that the limit system (which is surprisingly globally parabolic as proved by I. Gallagher in [61]) in that case is globally well-posed. Using that, R. Danchin proved in [42] the analog of Theorem 6.2, page 119.

For Ekman boundary layers, the pioneering mathematical work is the work by N. Masmoudi and E. Grenier (see [72]) where Theorem 7.1, page 156, is proved. This corresponds to the case of well-prepared data. The case of ill-prepared data with horizontal periodic boundary conditions was investigated by N. Masmoudi in [95]. Theorem 7.2, page 157, was proved by the authors in [34].

PART IV

Perspectives

The aim of this last part is to present some open questions related to the stability of Ekman boundary layers, or to other types of boundary layers.

In Chapter 9 we first discuss the stability of Ekman boundary layers, define the notion of critical Reynolds number, give some hints to compute it, and some related results (instability as well as more recent stability results). This leads us to discuss basic ideas on the transition between a laminar and a turbulent regime. In Chapter 10 we review boundary layer effects in magnetohydrodynamics and quasigeostrophic equations, which are very close to genuine Ekman layers. In Chapter 11 we then introduce the boundary layers which appear near vertical walls and formally link them with the classical Prandlt equations, and in the last chapter, we introduce spherical layers, whose study is completely open.

9

Stability of horizontal boundary layers

Let us now detail the stability properties of an Ekman layer introduced in Part I, page 11. First we will recall how to compute the critical Reynolds number. Then we will describe briefly what happens at larger Reynolds numbers.

9.1 Critical Reynolds number

The first step in the study of the stability of the Ekman layer is to consider the linear stability of a pure Ekman spiral of the form

$$u^E(t, x_1, x_2, \zeta) = U_\infty \begin{pmatrix} 1 - e^{-\zeta/\sqrt{2}} \cos \frac{\zeta}{\sqrt{2}} \\ -e^{-\zeta/\sqrt{2}} \sin \frac{\zeta}{\sqrt{2}} \end{pmatrix}, \tag{9.1.1}$$

where U_∞ is the velocity away from the layer and ζ is the rescaled vertical component $\zeta = x_3/\sqrt{\varepsilon\nu}$. The corresponding Reynolds number is

$$Re = U_\infty \sqrt{\frac{\varepsilon}{\nu}}.$$

Let us consider the Navier–Stokes–Coriolis equations, linearized around u^E

$$(\mathrm{LNSC}_\varepsilon) \begin{cases} \partial_t u + u^E \cdot \nabla u + u \cdot \nabla u^E - \nu \Delta u + \dfrac{e^3 \times u}{\varepsilon} + \dfrac{\nabla q}{\varepsilon} = 0 \\ \operatorname{div} u = 0. \end{cases}$$

The problem is now to study the (linear) stability of the 0 solution of the system ($\mathrm{LNSC}_\varepsilon$). If $u = 0$ is stable we say that u^E is linearly stable, if not we say that it is linearly unstable. Numerical results show that $u = 0$ is stable if and only if $Re < Re_c$ where Re_c can be evaluated numerically. Up to now there is no mathematical proof of this fact, and it is only possible to prove that 0 is linearly stable for $Re < Re_1$ and unstable for $Re > Re_2$ with $Re_1 < Re_c < Re_2$, Re_1 being obtained by energy estimates and Re_2 by a perturbative analysis of the case $Re = \infty$. We would like to emphasize that the numerical results are very reliable and can be considered as definitive results, since as we will see below, the stability analysis can be reduced to the study of a system of ordinary differential equations posed on the half-space, with boundary conditions on both ends, a system which can be studied arbitrarily precisely, even on desktop computers (first computations were done in the 1960s by Lilly [89]).

To reduce the system to ordinary differential equations, we take the Fourier transform in the (x_1, x_2) plane by introducing a wavenumber $k \in \mathbf{R}^2$, and we take the Laplace transform in time, and look for a solution u of $(\mathrm{LNSC}_\varepsilon)$ of the form $\exp(ik \cdot (x_1, x_2) - i\|k\|ct)v_0(x_3)$ where v_0 is a vector valued function. Note the $-i\|k\|c$ factor, which is traditional in fluid mechanics. To simplify the equations we first make a change of variables in the (x_1, x_2) plane and take

$$(-k^\perp/\|k\|, k/\|k\|)$$

as a new frame, and $(\tilde{x}_1, \tilde{x}_2)$ as new coordinates. In these new coordinates, k is parallel to the second vector of the basis. This change is equivalent to a rotation in u^E which becomes (dropping the tildes for convenience)

$$u^E(t, x_1, x_2, x_3) = -U_\infty \begin{pmatrix} \cos\gamma - e^{-\zeta/\sqrt{2}} \cos\left(\dfrac{\zeta}{\sqrt{2}} + \gamma\right) \\ -\sin\gamma + e^{-\zeta/\sqrt{2}} \sin\left(\dfrac{\zeta}{\sqrt{2}} + \gamma\right) \\ 0 \end{pmatrix},$$

where γ is the angle of rotation of the frame. Now u is of the form $\exp(ikx_2 - ikct)v_1(x_3)$ where $k \in \mathbf{R}$. However, as u does not depend on x_1, we can introduce a stream function Ψ and look for u in the form

$$u(t, x_1, x_2, \zeta) = \exp(ikx_2 - ikct) \begin{pmatrix} U(\zeta) \\ \Psi'(\zeta) \\ -ik\Psi(\zeta) \end{pmatrix}.$$

System $(\mathrm{LNSC}_\varepsilon)$ then reduces to the following 4×4 system on the two functions (U, Ψ)

$$\partial_\zeta^2 U - k^2 U + 2\partial_\zeta \Psi = ikRe\left((v_l - c)U - \Psi \partial_\zeta u_l\right), \tag{9.1.2}$$

$$\left(\partial_\zeta^2 - k^2\right)^2 \Psi - ikRe\left((v_l - c)(\partial_\zeta^2 - k^2)\Psi - \Psi \partial_\zeta^2 v_l\right) - 2\partial_\zeta U = 0 \tag{9.1.3}$$

where

$$u_l = \cos\tilde{\gamma} - \exp\left(-\zeta\sqrt{\frac{\beta}{2}}\right) \cos\left(\tilde{\gamma} + \zeta\sqrt{\frac{\beta}{2}}\right) \quad \text{and}$$

$$v_l = -\left(\sin\tilde{\gamma} - \exp\left(-\zeta\sqrt{\frac{\beta}{2}}\right) \sin\left(\tilde{\gamma} + \zeta\sqrt{\frac{\beta}{2}}\right)\right),$$

and where $\tilde{\gamma}$ is the angle between the direction of the flow outside the boundary layer and the direction of k, with boundary conditions

$$U(0) = 0, \quad \Psi(0) = \partial_\zeta \Psi(0) = 0$$

on $\zeta = 0$ and

$$\partial_\zeta U = \partial_\zeta^2 \Psi = 0$$

at infinity.

Note that as $Re \to \infty$, system (9.1.2, 9.1.3) degenerates into

$$(v_l - c)U - \Psi \partial_\zeta u_l = 0, \tag{9.1.4}$$

$$(v_l - c)(\partial_\zeta^2 - k^2)\Psi - \Psi \partial_\zeta^2 v_l = 0, \tag{9.1.5}$$

and (9.1.5) is exactly Rayleigh's equation governing the stability of v_l for linearized Euler equations (ignoring rotation and viscosity). Moreover (9.1.4) and (9.1.5) are completely decoupled, (9.1.4) being easy to solve. Therefore, for high Reynolds numbers, the stability of v_l is the same as its stability for the Euler equations. For this latter equation, stability is mainly controlled by possible inflexion points in the tangential velocity profile. As v_l has many inflexion points, we can expect the flow to be unstable for Euler equations. This is indeed the case, and there exist solutions (Ψ, c) of (9.1.5) with $\Im m\, c > 0$, which give exponentially increasing solutions of linearized Euler equations. A perturbative analysis then shows that for sufficiently large Re, there exist solutions (Ψ, U, c) of (9.1.2) and (9.1.3) with $\Im m\, c > 0$. Hence for sufficiently large Reynolds numbers there exist exponentially increasing solutions of (LNSC$_\varepsilon$), and u_E is linearly unstable.

Once we have a linear instability, we can get a non-linear instability by using the techniques of [49] and prove the following theorem, showing the nonlinear instability of Ekman layers at supercritical Reynolds numbers.

Theorem 9.1 *Let u_E be given by (9.1.1). Then u_E is non-linearly unstable provided $Re > Re_c$ in the following sense. For every arbitrary large s, there exists a constant C_0 such that for every $\eta > 0$ there exists a solution u^η with*

$$\|u^\eta(0, \cdot) - u_E\|_{H^s} \le \eta,$$

$$\|u^\eta(T^\eta, \cdot) - u_E\|_{L^2} \ge C_0,$$

$$\|u^\eta(T^\eta, \cdot) - u_E\|_{L^\infty} \ge C_0,$$

for some time $T^\eta \le C_s \log \eta^{-1} + C_s$.

We refer to [49] for more precise results.

Let us now detail a little the numerical study of (9.1.2) and (9.1.3). It is an eigenvalue problem since we have to find solution (Ψ, U, c) of (9.1.2) and (9.1.3), with $\Im m\, c > 0$. But (9.1.2) and (9.1.3) can be seen as elliptic partial differential equations of order at most 4. We can discretize them in a classical fashion (using for instance the classical three-point approximation of the second derivative), using the values at N different points, and taking into account the boundary conditions. This gives large matrices and (9.1.2) and (9.1.3) is of the form

$$Ax = cBx$$

where x contains the discretized values of Ψ and U. It remains to invert B and to compute the spectrum of $B^{-1}A$.

Therefore if k, Re, and γ are given we can compute the spectrum of (9.1.2) and (9.1.3) with an arbitrary solution. For a given N, we get of course only a finite number of eigenvalues, but as N increases, a part of the eigenvalues concentrates in a continuous spectrum and a part of the eigenvalues remains isolated and "true eigenvalues". If there exists an eigenvalue with positive imaginary part, the flow is unstable, for these parameters, if not it is stable. To get the critical Reynolds number it remains to find the smallest Reynolds number Re for which there exists parameters k and γ and a corresponding eigenvalue c with $\Im m\, c > 0$. Lilly has found in [89] a critical Reynolds number $Re_c \sim 55$ and has computed the most unstable mode for various values of k and Re. Note that as Re_c is moderately high, we do not need many points N to discretize correctly the solution (N of order 50 to 100 is sufficient).

9.2 Energy of a small perturbation

The aim of this section is to discuss the evolution of the energy of a perturbation u, a solution of (LNSC$_\varepsilon$).

For $Re < Re_1$ as seen at the end of Part I, the energy is decreasing. As the flow is stable, it goes to 0.

For $Re > Re_c$ there exist exponentially increasing modes, therefore, in general the energy of the perturbation will increase exponentially.

In the range $[Re_1, Re_c]$ the situation is slightly different. First as the Reynolds number is subcritical the energy of an arbitrary perturbation tends to 0 as time goes to $+\infty$. The question is then to know whether the energy is decreasing or if it begins to increase before ultimately decreasing. Let Re_3 be the supremum of the Reynolds numbers such that for every $Re > Re_3$, the energy of any perturbation decreases continuously. It is possible to compute numerically Re_3, which is of order 8. The final picture is the following:

- $Re < Re_3$: the energy of any perturbation goes to 0 in a monotonic way;
- $Re_3 < Re < Re_c$: the energy of any perturbation tends to 0 as time goes to $+\infty$, but may begin to increase, before its decay;
- $Re > Re_c$: the energy of a general perturbation goes to $+\infty$ as time goes to $+\infty$ (not always in a monotonic way).

The main consequence is that linear stability cannot be proved by energy estimates in the range $[Re_3, Re_c]$ since in this area we have only energy estimates of the form $\partial_t \|u\|_{L^2}^2 \leq C\|u\|_{L^2}^2/\varepsilon$, which are useless in the limit $\varepsilon \to 0$. In this range linear stability can only be proved by spectral arguments, using refined pseudodifferential techniques. This has been done by G. Métivier and K. Zumbrun [97] in the case of the vanishing viscosity limit of parabolic systems, leading to hyperbolic systems of conservation laws. As in rotating fluids, boundary layers appear, which are stable under a smallness criterion. Simple energy estimates give stability of such layers with a stability threshold much smaller than the optimal one. The proof that the spectral stability of the boundary layer implies the stability of the complete solution, and the justification of the classical formal Ansatz is very technical and difficult, involving careful pseudodifferential analysis.

Very recently, similar work has been done on Ekman layers, for well-prepared initial data, by F. Rousset [73] who proved that sequences of solutions of Navier–Stokes–Coriolis system (in the well-prepared case) converge to a solution of damped Euler equations, provided the Reynolds number of the limit solution always remains smaller than the critical Reynolds number Re_c.

9.3 Rolls and turbulence

When $Re > Re_c$ the Ekman layer is linearly unstable, and more precisely as Re crosses Re_c the layer exhibits a Hopf bifurcation [76] since two isolated eigenvalues cross the imaginary axis. The flow is no longer laminar and gets organized into rolls of typical size $\sqrt{\varepsilon\nu} \times \sqrt{\varepsilon\nu}$, which make a given angle with the direction of the flow at infinity, U_∞. Their size is proportional to $\sqrt{Re - Re_c}$ which is the classical behavior in Hopf bifurcations. Moreover the rolls move with a fixed velocity. The question of the influence of these rolls on the interior behavior (Ekman pumping, energy balance, and so on) is widely open (and completely open from a mathematical point of view). In particular it is not known whether for $Re < 100$ or even for slightly supercritical Reynolds numbers, flows of highly rotating fluids consist of a two-dimensional incompressible flow in the interior of the domain, bounded by two layers consisting of rolls (or of Ekman's layers), and whether the limit two-dimensional flow satisfies a damped Euler equation or not.

These rolls are the first step towards turbulent behavior. They are destabilized at some higher Reynolds (of order $120 \sim 150$) and "burst" into a real three-dimensional turbulent boundary layer. The mathematical viewpoint is of course open.

Note that it is not clear whether turbulent boundary layers separate from the boundary and enter the domain or whether they remain close to the boundary and only affect the Ekman pumping term and the local energy dissipation. It seems that turbulent layers dissipate less energy than laminar layers (as for Prandtl-type layers). It is almost impossible to make computations precise enough to answer this question, which makes theoretical studies all the more interesting.

10

Other systems

The methods developed in this book can be applied to various physical systems. We will not detail all the possible applications and will only quote three systems arising in magnetohydrodynamics (MHD) and meteorology, namely conducting fluids in a strong external "large scale" magnetic field, a classical MHD system with high rotation, and the quasigeostrophic limit. The main theorems of this book can be extended to these situations.

10.1 Large magnetic fields

The theory of rotating fluids is very close to the theory of conducting fluids in a strong magnetic field. Namely the Lorenz force and the Coriolis force have almost the same form, up to Ohm's law. The common feature is that these phenomena appear as singular perturbation skew-symmetric operators. The simplest equations in MHD are Navier–Stokes equations coupled with Ohm's law and the Lorenz force

$$\begin{aligned} \partial_t u + (u \cdot \nabla)u - \nu\Delta u + \nabla p + \frac{j \times e}{\varepsilon} &= 0, \\ \nabla \cdot u &= 0, \\ j &= -\nabla\phi + u \times e, \\ \nabla \cdot j &= 0, \end{aligned} \tag{10.1.1}$$

where $\nabla\phi$ is the electric field, j the current, and e the direction of the imposed magnetic field. In this case ε is called the Hartmann number. In physical situations, like the geodynamo (study of the magnetic field of the Earth), it is really small, of order 10^{-5}–10^{-10}, much smaller than the Rossby number.

These equations are the simplest model in geomagnetism and in particular in the geodynamo. As $\varepsilon \to 0$ the flow tends to become independent of x_3. This is not valid near boundaries. For horizontal boundaries, Hartmann layers play the role of Ekman layers and in the layer the velocity is given by

$$u(t, x_1, x_2, x_3) = u_\infty(t, x_1, x_2)\Big(1 - \exp(-x_3/\sqrt{\varepsilon\nu})\Big). \tag{10.1.2}$$

The critical Reynolds number for linear instability is very high, of order $Re_c \sim 10^4$. The main reason is that there is no inflexion point in the boundary layer profile (10.1.2), therefore it is harder to destabilize than the Ekman layer since the Hartmann profile is linearly stable for the inviscid model associated with (10.1.1). As for Ekman layers, Hartmann layers are stable for $Re < Re_c$ and unstable for $Re > Re_c$. There is also something similar to Ekman pumping, which is responsible for friction and energy dissipation. Vertical layers are simpler than for rotating fluids since there is only one layer, of size $(\varepsilon\nu)^{1/4}$. We refer to [109], [110] for physical studies.

10.2 A rotating MHD system

Currently the persistence of the magnetic field of the Earth is not explained and much work has been done, from a numerical, experimental or mathematical point of view to try to explain why the Earth has a non-zero large-scale magnetic field whose polarity turns out to invert over several hundred centuries. One possible model is the following:

$$\begin{aligned} \partial_t u + u\cdot\nabla u + \nabla p - \nu\Delta u + 2\Omega e \times u - (\nabla\times B)\times B &= 0,\\ \operatorname{div} u &= 0,\\ \partial_t B - \nabla\times(u\times B) - \eta\Delta B &= 0,\\ \operatorname{div} B &= 0, \end{aligned}$$

where u is the velocity field, B the magnetic field, ν the viscosity, η the diffusion of the magnetic field and Ω the rotation speed. There are many different interesting scalings, but usually ν and ε go to 0, Ω goes to $+\infty$ and $\eta\to\infty$ as well (large diffusion of the magnetic field). It is in particular interesting to enforce a strong, large-scale external magnetic field, in the e-direction to simplify and to consider perturbations of it: $B=e+b$. We refer to [51], [44],[45] and to the references therein.

10.3 Quasigeostrophic limit

Let us go back to meteorology and oceanography. A first attempt to include more complex effects to the (NSC$_\varepsilon$) system is to add a temperature equation and to couple it with the vertical motion (see [103],[31]). This is done in the following quasigeostrophic system

$$\begin{aligned} \partial_t u + u\cdot\nabla u + \nabla p - \nu\Delta u &= \frac{1}{\varepsilon}\begin{pmatrix} u_2\\ -u_1\\ \theta\end{pmatrix},\\ \partial_t\theta + u\cdot\nabla\theta - \nu'\Delta\theta &= -\frac{u_3}{\varepsilon},\\ \operatorname{div} u &= 0, \end{aligned}$$

where θ denotes the temperature, ν the viscosity, and ν' the thermal diffusivity.

As ε goes to 0, the only way to control the right-hand side is to absorb it in the pressure term, which gives

$$\partial_1 p = u_2, \quad \partial_2 p = -u_1, \quad \partial_3 p = \theta, \quad 0 = -u_3.$$

As for rotating fluids, the limit flow is a divergence-free two-dimensional vector field, with $\theta = u_3 = 0$ which satisfies two-dimensional Euler or Navier–Stokes equations (depending on whether $\nu \to 0$ or not).

11

Vertical layers

11.1 Introduction

From a physical point of view, as well as from a mathematical point of view, horizontal layers (Ekman layers) are now well understood. This is not the case for vertical layers which are much more complicated, from a physical, analytical and mathematical point of view, and many open questions in all these directions remain open. Let us, in this section, consider a domain Ω with vertical boundaries. Namely, let Ω_h be a domain of $\mathbf{R}^2$ and let $\Omega = \Omega_h \times [0, 1]$. This domain has two types of boundaries:

- horizontal boundaries $\Omega_h \times \{0\}$ (bottom) and $\Omega_h \times \{1\}$ (top) where Ekman layers are designed to enforce Dirichlet boundary conditions;
- vertical boundaries $\partial\Omega_h \times [0, 1]$ where again a boundary layer is needed to ensure Dirichlet boundary conditions. These layers, however, are not of Ekman type, since r is now parallel to the boundary.

Vertical layers are quite complicated. They in fact split into two sublayers: one of size $E^{1/3}$ and another of size $E^{1/4}$ where $E = \nu\varepsilon$ denotes the Ekman number. This was discovered and studied analytically by Stewartson and Proudman [105], [115]. Vertical layers can be easily observed in experiments (at least the $E^{1/4}$ layer, the second one being too thin) but do not seem to be relevant in meteorology or oceanography, where near continents, effects of shores, density stratification, temperature, salinity, or simply topography are overwhelming and completely mistreated by rotating Navier–Stokes equations. In MHD, however, and in particular in the case of rotating concentric spheres, they are much more important. Numerically, they are easily observed, at large Ekman numbers E (small Ekman numbers being much more difficult to obtain).

The aim of this section is to provide an introduction to the study of these layers, a study mainly open from a mathematical point of view. First we will derive the equation of the $E^{1/3}$ layer. Second we will investigate the $E^{1/4}$ layer and underline its similarity with Prandtl's equations. In particular, we conjecture that $E^{1/4}$ is always linearly and nonlinearly unstable. We will not prove this latter fact, which would require careful study of what happens at the corners of the domain, a widely open problem.

11.2 $E^{1/3}$ layer

Let us consider the Stokes–Coriolis equations

$$(\mathrm{SC}_\varepsilon) \qquad \begin{cases} \dfrac{e^3 \times u}{\varepsilon} - \nu \Delta u + \dfrac{\nabla p}{\varepsilon} = 0 \\ \operatorname{div} u = 0 \end{cases}$$

and let $E = \nu\varepsilon$. We have

$$-u_2 - E^2\Delta^2 u_2 + \partial_1 p + E\Delta\partial_2 p = 0,$$

$$-u_1 - E^2\Delta^2 u_1 - \partial_2 p + E\Delta\partial_1 p = 0,$$

$$-\Delta u_3 + E^{-1}\partial_3 p - E^2\Delta^3 u_3 + E\Delta^2\partial_3 p = 0,$$

hence, using the divergence-free equation,

$$\partial_{33} p + E^2\Delta^3 p = 0. \tag{11.2.1}$$

Let us go back to horizontal boundary layers. Ekman layers are in fact stationary solutions of rotating Stokes equations (since in their derivation we drop the non-linear transport term). Let us recover their size λ with the help of (11.2.1). Vertical derivatives are of order $O(\lambda^{-1})$ and horizontal derivatives of order $O(1)$ in the layer. Hence in (11.2.1), $\partial_{33}p$ is of order $O(\lambda^{-2})$ and $E^2\Delta^3 p$ of order $E^2\lambda^{-6}$, and equals $E^2\partial_3^6$ up to smaller order terms ($E^2\lambda^{-4}$). Therefore $\lambda^{-2} \sim E^2\lambda^{-6}$, and $\lambda \sim E^{1/2} \sim \sqrt{\varepsilon\nu}$. We can even derive the equation of Ekman layers, namely $\partial_{33}p + E^2\partial_3^6 p = 0$.

Let us repeat the same procedure for horizontal layers, say in the x_1-direction. Derivatives in the x_1-direction are of size $O(\lambda^{-1})$, in the x_2- and x_3-directions of size $O(1)$. Hence $\partial_{33}p$ is of order $O(1)$ and $E^2\Delta^3 p \sim E^2\partial_1^6 p$ is of order $E^2\lambda^{-6}$. Therefore $\lambda \sim E^{1/3}$. Let $X = x_1/\lambda$. The equation of the $E^{1/3}$ layer is therefore

$$\partial_{33} p + \partial_1^6 p = 0. \tag{11.2.2}$$

This is no longer an ordinary differential equation in the normal variable, but a partial differential equation of elliptic type in x_1, x_3 space, with different orders in the x_1- and x_3-directions. Equation (11.2.2) must be supplemented with boundary conditions on the velocity field, namely $u = 0$ on the boundary and the convergence of u to some constant (x_3 independent) vector for $x_1 \to +\infty$. We refer to [105] and [115] for the explicit resolution of (11.2.2) the cylindrical case, using Bessel functions.

11.3 $E^{1/4}$ layer

The derivation of this layer is more subtle than the previous one. Namely this layer would not exist if there were no Ekman pumping. Its role is to fit the Ekman pumping (at the horizontal boundaries) near the vertical boundaries. There is no $E^{1/4}$ layer if we assume periodicity in the x_3-direction (Ω of the form $\Omega_h \times \mathbf{T}$ where Ω_h is a two-dimensional open set).

Let us derive the equations in the case $\mathbf{R}^+ \times \mathbf{R} \times [0,1]$ (the case of a general cylinder $\Omega_h \times [0,1]$ being similar). Let (u, v, w) be the velocity field, where u is the radial velocity, v the orthoradial part, and w the normal velocity. In the layer, the orthoradial velocity is of order $O(1)$, the other parts being of order $O(E^{1/4})$. It is natural that the normal velocity if small, since the boundary layer cannot absorb a large flux. On the other hand, the vertical velocity is small since it is absorbed in horizontal Ekman layers. This $E^{1/4}$ vertical flow will close the global circulation of the fluid in the domain, since through Ekman pumping, flow enters or leaves the top and bottom Ekman boundary layers, a global circulation which is closed through flow near the vertical walls. Let us emphasize that $E^{1/4}$ layers would not exist without Ekman pumping (in particular they do not exist if x_3 is a periodic variable).

This leads to asymptotic expansions of the form

$$u(t,x_1,x_2,x_3) = E^{1/4}u_0 + E^{1/2}u_1 + \cdots,$$

$$v(t,x_1,x_2,x_3) = v_0 + E^{1/4}v_1 + \cdots,$$

$$w(t,x_1,x_2,x_3) = E^{1/4}w_0 + E^{1/2}w_1 + \cdots,$$

$$p(t,x_1,x_2,x_3) = p_0 + E^{1/4}p_1.$$

Now looking at $(\mathrm{SC}_\varepsilon)$ at order $E^{-3/4}$ gives

$$\partial_1 p_0 = 0, \tag{11.3.3}$$

at order $E^{-1/2}$

$$-v_0 + \partial_1 p_1 = 0, \tag{11.3.4}$$

$$\partial_2 p_0 = 0, \tag{11.3.5}$$

$$\partial_3 p_0 = 0. \tag{11.3.6}$$

At order $E^{-1/4}$ we get

$$-v_1 + \partial_1 p_2 = 0, \tag{11.3.7}$$

$$u_0 + \partial_2 p_1 = 0, \tag{11.3.8}$$

$$\partial_3 p_1 = 0, \tag{11.3.9}$$

at order E^0,

$$\partial_t v_0 + u_0\partial_1 v_0 + v_0\partial_2 v_0 + u_1 + \partial_2 p_2 - \partial_{11} v_0 = 0, \tag{11.3.10}$$

$$\partial_1 u_0 + \partial_2 v_0 = 0,$$

and at order $E^{1/4}$,

$$\partial_1 u_1 + \partial_2 v_1 + \partial_3 w_0 = 0. \tag{11.3.11}$$

Combining (11.3.7), (11.3.10), and (11.3.11) gives

$$\partial_1^3 v_0 - \partial_t\partial_1 v_0 - \partial_1(u_0\partial_1 v_0 + v_0\partial_2 v_0) + \partial_3 w_0 = 0. \tag{11.3.12}$$

Note that by (11.3.3)–(11.3.5), $p_0 = 0$ and that by (11.3.9), p_1 is independent of x_3. Using (11.3.4), (11.3.8) we deduce that u_0 and v_0 are independent of x_3, and using (11.3.12) we get that $\partial_1 w_0$ is independent of x_3.

As a first approximation, as the Ekman layer is much smaller than the $E^{1/4}$ layer, the vertical velocity at $x_3 = 1$ and $x_3 = 0$ is given by the vertical velocity just outside the Ekman layer, and hence (in the original spatial variables),

$$w = \pm \frac{1}{\sqrt{2}} E^{1/2} \operatorname{curl}(u, v)$$

with the minus sign at $x_3 = 1$ and the plus sign at $x_3 = 0$. After rescaling, this gives

$$w_0 = -\frac{1}{\sqrt{2}} \partial_1 v_0 \quad \text{for } x_3 = 1, \quad \text{and}$$

$$w_0 = \frac{1}{\sqrt{2}} \partial_1 v_0 \quad \text{for } x_3 = 0.$$

As w_0 is affine, this gives

$$\partial_3 w_0 = -\sqrt{2} \partial_1 v_0.$$

So we can integrate (11.3.12) to get

$$\partial_t v_0 + u_0 \partial_1 v_0 + v_0 \partial_2 v_0 - \partial_{11} v_0 + \sqrt{2} v_0 = f(x_1) \tag{11.3.13}$$

where f is some integrating constant, which only depends on x_1 (since the left of (11.3.13) is independent of x_3) and is given by the flow outside the layer. This equation must be supplemented with

$$\partial_1 u_0 + \partial_2 v_0 = 0, \tag{11.3.14}$$

$$u_0 = v_0 = 0 \quad \text{on} \quad x_1 = 0. \tag{11.3.15}$$

System (11.3.13)–(11.3.15) is very similar to Prandtl's equations. We recall that Prandtl's layer appears in the inviscid limit of Navier–Stokes equations near a wall and describes the transition between Dirichlet boundary condition and the flow away from the boundary, described at least formally by Euler's equations. Prandtl's system is exactly (11.3.13)–(11.3.15) except for the term $\sqrt{2} v_0$ which does not appear.

This is very bad news since Prandtl's system behaves very badly from a mathematical point of view, and little is known of the existence of solutions. Up to now we just have existence in small time of analytic type solutions for analytic initial data [1], [22], which is an important result. In the case of monotonic profiles, existence is established locally in time and space [99] and we are far from small-time existence of strong solutions, technically because of a lack of high-order estimates, but also because there are physical underlying instability phenomena that we will now detail.

Prandtl's layers try to describe boundary layers of Navier–Stokes equations in the regime of high Reynolds number ($\nu \to 0$), with a size $\sqrt{\nu}$. But it is well known, since the work of Tollmien in particular, that *any* shear layer profile is linearly unstable for Navier–Stokes equations provided the viscosity is small enough [52], [111]. More precisely, let $V = (0, V_0(x_1/\sqrt{\nu}))$ be some smooth shear layer profile and let us consider the linearized Navier-Stokes equation near V

$$\partial_t u + V \cdot \nabla u + u \cdot \nabla V - \nu \Delta u + \nabla p = 0, \tag{11.3.16}$$

$$\nabla \cdot v = 0. \tag{11.3.17}$$

The main result (see [52], [90], [111]) is that for any profile V_0, there exists a solution of (11.3.16), (11.3.17) of the form $u_0 \exp(\lambda t)$ with $\Re e\, \lambda > 0$, provided ν is small enough. This is quite natural if V_0 has an inflection point and is unstable for the limit system $\nu = 0$ (Euler equations) since the viscosity can not kill an inviscid instability if it is too small. It is, however, more surprising if V_0 has no inflection point and is stable at $\nu = 0$, since the viscosity then has a destabilizing role (which is less intuitive).

The only difference between Prandtl's equation and (11.3.13)–(11.3.15) is the damping term $\sqrt{2}u$. However this damping term improves the energy estimates only slightly and cannot be used to improve the existence results on Prandtl's equations. It is also not sufficient to kill the linear instabilities which arise at low viscosity, and therefore the equations of the $E^{1/4}$ layer behave like Prandtl's equations.

Therefore, we can only expect existence of analytic solutions local in time for analytic initial data, which is a technically difficult result, but not so interesting from a physical point of view. The second consequence is that "non-derivation" results of Prandtl's equation [71] can be extended to our system. In particular the flow is not as simple as an interior inviscid flow combined with a laminar boundary layer flow. Note that it is possible formally to construct a boundary layer, but that this boundary layer is not stable and therefore is not relevant in the analysis. This is an important difference in the mathematical and physical treatment of vertical and horizontal boundary layers.

11.4 Mathematical problems

Let us review some mathematical open problems in the direction of vertical layers.

First in the x_3 periodic case, $E^{1/4}$ does not appear. It seems in this case possible to handle rigorously the limit $\varepsilon, \nu \to 0$. In the interior the limit flow satisfies Euler (or Navier–Stokes) equations (without damping), and a boundary layer appears, of size $E^{1/3}$ near the boundary. The boundary layer is stable provided the limit interior flow is small enough near the boundary (stability under a Reynolds condition, like for Ekman layers), else it may be unstable. Linear and nonlinear instability of this layer seems open (critical Reynolds number, behavior of the unstable modes, and so on).

When $0 \le x_3 \le 1$, one important problem is the corners $\partial\Omega \times \{0\}$ and $\partial\Omega \times \{1\}$. Their structure is not clear, despite various studies [120]. Second, in a cylinder, exact solutions have been computed [115], but little is known in the general case. We conjecture an instability result of the type [71] for these layers. This result is probably very technical to obtain, since we must control the $E^{1/3}$ layer and the flow near the corners.

12

Other layers

Note that $\Omega = \Omega_h \times [0,1]$ is a particular case where the boundary layers are purely horizontal or purely vertical. In the general case of an open domain Ω, the boundaries have various orientations. As long as the tangential plane $\partial\Omega$ is not vertical, the boundary layers are of Ekman type, with a size of order $\sqrt{\nu\varepsilon/|\nu.r|}$ where ν is the normal of the tangential plane. When $\nu.r \to 0$, namely when the tangential plane becomes vertical, Ekman layers become larger and larger, and degenerate for $\nu.r = 0$ in another type of boundary layer, called equatorial degeneracy of the Ekman layer. We will now detail this phenomenon in the particular case of a rotating sphere. Mathematically, almost everything is widely open!

12.1 Sphere

Let $\Omega = B(0,R)$ be a ball. Let θ be the latitude in spherical coordinates. The equatorial degeneracy of the Ekman layer is difficult to study. We will just give the conclusions of the analytical studies of [105], [115]. The Ekman layer is a good approximation of the boundary layer as long as $|\theta| \gg E^{1/5}$. For $|\theta| \ll E^{1/5}$ the Ekman layer degenerates into a layer of size $E^{2/5}$.

The structure of the boundary layer is therefore the following:

- for $|\theta| \gg (\varepsilon\nu)^{1/5}$, Ekman layer of size $\sqrt{\varepsilon\nu/\sin(\theta)}$;
- for $|\theta|$ of order $(\varepsilon\nu)^{1/5}$, degeneracy of the Ekman layer into a layer of size $(\varepsilon\nu)^{1/5}$ in depth and $(\varepsilon\nu)^{2/5}$ in latitude.

12.2 Spherical shell

Let us now concentrate on the motion between two concentric rotating spheres, the speed of rotation of the spheres being the same. In this case, $\Omega = B(0,R) - B(0,r)$ where $0 < r < R$. Keeping in mind meteorology, the interesting case arises when $R - r \ll R$: the two spheres have almost equal radius. Let us study the fluid at some latitude θ. If $\theta \neq 0$, locally, the space between the two spheres can be considered as flat and treated as a domain between two nearby plates. The conclusions of the previous paragraphs can be applied. Two Ekman layers are created, one near the inner sphere and the other one near the outer sphere. The size of the layer is, however, different, since it is of order

$$\sqrt{\frac{\varepsilon\nu}{\sin\theta}},$$

and in particular it goes to infinity as θ goes to 0. The Ekman layer gets thicker and thicker as one approaches the equator, where it degenerates and is no longer a valid approximation.

This phenomenon can be observed in experiments where one can see for instance large currents leaving the equatorial area of a highly rotating sphere. In oceanography and meteorology, this degeneracy is also important, since it indicates that the behavior of oceans is very different near the equator. We must keep in mind however that other phenomena, not included in this toy model, have a prominent influence near the equator, like winds, global atmospheric currents (including very important vertical currents created by heating). Therefore the precise study of the degeneracy must not be taken too seriously in this context, and is completely overwhelmed by other physical phenomena.

12.3 Layer between two differentially rotating spheres

Let us turn to the case when Ω is the domain limited by two spheres of different radius R and r $(0 < r < R)$, rotating with different velocities $\Omega\varepsilon^{-1}$ and $\Omega\varepsilon^{-1} + \Omega'$, Ω' being of order 1 (a slight difference in the rotation speed). This situation is often studied in MHD where this rotation difference is considered as a possible source of the Earth's magnetic inner core field. This geometry is very rich and fascinating, and still far from being completely understood [105], [115], since there is a conflict between the spherical symmetry enforced by the domain, and the cylindrical geometry imposed by the large Coriolis force, which penalizes motions depending on the vertical direction.

For $x_1^2 + x_2^2 > r^2$, by the Taylor–Proudman theorem, the fluid moves with the speed of the outer sphere, namely $\Omega\varepsilon^{-1}$. For $x_1^2 + x_2^2 < r^2$, however, there is competition between the velocity imposed by the inner sphere and the velocity imposed by the outer sphere. It can be shown that the fluid then has an averaged velocity, and that two Ekman layers appear at the surfaces of the spheres to fit Dirichlet boundary conditions. The Ekman pumping velocity then creates a global circulation in the area $x_1^2 + x_2^2 < r^2$, fluid leaving the upper sphere to fall vertically on the inner sphere. At the equator $x_1^2 + x_2^2 = r^2$ of the inner sphere, however, Ekmàn layers degenerate. To close the global circulation, the fluid which by Ekman suction is absorbed by the Ekman layer goes to the equator, remaining in the Ekman layer. It then escapes the inner sphere and goes vertically to the outer sphere, in a vertical shear layer of size $E^{1/4}$. The situation would be simple without a whole bunch of other boundary layers created by the various singularities of the Ekman layer and of the vertical layers, new boundary layers of size $E^{1/5}$, $E^{2/5}$, $E^{1/3}$, $E^{7/12}$, $E^{1/28}$, and so on.

REFERENCES

[1] K. Asano, Zero-viscosity limit of the incompressible Navier-Stokes equation. II., *Mathematical analysis of fluid and plasma dynamics, I* (Kyoto, 1986), Sûrikaisekikenkyûsho Kôkyûroku No. **656**, (1988), pages 105–128.

[2] A. Babin, A.Mahalov, and B. Nicolaenko, Integrability and regularity of 3D Euler and equations for uniformly rotating fluids, *Comput. Math. Appl.* **31**, (1996), no. 9, pages 35–42

[3] A. Babin, A. Mahalov, and B. Nicolaenko, Global splitting, integrability and regularity of 3D Euler and Navier-Stokes equations for uniformly rotating fluids, *European J. Mech. B Fluids*, **15**, (1996), no. 3, pages 291–300.

[4] A. Babin, A. Mahalov, and B. Nicolaenko, Global splitting and regularity of rotating shallow-water equations, *European J. Mech. B Fluids*, **16**, (1997), no. 5, pages 725–754.

[5] A. Babin, A. Mahalov, B. Nicolaenko: Regularity and integrability of 3D Euler and Navier-Stokes equations for rotating fluids, *Asymptot. Anal.*, **15**, (1997), no. 2, pages 103–150.

[6] A. Babin, A. Mahalov, and B. Nicolaenko, Global regularity of 3D rotating Navier-Stokes equations for resonant domains, *Indiana Univ. Math. J.*, **48**, (1999), no. 3, pages 1133–1176.

[7] A. Babin, A. Mahalov, B. Nicolaenko, Fast singular oscillating limits and global regularity for the 3D primitive equations of geophysics, *Special issue for R. Temam's 60th birthday. M2AN Math. Model. Numer. Anal.*, **34**, (2000), no. 2, pages 201–222.

[8] H. Bahouri and J.-Y. Chemin, Équations d'ondes quasilinéaires et inégalités de Strichartz et équations d'ondes quasilinéaires, *American Journal of Mathematics*, **121**, (1999), pages 1337–1377.

[9] G.K. Batchelor, *An introduction to fluid dynamics*, Cambridge University Press, (1967).

[10] T. Beale and A. Bourgeois, Validity of the quasigeostrophic model for large–scale flow in the atmosphere and the ocean, SIAM Journal of Mathematical Analysis, **25**, (1994), pages 1023–1068.

[11] P. Bougeault and R. Sadourny: *Dynamique de l'atmosphère et de l'océan*, Éditions de l'École polytechnique (2001).

[12] Y. Brenier, Homogeneous hydrostatic flows with convex velocity profiles, *Nonlinearity,* **12**, (1999), no. 3, pages 495–512.

[13] P. Brenner, On $L^p - L^{p'}$ estimates for the wave-equation. *Math. Z.*, **145**, (1975), no. 3, pages 251–254.

[14] P. Brenner, On L^p-decay and scattering for nonlinear Klein-Gordon equations. *Math. Scand.*, **51**, (1982), pages 333–360.
[15] D. Bresch, J. Simon, On the effect of friction on wind driven shallow lakes, *J. Math. Fluid Mech.*, **3**, (2001), no. 3, pages 231–258
[16] D. Bresch, J. Lemoine, J. Simon, A geostrophic model with vertical diffusion, *Nonlinear Anal.* **43**, (2001), no. 4, *Ser. A: Theory Methods*, pages 449–470.
[17] D. Bresch, J. Lemoine, J. Simon, Nonstationary models for shallow lakes, *Asymptot. Anal.* **22**, (2000), no. 1, pages 15–38.
[18] D. Bresch, J. Lemoine, J. Simon, A vertical diffusion model for lakes, *SIAM J. Math. Anal.*, **30**, (1999), no. 3, pages 603–622
[19] D. Bresch, B. Desjardins, E. Grenier, C.-K. Lin, Low Mach number limit of viscous polytropic flows: formal asymptotics in the periodic case, *Stud. Appl. Math.*, **109**, (2002), no. 2, pages 125–149.
[20] F. Browder, On the spectral theory of elliptic differential operators, I, *Math. Annalen*, **142**, (1961) pages 22–130.
[21] L. Caffarelli, R. Kohn and L. Nirenberg, Partial regularity of suitable weak solutions of the Navier-Stokes equations. *Comm. Pure Appl. Math.*, **35**, (1982), no. 6, pages 771–831.
[22] R. Caflisch, A simplified version of the abstract Cauchy-Kowalewski theorem with weak singularities, *Bull. Amer. Math. Soc. (N.S.)* **23**, (1990), no. 2, pages 495–500.
[23] R. Caflisch, M. Sammartino: Zero viscosity limit for analytic solutions, of the Navier-Stokes equation on a half-space. I., II *Comm. Math. Phys.*, **192**, (1998), no. 2, pages 463–491.
[24] M. Cannone, Ondelettes, paraproduits et Navier–Stokes, Diderot éditeur, Arts et Sciences, 1995.
[25] M. Cannone, Y. Meyer and F. Planchon, Solutions auto-similaires des équations de Navier-Stokes, *Séminaire sur les Équations aux Dérivées Partielles*, 1993–1994, Exp. No. VIII, 12 pp., École polytechnique, Palaiseau, (1994).
[26] T. Cazenave, *Semilinear Schrödinger equations*, Courant Lecture Notes in Mathematics, 10. New York University, Courant Institute of Mathematical Sciences, New York; American Mathematical Society, Providence, RI, (2003).
[27] F. Charve, Convergence de solutions faibles du système primitif des équations quasigéostrophique, *Asymptotic Analysis*, **42**, (2005), pages 173–209.
[28] F. Charve, Global wellposedness and asymptotics for a geophysical fluid system, *Communications in Partial and Differential Equations*, **29**, (2004), pages 1919–1940.
[29] J.-Y. Chemin, Perfect incompressible fluids. *Translated from the 1995 French original by I. Gallagher and D. Iftimie. Oxford Lecture Series*

in Mathematics and its Applications, **14**, The Clarendon Press, Oxford University Press, New York, (1998).

[30] J.-Y. Chemin, Localization in Fourier space and Navier-Stokes system *Phase space analysis of Partial Differential Equations*, Vol I, page 53–135, Pubbl. Cert. Ric. Mat. Ennio de Gorg *Scuola Norma. Sup.* Pisa, 2004.

[31] J.-Y. Chemin, Système primitif de l'océan-atmosphère et limite quasi-géostrophique, *Séminaire sur les équations aux dérivées partielles*, 1995–1996, Exp. No. VII, 15 pp., École polytechnique, Palaiseau, (1996).

[32] J.-Y. Chemin, B. Desjardins, I. Gallagher and E. Grenier, Anisotropy and dispersion in rotating fluids Non linear Partial Differential Equations and their Applications, *Studies in Mathematics and its applications*, North Holland, 2002, pages 171–192.

[33] J.-Y. Chemin, B. Desjardins, I. Gallagher and E. Grenier, Fluids with anisotropic viscosity, *Special issue for R. Temam's 60th birthday. M2AN Math. Model. Numer. Anal.*, **34**, (2000), no. 2, pages 315–335.

[34] J.-.Y Chemin, B. Desjardins, I. Gallagher and E. Grenier, Ekman boundary layers in rotating fluids, *Control, Optimisation and Calculus of Variations*, **8**, (2002), pages 441–466.

[35] J.-Y. Chemin and C.-J. Xu: Inclusions de Sobolev en calcul de Weyl-Hörmander et champs de vecteurs sous-elliptiques, *Ann. Sci. École Norm. Sup.*, **30**, (1997), no. 6, pages 719–751.

[36] J. Colliander, M. Keel, G. Staffilani, H. Takaoka, T. Tao, Global existence and scattering for rough solutions of a nonlinear Schrödinger equation on $\mathbf{R}^3$. *Comm. Pure Appl. Math.*, **57**, (2004), no. 8, pages 987–1014.

[37] T. Colin and P. Fabrie, Rotating fluid at high Rossby number driven by a surface stress : existence and convergence, *Advances in Differential Equations*, **2**, (1997), pages 715–751.

[38] P. Constantin and C. Foias: *Navier–Stokes Equations*, Chicago University Press, (1988).

[39] P. Constantin, A. Majda, E. Tabak, Formation of strong fronts in the 2-D quasigeostrophic thermal active scalar, *Nonlinearity*, **7**, (1994), no. 6, pages 1495–1533.

[40] B. Cushman-Roisin, *Introduction to geophysical fluid dynamics*, Prentice-Hall (1994).

[41] R. Danchin, Zero Mach number limit in critical spaces for compressible Navier-Stokes equations, *Annales Scientifiques de l'École Normale Supérieure*, **35**, (2002), pages 27–75.

[42] R. Danchin, Zero Mach number limit for compressible flows with periodic boundary conditions, *Amer. J. of Mathematics*, **124**, (2002), no. 6, pages 1153–1219.

[43] R. Dautray, J.-L. Lions, Mathematical analysis and numerical methods for science and technology, *Translated from the French by John C. Amson.* Springer-Verlag, Berlin, (1990).

[44] B. Desjardins, E. Dormy, E. Grenier, Stability of mixed Ekman-Hartmann boundary layers, *Nonlinearity*, **12**, (1999), no. 2, pages 181–199.

[45] B. Desjardins, E. Dormy, E. Grenier, Instability of Ekman-Hartmann boundary layers, with application to the fluid flow near the core-mantle boundary, *Physics of the Earth and Planetary Interiors*, **123-1**, (2001), pages 15–27.

[46] B. Desjardins, E. Grenier, Derivation of quasi-geostrophic potential vorticity equations, *Adv. Differential Equations*, **3**, (1998), no. 5, pages 715–752.

[47] B. Desjardins, E. Grenier, Low Mach number limit of viscous compressible flows in the whole space, *R. Soc. Lond. Proc. Ser. A Math. Phys. Eng. Sci.*, **455**, (1999), no. 1986, pages 2271–2279.

[48] B. Desjardins, E. Grenier, On the homogeneous model of wind-driven ocean circulation, *SIAM J. Appl. Math.*, **60**, (2000), no. 1, pages 43–60.

[49] B. Desjardins, E. Grenier, Linear instability implies nonlinear instability for various boundary layers, *Ann. Inst. Henri Poincaré Anal. non Linéaire*, **20**, (2003), no. 1, pages 87–106.

[50] B. Desjardins, E. Grenier, P.-L. Lions, N. Masmoudi, Incompressible limit for solutions of the isentropic Navier-Stokes equations with Dirichlet boundary conditions, *J. Math. Pures Appl.*, **78**, (1999), no. 5, pages 461–471.

[51] E. Dormy, D. Jault, A. Soward, A super rotating shear layer in magnetohydrodynamic spherical Couette flow, *Journal of Fluid Mechanics*, **452**, (2002), pages 263-291.

[52] P.G. Drazin, W.H. Reid, Hydrodynamic stability, *Cambridge monographs on mechanics and applied mathematics, Cambridge University Press*. (1981).

[53] W. E, B. Engquist, Blowup of solutions of the unsteady Prandtl's equation, *Comm. Pure Appl. Math.* **50**, (1997), no. 12, pages 1287–1293.

[54] P. Embid, A. Majda. Averaging over fast gravity waves for geophysical flows with arbitrary potential vorticity, *Comm. Partial Differential Equations*. **21**, (1996), no. 3-4, pages 619–658.

[55] P. Embid, A. Majda, Low Froude number limiting dynamics for stably stratified flow with small or finite Rossby numbers, *Geophys. Astrophys. Fluid Dynam.*, **87**, (1998), no. 1-2, pages 1–50.

[56] R. Fjortoft, Application of integral theorems in deriving criteria of stability for laminar flows and for the baroclinic circular vortex, *Geofys. Publ., Oslo*, **17**, (1950), pages 1–52.

[57] H. Fujita and T. Kato, On the Navier-Stokes initial value problem I, *Archiv for Rationnal Mechanic Analysis*, **16**, (1964), pages 269–315.

[58] I. Gallagher, The tridimensional Navier–Stokes equations with almost bidimensional data: stability, uniqueness and life span, *International Mathematical Research Notices*, **18**, (1997), pages 919–935.

[59] I. Gallagher, Asymptotic of the solutions of hyperbolic equations with a skew-symmetric perturbation, *J. Differential Equations*, **150**, (1998), no. 2, pages 363–384.

[60] I. Gallagher, Applications of Schochet's methods to parabolic equations, *Journal de Mathématiques Pures et Appliquées*, **77**, (1998), pages 989–1054.

[61] I. Gallagher, A Remark on smooth solutions of the weakly compressible Navier–Stokes equations, *Journal of Mathematics of Kyoto University*, **40**, (2000), 3, pages 525–540.

[62] I. Gallagher and F. Planchon, On global infinite energy solutions to the Navier–Stokes equations in two dimensions, *Archive for Rational Mechanics and Analysis*, **161**, (2002), pages 307–337.

[63] M. Ghil and S. Childress, *Topics in Geophysical Fluid Dynamics: Atmospheric Dynamics, Dynamo Theory, and Climate Dynamics, Applied Mathematical Sciences*, Springer Verlag, New–York, **60**, (1987).

[64] A.D. Gilbert, A.M. Soward, S. Childress, A fast dynamo of $\alpha\omega$-type. *Geophys. Astrophys. Fluid Dynam.*, **85**, (1997), no. 3-4, pages 279–314.

[65] A.E. Gill, *Atmosphere - ocean dynamics*, International geophysics series, vol 30, Academic Press, (1982).

[66] J. Ginibre and G. Velo, Generalized Strichartz inequalities for the wave equation, *J. Funct. Anal.* ,**133**, (1995), pages 50–68.

[67] H.P. Greenspan, *The theory of rotating fluids*. Reprint of the 1968 original. *Cambridge Monographs on Mechanics and Applied Mathematics*. Cambridge University Press, Cambridge-New York, (1980).

[68] E. Grenier, On the nonlinear instability of Euler and Prandtl equations., *Comm. Pure Appl. Math.*, **53**, (2000), no. 9, pages 1067–1091.

[69] E. Grenier, Oscillatory perturbations of the Navier-Stokes equations. *J. Math. Pures Appl.*, **76**, (1997), no. 6, pages 477–498.

[70] E. Grenier, On the derivation of homogeneous hydrostatic equations. *M2AN Math. Model. Numer. Anal.*, **33**, (1999), no. 5, pages 965–970.

[71] E. Grenier, Non dérivation des équations de Prandtl, *Séminaire sur les Équations aux Dérivées Partielles*, 1997 – 1998, Exp. No. XVIII, 12 pp., École polytechnique, Palaiseau, 1998.

[72] E. Grenier, N. Masmoudi, Ekman layers of rotating fluids, the case of well prepared initial data, *Comm. Partial Differential Equations*, **22**, (1997), no. 5 – 6, pages 953–975.

[73] E. Grenier, F. Rousset, Stability of one dimensional boundary layers using Green's functions, *Comm. Pure Appl. Math.*, **54**, (2001), no. 11, pages 1343–1385.

[74] E. Hopf, Ein allgemeiner Endlichkeitssatz der Hydrodynamik. *Math. Ann*, **117**, (1941), pages 764–775.

[75] L. Hörmander, The Analysis of Partial Differential Operators. Vol I, *Springer Study Edition*, 2nd Edition, (1990).

[76] G. Iooss, Existence et stabilité de la solution périodique secondaire intervenant dans les problèmes d'évolution du type Navier Stokes, *Arch. Rat. Mech. Anal.*, **47**, (1972) pages 301-329.

[77] J.-L. Joly, G. Métivier and J. Rauch, Generic rigorous asymptotic expansions for weakly nonlinear multidimensional oscillatory waves, *Duke Mathematical Journal*, **70**, (1993), pages 373–404.

[78] J.-L. Joly, G. Métivier and J. Rauch: Coherent and focusing multidimensional nonlinear geometric optics, *Annales Scientifiques de l'ENS Paris*, **28**, (1995), pages 51–113.

[79] T. Kato, Strong L^p solutions of the Navier–Stokes equations in $\mathbf{R}^m$ with applications to weak solutions, *Mathematische Zeitschrift*, **187**, (1984), pages 471–480.

[80] M. Keel, T. Tao, Endpoint Strichartz estimates, *Amer. J. Math.* **120**, (1998), no. 5, pages 955–980.

[81] S. Klainerman, A commuting vectorfields approach to Strichartz-type inequalities and applications to quasi-linear wave equations, *Internat. Math. Res. Notices*, **2001**, no. 5, pages 221–274.

[82] S. Klainerman, A. Majda: Compressible and incompressible fluids, *Comm. Pure Appl. Math.*, **35** , (1982), no. 5, pages 629–651.

[83] S. Klainerman and I. Rodnianski, Improved local well-posedness for quasilinear wave equations in dimension three, *Duke Math. J.*, **117**, (2003), pages 1–124.

[84] H. Koch and D. Tataru: Well–posedness for the Navier–Stokes equations, *Advances in Mathematics*, **157**, (2001), pages 22–35.

[85] H. Lamb, Hydrodynamics, *Cambridge University Press*, (1932).

[86] P.-G. Lemarié-Rieusset, *Recent Developments in the Navier-Stokes problem*, Chapman-Hall, (2003).

[87] J. Leray, Essai sur le mouvement d'un liquide visqueux emplissant l'espace, *Acta Mathematica*, **63**, (1933), pages 193–248.

[88] R. Lewandowski, *Analyse Mathématique et Océanographie*. Masson, (1997).

[89] D.K. Lilly, On the instability of the Ekman boundary layer, *J. Atmos. Sci.*, **23** (1966), pages 481–494.

[90] C.C. Lin, The theory of hydrodynamic stability, *Cambridge University Press*, (1955).

[91] J.-L. Lions, G. Prodi: Un théorème d'existence et unicité dans les équations de Navier-Stokes en dimension 2, *C. R. Acad. Sci. Paris* **248**, (1959), pages 3519–3521.

[92] P.-L. Lions, Mathematical topics in fluid mechanics. Vol. 1 and 2. *Oxford Lecture Series in Mathematics and its Applications,* 3. Oxford Science Publications. The Clarendon Press, Oxford University Press, New York, (1996).

[93] P.-L. Lions, N. Masmoudi, Incompressible limit for a viscous compressible fluid, *J. Math. Pures Appl.*, **77**, (1998), no. 6, pages 585–627.

[94] A. Majda, *Introduction to PDEs and Waves for the Atmosphere and Ocean*, Courant Lecture Notes in Mathematics (2003).

[95] N. Masmoudi, Ekman layers of rotating fluids: the case of general initial data, *Comm. Pure Appl. Math.*, **53**, (2000), no. 4, pages 432–483.

[96] N. Masmoudi, Incompressible, inviscid limit of the compressible Navier–Stokes system, *Annales de l'Institut Henri Poincaré, Analyse Non Linéaire*, **18**, (2001), pages 199–224.

[97] G. Métivier and K. Zumbrun, Stabilité de couches limites multi-dimensionnelles, *Séminaire Équations aux Dérivées Partielles, École polytechnique*, (2002).

[98] Y. Meyer, *Wavelets, Paraproducts and Navier–Stokes*, Current Developments in Mathematics, International Press, Cambridge, Massachussets, (1996).

[99] O. Oleinik, On the system of Prandtl equations in boundary-layer theory, *Dokl. Akad. Nauk SSSR*, **150**, (1963), pages 28–31.

[100] C. Oseen, Sur les formules de Green généralisées qui se présentent dans l'hydrodynamique et sur quelques unes de leurs applications, (seconde partie) *Acta Mathematica*, **35**, (1912), pages 97–192.

[101] M. Paicu, Étude asymptotique pour les fluides anisotropes en rotation rapide dans le cas periodique, *J. Math. Pures Appl.*, no. **83**, (2004), pages 163–242.

[102] H. Pecher, Das Verhalten globaler Lösungen nichtlinearer Wellengleichungen für große Zeiten, *Math. Z.*, **136**, (1974), pages 67–92.

[103] J. Pedlovsky, Geophysical Fluid dynamics, *Springer Verlag*, (1987).

[104] G. Ponce and T. Sideris, Local regularity of non linear wave equations in three space dimensions, *Communications in Partial Differential Equations*, **18**, (1993), pages 169–177.

[105] I. Proudman, The almost rigid rotation of viscous fluid between concentric spheres, *J. Fluid Mech.*, **1**, pages 505–516.

[106] Lord Rayleigh, On the stability, or instability, of certain fluid motions, *Proc. London Math. Soc.*, **11**, (1880), pages 57–70.

[107] Lord Rayleigh, Scientific papers, vol I to VI, *Cambridge University Press.*

[108] M. Reed, B. Simon, *Methods of modern mathematical physics.* Academic Press [Harcourt Brace Jovanovich, Publishers], New York-London, (1978).

[109] P.H. Roberts, An introduction to magnetohydrodynamics, *Longmans*, (1967), London.

[110] P.H. Roberts, Singularities of Hartmann layers, *Proc. Roy. Soc. Lond. A* **300**, pages 94–107.

[111] H. Schlichting, Boundary layer theory, *McGraw Hill, New York*, (1968).

[112] S. Schochet, Fast singular limits of hyperbolic PDEs, Journal of Differential Equations, **114**, (1994), pages 476–512.

[113] J. Serrin, On the stability of viscous fluid motions, *Arch. Rat. Mech. Anal.*, **3**, (1959), pages 1–13.

[114] H.B. Squire, On the stability of three-dimensional disturbances of viscous flow between parallel walls, *Proc. Roy. Soc. A*, **142**, (1933), pages 129–155.

[115] K. Stewartson, On almost rigid rotations, Part 2, *J. Fluid Mech.*, **3**, (1957), pages 17–26.

[116] R. Strichartz, Restriction Fourier transform of quadratic surfaces and decay of solutions of the wave equations, *Duke Mathematical Journal*, **44**, (1977), pages 705–714.

[117] L. Tartar, Topics in non linear analysis, *Publications Mathématiques d'Orsay*, **78.13**.

[118] R. Temam, *Navier–Stokes Equations, Theory and Numerical Analysis*, North Holland, Amsterdam, (1984) (reprint AMS Chelsea Publishing, Providence, RI, 2001).

[119] S. Ukai, The incompressible limit and the initial layer of the compressible Euler equation. *J. Math. Kyoto Univ.*, **26**, no. 2, (1986), pages 323-331.

[120] A.I. Van de Vooren, The connection between Ekman and Stewartson layers for a rotating disk, *J. of Engineering Mathematics*, **27**, (1993), pages 189–207.

[121] W. von Wahl: *The Equations of Navier-Stokes and Abstract Parabolic Equation,* Aspects of Mathematics, Friedr. Vieweglsohn Braunschweig, (1985).

LIST OF NOTATIONS

INDEX

Printed in the United States
1377LVUK00006B/49/P